T0173013

Molecular Dosimetry and Human Cancer: ANALYTICAL, EPIDEMIOLOGICAL, AND SOCIAL CONSIDERATIONS

Molecular Dosimetry and Human Cancer: Analytical, Epidemiological, and Social Considerations

Edited by

John D. Groopman
Johns Hopkins University
School of Hygiene and Public Health
Baltimore, MD

and

Paul L. Skipper
Massachusetts Institute of Technology
Division of Toxicology
Cambridge, MA

CRC Press
Boca Raton   Ann Arbor   Boston

Library of Congress Cataloging-in-Publication Data

Molecular dosimetry and human cancer : analytical, epidemiological, and social
considerations / edited by John D. Groopman and Paul L. Skipper.
    p.   cm.
  Based on a joint meeting of the American Society for Pharmacology and Experimental
Therapeutics and the Society of Toxicology held in Baltimore, MD, Aug. 17–21, 1986.
  Includes bibliographical references and index.
  ISBN 0-8493-8800-7
  1. Carcinogenesis—Congresses. 2. Cancer—Epidemilogy—
Congresses. 3. Pathology, Molecular—Congresses. 4. Cancer—Etiology—
Congresses. I. Groopman, John D. II. Skipper, Paul L. III. American Society of
Pharmacology and Experimental Therapeutics. IV. Society of Toxicology.
  [DNLM: 1. Carcinogens—analysis—congresses. 2. DNA—metabolism—
congresses 3. Environmental Exposure—congresses. 4. Risk—congresses.
QZ 202 M7184]
RC268.5.M635   1991
616.99'4071—dc20
DNLM/DLC
for Library of Congress   91-13544
                                                             CIP

Developed by Telford Press

Acknowledgment

The origins of this book can be traced to a joint meeting of the American Society for Pharmacology and Experimental Therapeutics and the Society of Toxicology held in Baltimore, MD August 17–21, 1986. Dr. Francis J. Koschier was the organizer of one of the symposia, entitled "Quantification/ Assessment of Mutagens and Carcinogens in Man", in which we were participants. Afterwards, he brought up the idea that a book should be written which would reflect the contents of that symposium and expand upon it. We agreed, and although he did not participate in the project to its completion, without his initial energy and persistance we would never have progressed beyond the discussion stage. We thus wish gratefully to acknowledge Francis's essential contribution.

PREFACE

The molecular dosimetry field has developed in response to the decreasing success of existing techniques to clarify the etiology of human cancers or to provide satisfactory estimates of the hazard associated with exposure to known or suspected etiological agents. Epidemiology has had notable successes in the first endeavor, but its applicability is limited. Epidemiology grew out of the pioneering studies of John Snow, who identified the cause of the cholera epidemic in London in the mid-nineteenth century, and Alice Hamilton, who was one of the first pathologists to study occupational diseases. The power of these early epidemiological studies is remarkable; they were successful in identifying causes and describing disease outcomes in the absence of knowledge of the mechanisms of the disease process. Epidemiology relies on population-based statistics, and these early studies were favored with high disease incidences in large populations. Morever, infectious and acute diseases typically bear a close temporal relationship with exposure. Unfortunately for the epidemiologist, high incidence and temporal proximity to exposure are atypical of human cancer.

Risk estimation is partly a statistical process which, like epidemiology, becomes increasingly uncertain as the frequency of a given outcome diminishes and as the outcome is removed in time from its cause. Risk estimation also depends heavily on exposure assessment. Quantitative accuracy is of course essential but exposure assessment also has a qualitative aspect to it. Not all individuals respond equally to the same exposure, nor do they necessarily respond equally when one agent is presented as part of a complex mixture. Statistical approaches are powerless to address these ancillary aspects of risk estimation.

Clearly a need has arisen for alternatives to or enhancements of the statistical processes involved in epidemiology and risk estimation. The emergence of molecular dosimetry is the primary response to that need. When the term was introduced it was used to mean determination of the concentration of a genotoxin at its biological target site. As such it would be a vastly superior form of exposure assessment than environmental sampling, thereby increasing the accuracy of risk estimation insofar as risk estimates depend on exposure assessment. Molecular dosimetry is also an obvious means for increasing the power of epidemiological investigations; at a minimum it would reduce misclassification of subjects as exposed or not exposed.

Beyond these relatively simple goals lies a more profound hope, namely, that molecular dosimetry will allow us to understand more fully the reasons for interindividual differences in susceptibility to cancer. This hope is based on the belief that the apparently stochastic nature of cancer incidence may be revealed as a more deterministic process if the molecular mechanisms are more apparent. Molecular dosimetry is seen as a means for elucidating those molecular mechanisms.

The working definition of molecular dosimetry has broadened somewhat in the years since the concept was introduced. Measurement of the concentration of a reactive metabolite in a target tissue is not a realistic goal, so some type of proxy measure has always been required. Since the ultimate carcinogenic form of a chemical carcinogen is presumed to be the species that reacts with the bases in DNA, adducts formed in DNA were the first dosimeters to be proposed.

DNA adducts in target tissue are problematical, however, primarily for reasons of practicality. Thus other alternatives have emerged as the field progresses. These include DNA adducts in peripheral blood leukocytes, DNA adducts excreted in urine, adducts formed by blood proteins, and other excreted products such as metabolites or phase II conjugates of the carcinogen or of a metabolite.

Preliminary DNA- and protein-adduct studies in human samples are promising. It appears that these adducts are useful as indicators of environmental exposures and as such may reveal populations at elevated risk that can subsequently be investigated further. The studies described in this book have concurrently attempted to both validate methodologies as well as assess human exposure status. At the present stage of development, the field of biomarkers is beginning to provide the foundation for routine molecular dosimetry, but much remains to be done.

It should be emphasized that none of the biomarkers discussed has reached the status of a routine clinical chemistry method. The techniques and technologies may be similar to standard clinical chemistry analyses but there are critical differences. One is that the biomarker field has as yet no ''normal'' values to refer to when comparing one human sample to another. Indeed, part of the function of this field has been to establish databases on background levels of carcinogen-macromolecular adducts and genetic changes in the human population. This background information will be essential to identify appropriate control cohorts for epidemiological studies. Another difference is that the analytical sensitivity needed to detect carcinogen lesions is often several orders of magnitude lower than the levels of normal biochemicals measured in clinical assays. The human samples analyzed in the investigations described in this volume often contain adducts at or near the limit of detectability of the assay. For this reason the signal-to-noise ratio of many of the reported analyses is low. Nonetheless, detectable levels of environmental chemical damage can be found in human samples, clearly warranting further investigations to improve our interpretation of the data.

Many of the studies reported here only became possible in the last few years as the result of the development of certain analytical methodologies. Some of these were developed specifically for the detection of carcinogen–

macromolecule adducts. Others developed quite independently, but have been adapted by researchers in molecular dosimetry. Several methods in particular have facilitated and stimulated the investigation of biomarkers in people: (a) detection of small molecules and modified macromolecules by immunoassay; (b) detection of carcinogen DNA adducts by ^{32}P-postlabeling; (c) bonded phase capillary column gas chromatography–mass spectrometry; (d) high-performance liquid chromatography; (e) immunoaffinity chromatography using monoclonal antibodies.

The scientific community has long recognized the role of environmental contaminants in human disease, even when it has lacked the technical means to fully elucidate the mechanisms involved. In recent years the general public has become increasingly aware of the problems associated with environmental contamination as attention has been focused on issues such as hazardous waste in water supplies, pesticides in foods, and health problems associated with passive smoking. This awareness has resulted in a challenge to develop methods that can be used not only to assess individual exposure status but also to quantify the risk of developing diseases such as cancer as the result of exposure to environmental toxins.

Public awareness brings with it a second challenge: to use the results in as near a legally, ethically, and morally acceptable manner as possible. Many scientists are not fully aware of the social ramifications of the increased power to determine exposure and estimate risk that molecular dosimetry may bring. The data generated in future studies will likely be used in litigation involving community groups, worker compensation cases, and in tort cases, to mention a few. Questions of confidentiality and the right of access will be raised as well as other ethical and economic questions such as hiring and firing workers. We have ended this book with several chapters that address some of these issues in an attempt to stimulate awareness among those of us involved in the research about the social considerations of molecular dosimetry studies.

We hope this book will provide the reader with a broad view of the field of molecular dosimetry, from its origins and the concepts underlying it, to a wide range of approaches to making the measurements, and finally, to some thoughts on how to make the best use of the results.

John D. Groopman and Paul L. Skipper

CONTRIBUTORS

George J. Annas
Boston University Schools of
  Medicine and Public Health
Boston, Massachusetts

Nicholas A. Ashford
Massachusetts Institute
  of Technology
Cambridge, Massachusetts

Helmut Bartsch
Unit of Environmental
  Carcinogens and Host Factors
International Agency for Research
  on Cancer
Lyon, France

Leslie I. Boden
Environmental Health Section
Boston University School of
  Public Health
Boston, Massachusetts

Peter B. Farmer
MRC Toxicology Unit
Carshalton, Surrey
United Kingdom

Norbert Fedtke
Chemical Industry Institute
  of Toxicology
Department of Biochemical
  Toxicology and Pathobiology
Research Triangle Park
North Carolina

Siv Osterman-Golkar
Department of Radiobiology
Arrhenius Laboratories for
  Natural Sciences
Stockholm University
Stockholm, Sweden

John D. Groopman
The Johns Hopkins University
School of Hygiene and
  Public Health
Department of Environmental
  Health Sciences
Baltimore, Maryland

F. Peter Guengerich
Department of Biochemistry and
  Center in Molecular Toxicology
Vanderbilt University
Nashville, Tennessee

Nancy J. Haley
American Health Foundation
Valhalla, New York

Curtis C. Harris
Laboratory of Human
  Carcinogenesis
National Cancer Institute
Bethesda, Maryland

Stephen S. Hecht
American Health Foundation
Valhalla, New York

Dietrich Hoffmann
American Health Foundation
Valhalla, New York

A. M. Jeffrey
Division of Environmental Science
Institute of Cancer Research and
  Department of Pharmacology
Columbia University
New York, New York

Ruggero Montesano
Unit of Mechanisms of
  Carcinogenesis
International Agency for Research
  on Cancer
Lyon, France

H.-G. Neumann
Institute of Pharmacology and
* Toxicology*
University of Würzburg
Würzburg, Germany

Hiroshi Ohshima
Unit of Environmental
* Carcinogens and Host Factors*
International Agency for Research
* on Cancer*
Lyon, France

David M. Ozonoff
Department of Environmental
* Health Sciences*
Boston University School of
* Public Health*
Boston, Massachusetts

Frederica P. Perera
Columbia University School of
* Public Health*
Division of Environmental
* Sciences*
New York, New York

Michael J. Paolino
Department of Pathology
Medical College of Ohio
Toledo, Ohio

David H. Phillips
Chester Beatty Laboratories
Institute of Cancer Research
London, England

Miriam C. Poirier
National Cancer Institute
National Institutes of Health
Bethesda, Maryland

Erika Randerath
Department of Pharmacology
Baylor College of Medicine
Houston, Texas

Kurt Randerath
Department of Pharmacology
Baylor College of Medicine
Houston, Texas

Gabriele Sabbioni
Institute of Pharmacology and
* Toxicology*
University of Würzburg
Würzburg, Germany

Regina M. Santella
Comprehensive Cancer Center and
* Division of Environmental*
* Sciences*
School of Public Health
Columbia University
New York, New York

David E. G. Shuker
Unit of Environmental
* Carcinogens and Host Factors*
International Agency for Research
* on Cancer*
Lyon, France

Gary D. Stoner
Department of Pathology
Medical College of Ohio
Toledo, Ohio

James A. Swenberg
Chemical Industry Institute
* of Toxicology*
Department of Biochemical
* Toxicology and Pathobiology*
Research Triangle Park
North Carolina

Margareta Törnqvist
Department of Radiobiology
Arrhenius Laboratories for
* Natural Sciences*
Stockholm University
Stockholm, Sweden

Christopher P. Wild
Unit of Mechanisms of
 Carcinogenesis
International Agency for Research
 on Cancer
Lyon, France

W. P. Watson
Shell Research Limited
Sittingbourne Research Centre
Sittingbourne, Kent
United Kingdom

A. S. Wright
Shell Research Limited
Sittingbourne Research Centre
Sittingbourne, Kent
United Kingdom

Daniel Wartenberg
Department of Environmental and
 Community Medicine
Robert Wood Johnson
 Medical School
University of Medicine and
 Dentistry of New Jersey
Piscataway, New Jersey

CONTENTS

Chapter 9
DNA-ADDUCT ANALYSIS BY ^{32}P-POSTLABELING IN THE
STUDY OF HUMAN EXPOSURE TO CARCINOGENS   151
David H. Phillips

Chapter 10
QUANTITATIVE ANALYSIS OF DNA ADDUCTS: THE
POTENTIAL FOR MASS SPECTROMETRIC TECHNIQUES   171
Norbert Fedtke and James A. Swenberg

Chapter 11
ANALYTICAL APPROACHES FOR THE DETERMINATION
OF PROTEIN-CARCINOGEN ADDUCTS USING
MASS SPECTROMETRY

Peter B. Farmer

Chapter 12
DEVELOPMENT OF IMMUNOASSAYS FOR THE DETECTION
OF CARCINOGEN-DNA ADDUCTS

Miriam C. Poirier

PART III: ETHICAL, ECONOMIC AND LEGAL CONSIDERATIONS

Part

I

Basic Principles
and Concepts

The Direct Determination of Human Carcinogens and Assessment of Low-level Risks

A. S. Wright and W. P. Watson
Shell Research Limited
Sittingbourne Research Centre
Sittingbourne, Kent, United Kingdom

1. INTRODUCTION

During recent years, advances in our understanding of the mechanisms of the action of chemical carcinogens have been applied to develop radical new methods for the detection and identification of chemical carcinogens and to improve the quality of cancer risk assessment (for a review of see Wright *et al.*, 1988). The main focus has been on qualitative applications, i.e., procedures for the detection and identification of genotoxic carcinogens. The emerging methods are designed to determine primary DNA damage and mutation at the molecular level by direct analysis of DNA or by analysis of appropriate surrogates and are characterized by high precision, sensitivity, and resolving power. These methods have the additional important advantage that they can be applied directly to man.

Although the detection and identification of hazards are the key steps in cancer prevention, the importance of the more difficult problem of assessing the attendant health risks should not be underestimated. However, with certain notable exceptions, few groups have applied modern mechanistic knowledge to the problem of quantitative risk assessment. Nevertheless, this quantitative aspect needs to be developed if the potential of the emerging technology is to be fully realized. Valid quantitative approaches are also essential for the development of a rational scientific basis for the management of cancer risks.

2. THE MANAGEMENT OF CANCER RISKS

The management of toxicological risks implies a capacity to control exposures to toxic chemicals within acceptable safety limits. Effective control is, therefore, dependent not only on the qualitative determination of chemical hazards but also on a capacity to determine exposure and evaluate the health risks. This last requirement necessitates knowledge of quantitative human dose–response relationships.

The concept of acceptable safety limits is readily accepted when applied to the many classes of toxic chemicals that display thresholded dose–response curves, indicative of very high or even absolute margins of safety at sub-threshold doses. Absolute safety margins cannot be guaranteed in the case of exposures to genotoxic chemicals, particularly when viewed in the context of linear target dose-response curves for the induction of mutation by low doses of genotoxic chemicals *vide infra* (for a discussion of target dose). The absence of a threshold leads to conservative regulatory measures, and in practical terms, even a purely qualitative indication of genotoxicity can act as an absolute deterrent to the development of new products.

It may be argued that knowledge of quantitative human dose–response relationships is not a key requirement for the effective management of cancer risks. Thus it is generally accepted that human contact with carcinogens should be minimized. Purely qualitative identification of the hazard would permit the design of measures to limit exposure and therefore minimize carcinogenic impact. However, such a view fails to take account of the serious limitations of current methods for the detection and identification of human carcinogens or of the high sensitivity of emerging methods. The detection limits of these new molecular approaches correspond to very low levels of risk, and current indications are that they may be so low as to be considered acceptable (Silbergeld *et al.*, 1987). Clearly, positive findings obtained using such sensitive methods necessitate quantitative risk perspective in order to determine whether the risk is acceptable or unacceptable. It is equally clear that in order to achieve this goal, the focus must be on the assessment of low-level risks.

3. CURRENT APPROACHES TO DETECT AND IDENTIFY HUMAN CARCINOGENS

3.1. Environmental Carcinogens

The first stage in seeking to reduce chemical-induced cancer is to identify the causative agents or at least their sources, e.g., cigarette smoking. Until very recently, epidemiological methods to detect and identify environmental carcinogens were based, almost exclusively, on the analysis of tumor incidence and chromosome aberrations in human populations. However, these methods *per se* lack the resolving power required to discriminate between different contributory factors, and it is only in instances of specific, high, and often localized exposures that the individual causative agents or their sources have been identified. Nevertheless, retrospective epidemiological studies indicate that as-yet unidentified chemicals, which may include both natural and human-made compounds, in food or drink or in the local or general environment, play a major and broad role in the etiology of human cancer. The identification of these chemical factors is, therefore, a major goal in cancer prevention. Even when a specific carcinogenic hazard has been identified, it is difficult to estimate the risk posed by low exposures. Indeed, information on the potency (dose–response relationships) of chemical carcinogens is almost completely lacking in humans—particularly in the low-dose region.

3.2. New Products

In addition to the need for identifying environmental carcinogens and their sources, cancer prevention also necessitates the effective prescreening of new chemical products. Until the mid-1970s, cancer studies in laboratory animals were regarded as providing the only valid basis for the prospective determination of human carcinogens. The main strength of this approach is that it is generic, i.e., detection of carcinogenic action is independent of mechanism. However, the interpretation of the results of experimental cancer studies posed severe problems due primarily to the poor sensitivity of the methods and the need to extrapolate to humans.

The detection limits of experimental cancer studies rarely approaches 10^{-2}, which is 3 to 4 orders of magnitude higher than a level of risk that may be considered acceptable from a sociopolitical standpoint (Silbergeld *et al.*, 1987). This shortfall suggests that cancer bioassays may fail to detect some significant carcinogens. High doses are often used to compensate for this poor sensitivity. However, a high-dose strategy may fail in cases where there is a threshold limit on metabolic activation. In such instances, an increase in dose will not lead to an increase in the systemic concentrations of the ultimate carcinogen. Conversely, the administration of high doses may increase tumor incidence by mechanisms, e.g., induction of cell proliferation, that do not operate at low doses and do not, therefore, have any relevance to an assessment of the risks posed by low doses or exposures. Moreover, the determination of high-level risks *per se* has, at best, only limited value in the management of genotoxic risks. The concerns are focused on low-level risks associated with the low exposures encountered in the general, domestic, or occupational environments.

The estimation of the low-level risks to humans necessitates two extrapolative steps. The first step involves extrapolation from the high-dose range, i.e., the measurable effect range, to the low-dose (risk) range of interest. A linear projection is usually employed, and this may give rise to a gross underestimation or overestimation of the true risk for reasons of the type outlined above (Ehrenberg *et al.*, 1983). The second stage, i.e., the extrapolation of the "experimental risk estimate" to man, leads to additional uncertainties and errors. Certainly, the frequent occurrence of major species differences in response to chemical carcinogens cautions against direct extrapolation and indicates a need for additional or supplementary information to improve the evaluation of low-level human risks.

3.3. Short-term Tests

The serious limitations of the classical epidemiological and prospective approaches to detect and identify human carcinogens and evaluate the low-level risks to human populations focused attention on the need to develop new approaches possessing greater sensitivity and resolving power. The development of new approaches hinged on advances in mechanistic knowledge. In this respect the major advance came with the identification of DNA as the critical target of genotoxic chemicals and the recognition that mutagenic properties are key to the action of probably all chemicals that induce cancer *de novo*. These advances in understanding led directly to the development of rapid *in vitro* genotoxicity tests for the prediction of carcinogenic (initiating) activity.

The introduction of these rapid assays permitted widespread applications in screening for environmental and occupational mutagens and also in screening new products, e.g., drugs and agrochemicals, for carcinogenic activity. The performance of these genotoxicity assays was gauged by reference to results obtained in long-term cancer studies, and it is not surprising that numerous qualitative discrepancies emerged (Tennant *et al.*, 1987). The reasons for these discrepancies in the correlation between the results of *in vitro* genotoxicity assays and long-term *in vivo* cancer studies have been widely discussed (Wright *et al.*, 1988; Butterworth, 1988). However, one obvious reason, which is not often discussed but which is relevant to this discussion, is the difference in sensitivity between *in vitro* genotoxicity assays and *in vivo* cancer studies. Thus the higher sensitivity of many genotoxicity assays permits carcinogenic (initiating) properties to be detected at exposures below the limit of detection of *in vivo* cancer studies and leads to apparent ''false''-positive results. However, the evidence that genotoxic action is critical to the initiation of carcinogens is overwhelming. When this evidence is considered in the context of the poor sensitivity of *in vivo* cancer studies, it seems clear that the so-called ''false''-positive results may in fact be signaling the detection of significant carcinogenic hazards. Furthermore, this reasoning casts doubt on the validity of *in vivo* carcinogenicity data as a reliable basis for gauging the performance of genotoxicity assays. These interpretative problems will become more severe as genotoxicity assays become increasingly prospective and increasingly sensitive.

4. STRATEGIES TO IMPROVE HAZARD DETECTION AND RISK ASSESSMENT

In addition to applications in prescreening new chemical products, *in vitro* genotoxicity assays provided the basis of the first practicable, systematic approach to identify environmental carcinogens. This approach has been successfully applied to identify a large number of genotoxic chemicals in the environment (Ames, 1989). The most abundant of these chemicals are natural products occurring in plants and products of cooking and combustion processes. However, this approach demands much time and effort for fractionating environmental samples and testing the individual components. Furthermore, the results of these tests are generally regarded to be of little value in providing information on the carcinogenic potency required to develop an estimate of the quantitative risk to humans. Long-term cancer studies continue to be preferred for this purpose. However, the latter approach necessitates major extrapolations to low doses and from experimental species to humans. Both of these extrapolative stages are subject to uncertainties and errors (*vide supra*).

Initially, research aimed at reducing these extrapolative errors sought to compensate for differences between prospective risk models and humans in the operation of systemic factors and any general extrinsic factors that determine the quantitative relationships between exposure dose and the biological effect or response. These factors include metabolism and related toxicokinetic factors; rates and fidelity of DNA repair and replicative processes; immune surveillance together with promoting or antipromoting factors; and other, as yet, poorly defined factors that influence the progression of initiated cells into malignant neoplasms (Wright, 1983). These factors strongly influence or determine each stage in the carcinogenic process and are primarily responsible for differences in response at the level of the cell, tissue, individual, strain, and species. Clearly, the estimation of the individual contributions of each of these factors is a complex and daunting task. However, advances in the understanding of the nature and mechanism(s) of the action of chemical carcinogens have permitted some simplifications that also address the problem of high-dose–low-dose extrapolation.

The identification of DNA as the primary and critical target of genotoxic carcinogens provided the key to these new developments. In general, genotoxic character is conferred by possession of a center(s) of electrophilic reactivity that permits the chemical to undergo covalent chemical reactions with nucleophilic centers in the target molecule (DNA). Many chemicals classified as genotoxic are in fact precursor genotoxic agents that require metabolic activation into reactive forms (electrophiles) in order to undergo covalent reaction with DNA. Primary products of the reactions with DNA are generally promutagenic (or lethal), and their occurrence signifies an increased risk of mutation and cancer. For this reason they may be described as key lesions (Wright, 1981). Among the primary reaction products, DNA adducts have been singled out for special attention, not only because of their mutagenic propensity but also because of the sensitivity and specificity of methods for their analysis. These new analytical techniques permit, for example, the estimation of target dose, i.e., the dose of the ultimate form(s) of a genotoxic carcinogen at its cellular target (DNA).

5. TARGET DOSE

The introduction of the concept of target dose by Ehrenberg in the early 1970s represented the first effective step in strategies to improve the prospective assessment of genotoxic risks (Ehrenberg *et al.*, 1974). This new dose concept was developed to provide a measure of the dose of ultimate genotoxic agent reaching the DNA in cells or tissues following exposure to a particular genotoxic chemical or a precursor. Target (DNA) dose (Lee, 1976) can be determined by measuring the primary products of reactions between the genotoxic chemical and DNA, e.g., DNA adducts. However, it must be noted that the rate constants of reactions leading to the formation of specific adducts and their biological half-lives must also be determined in order to transform values of the amounts of adducts into estimates of target dose. Indirect measurements of DNA dose may also be made using alternative monitors such as blood proteins, e.g., hemoglobin (Osterman-Golkar *et al.*, 1976).

The determination of target dose in experimental species and man serves two main purposes. First, such determinations automatically compensate for individual or species differences in metabolism or related toxicokinetic factors that control the quantitative relationships between the exposure dose and the dose of the ultimate toxicant delivered to the target. Second, because target dose can be measured at extremely low dose levels, such measurements may be applied to correct for any deviation from linearity in the relationships between the exposure dose and the dose of ultimate genotoxic agent at the target, of critical relevance to the estimation of low-level risks. Thus the concept may be applied to improve the quality of extrapolative risk models and may also lead to an improved definition of individual risk. The measurement of target dose in humans may, therefore, be viewed as an approach towards risk monitoring as well as an improved approach for exposure monitoring.

5.1. Assessment of Target Dose

Much of our present knowledge on the interactions between carcinogens and cellular macromolecules has been obtained by the use of radiolabeled compounds. In general, radiolabeled genotoxic agents may be used in experimental studies for determining amounts of specific adducts formed in the cells or tissues of laboratory organisms. Such direct methods cannot be applied in humans, and two general indirect approaches have been developed to monitor doses of genotoxic chemicals delivered to the DNA in inaccessible tissues, e.g., human tissues:

1. Determination of DNA adducts in more accessible tissues or body fluids, e.g., white blood cells, sperm, placenta, or skin.
2. Determination of the corresponding adducts in accessible proteins, e.g., plasma albumin or hemoglobin.

Such indirect approaches require validation. The quantitative relationships between the dose delivered to DNA in the target tissue and to the surrogate dose monitor can be readily determined in experimental models. However, these relationships cannot normally be determined in humans. Therefore reliance must be placed on experimentally determined coefficients relating doses delivered to the target and to the monitor. In order to justify applications to humans, it is necessary to demonstrate that these coefficients are not subject to significant variation between species.

6. DIRECT HUMAN RISK MONITORING

In addition to improving our ability to monitor low exposures to specified chemicals and interpret the biological consequences of such exposures, insights into mechanisms have also provided a basis for generic procedures, e.g., short-term genotoxicity assays, for the prospective detection of carcinogens operating through a common mechanism. The development of generic (and specific) procedures for the analysis of DNA and protein adducts represents a further step in the trend towards increasingly prospective, precise, and sensitive methods. The theoretical basis for such approaches is now firmly established. Thus a sequential mechanism links electrophilic reactivity with a capacity to induce primary damage in DNA, leading to an increased risk of mutation and cancer. It follows that the detection of DNA adducts in humans provides qualitative evidence of exposure to a genotoxic carcinogen. Of course, not all DNA adducts are equally promutagenic or procarcinogenic. However, since no intrinsically reactive electrophile displays absolute specificity for any particular nucleophile, the detection of even a weakly promutagenic adduct signals the formation of more strongly promutagenic adducts. The detection of covalent protein adducts may also furnish evidence of human exposure to a genotoxic hazard (Ehrenberg, 1980; Hemminki and Randerath, 1987).

The new analytical procedures for the determination of DNA and protein adducts provide selective approaches for the detection and quantification of specific primary reaction products. By virtue of their high resolving power and very high sensitivity, these methods are potentially of great value in assisting in the identification of agents responsible for the initiation of human cancers. Indeed these procedures are already finding applications in "molecular epidemiology" studies to detect and monitor human exposures to genotoxic chemicals. Furthermore, ^{32}P-postlabeling methods (Randerath *et al.*, 1981) have been successfully coupled with preparations of human tissue *in vitro*, first, to validate the metabolizing systems of these *in vitro* systems and, second, to screen new chemical products for genotoxic activity towards humans (Watson *et al.*, 1988).

The initial promise of the emerging technology appears to have been borne out by the detection of "background" exposures in human subjects (Randerath *et al.*, 1988). The direct determination of "background" DNA (and protein) adducts in humans illustrates the high sensitivity and resolving power of these analytical procedures and focuses attention on the causative chemicals as being likely to contribute to the initiation of human cancer. Such findings are much more informative than are the results obtained using extrapolative models to screen for genotoxic chemicals in the environment. Nevertheless, the results obtained with such extrapolative models, e.g., microbial mutation assays or *in vitro* ^{32}P-postlabeling assays, may assist in the characterization of human DNA adducts by identifying candidate causative agents. The characterization of the chemical structures of human DNA adducts would provide a basis for identifying the chemical initiators of human cancer. This is a key objective in cancer prevention and spurs further development and refinement of the emerging technology.

Although very high sensitivity and resolving power are essential for the direct detection of carcinogenic hazards in humans, it is probable that the sensitivity of the new methods will exceed that required for the determination of acceptable risks. In terms of sensitivity, therefore, the new methods for the determination of DNA and protein adducts would satisfy the criteria for a test to detect human carcinogens. However, it seems impracticable that tests which exceed the sensitivity required to determine acceptable risks can be applied in a purely qualitative mode—at least in a regulatory sense. With such sensitive assays it is essential to determine whether positive findings are indicative of a negligible risk or an unacceptably high risk. Thus as we progress to increasingly prospective and sensitive assays, the important issue shifts from the qualitative determination of whether or not a chemical is carcinogenic to the quantitative question: How carcinogenic is the chemical? Quantitation of the risk becomes essential.

7. EVALUATION OF DNA REPAIR AND PROMOTIONAL PRESSURE

The sensitivity of the new methods for determining DNA and protein adducts permits applications to assess tissue DNA doses of genotoxic chemicals at low environmental or occupational levels in humans, laboratory species, and environmental flora and fauna. The concept of target dose was introduced as a measure to compensate for cellular, individual, and species variations in metabolism and related systemic factors that determine the concentration of ultimate reactive genotoxic agents in the microenvironment of the target molecule. The concentration of the ultimate reactant at the target molecule is the prime determinant of the rate of formation of key chemical lesions, e.g., adducts, in DNA. By compensating for these toxicokinetic differences between extrapolative models and humans, the determination of target dose improves the translation of experimental risk data to humans and provides a better definition of individual risks posed by exposures to genotoxic chemicals. However, many problems remain. In order to further improve the quality of prospective risk assessment, it is necessary to develop procedures to correct for differences between the extrapolative models and humans in the operation of systemic and general extrinsic factors that determine the progression, first, of key lesions into cancer mutations and, second, of the initiated cells into malignant neoplasms.

The choice of the extrapolative model is very important. Experimental cancer studies are often viewed as the only realistic possibility because of the obvious relevance of the biological endpoint to the assessment of the human hazard. However, the important requirement is to assess low-level risks, and this necessitates extrapolations over several orders of magnitude.

Measurements of target dose can reduce the errors associated with such high-dose–low-dose extrapolations (*vide supra*). However, there is one critical proviso: the mechanism(s) giving rise to the tumors observed at the high doses required to produce measurable effects must also be the prime determinants of (theoretical) tumorigenic action at the low doses corresponding to acceptably low levels of risk. In this respect, conventional cancer studies are probably fundamentally unsound. Thus it is widely held that many nongenotoxic chemicals and genotoxic chemicals can cause cell proliferation and exhibit tumor-promoting activity at the high dose levels used in such studies (Ames, 1989). Such activity can have a profound influence on the occurrence or magnitude of the carcinogenic response. In considering low-level risks associated with low exposures to genotoxic chemicals, it is unlikely that such chemicals would exert significant promoting activity at low doses: initiating activity would be expected to predominate. Conventional cancer studies would therefore grossly overestimate low-level risks posed by low doses of genotoxic carcinogens or indeed any other "carcinogen" possessing significant promoting activity. It appears, therefore, that there is little scientific justification in employing models possessing a major promotive element to evaluate low-level genotoxic risks. Nevertheless, an assessment of the influence of general promotional pressure exerted on initiated cells in human populations, e.g., as a consequence of high or relatively high exposures to natural dietary components, must be made in order to evaluate the cancer risks posed by low exposures to genotoxic agents.

Based on these arguments, the fundamental requirements for the assessment of these low-level risks comprise, first, the determination of the quantitative relationships between low target doses of genotoxic chemicals and the induction of mutation, particularly cancer mutations (initiation), and, second, the development of procedures to evaluate the impact of promotional pressures and any other modulating factor on the progression of the initiated cells with tumors. High-precision experimental procedures required for the direct investigation of the quantitative relationships between specific DNA adducts and specific mutations at the molecular level in DNA are only just emerging. Such studies have fundamental relevance to cancer risk assessment, but the critical correlative data is unlikely to be available until well into the 1990s. In order to deal with the immediate problem, Ehrenberg and co-workers have developed a comprehensive strategy for the assessment of low-level cancer risks, known as the radiation-dose-equivalent approach, that is designed not only to compensate for the operation of factors, e.g., DNA repair, that influence the mutagenic effectiveness of low target doses of genotoxic chemicals in man but also to compensate for all extrinsic and systemic factors that influence the progression of initiated cells in human populations.

The risk model developed by Ehrenberg and co-workers (Ehrenberg *et al.*, 1974; Ehrenberg, 1979, 1980; Calleman *et al.*, 1978; Törnqvist and Osterman-Golkar, this volume) is based on current perceptions of the mechanism(s) of action of genotoxic carcinogens. Current research is providing increasingly refined and precise insights into these mechanisms. These insights and associated technical developments hold the promise of more refined and direct approaches to estimate genotoxic risks, e.g., by direct determination of the mutagenic (initiating) propensity of specific DNA adducts at the molecular level. However, at the present time, the radiation-dose-equivalent concept provides the only rational scientific approach to evaluate the cancer risks posed by low-level exposures to genotoxic chemicals.

REFERENCES

B. N. Ames, Environmental pollution and the causes of human cancer: Six errors. In: *Important Advances in Oncology,* eds. V. T. De Vita, Jr., S. Hellman, and S. A. Rosenberg, J. B. Lippincott, Philadelphia, PA pp. 237–247 (1989).

B. E. Butterworth, Strengths and limitations of short-term tests to identify potential human carcinogens and mutagens. In: *Management of Risk from Genotoxic Substances in the Environment. Proceedings of Symposium,* October 3–6, Stockholm, ed. L. Freij, Swedish National Chemicals Inspectorate Sweden, pp. 63–97 (1988).

C. J. Calleman, L. Ehrenberg, B. Jansson, S. Osterman-Golkar, D. Segerbäck, K. Svensson, and C. A. Wachtmeister, Monitoring and risk assessment by means of alkyl groups in hemoglobin in persons occupationally exposed to ethylene oxide, *J. Env. Path. Tox.,* **2,** 427–442 (1978).

L. Ehrenberg, Risk assessment of ethylene oxide and other compounds. In: *Assessing Chemical Mutagens: The Risk to Humans,* eds. V. K. McElheny and S. Abrahamson, Banbury Report 1, CSH Press, Cold Spring Harbor, NY, pp. 157–190 (1979).

L. Ehrenberg, K. D. Hiesche, S. Osterman-Golkar, and I. Wennberg, Evaluation of genetic risks of alkylating agents: Tissue doses in the mouse from air contaminated with ethylene oxide, *Mutat. Res.,* **24,** 83–103 (1974).

L. Ehrenberg, E. Moustacchi, S. Osterman-Golkar, and G. Ekman, Dosimetry of genotoxic agents and dose-response relationships of their effects, *Mutat. Res.,* **123,** 121–182 (1983).

K. Hemminki and K. Randerath, Detection of genetic interaction of chemicals by biochemical methods: Determination of DNA and protein adducts. In: *Mechanisms of Cell Injury: Implications for Human Health,* ed. B. A. Fowler, John Wiley, London, pp. 209–227 (1987).

A. Kolman, M. Naslund, S. Osterman-Golkar, G. P. Scalia-Tomba, and A. L. Meyer, Comparative studies of *in vitro* transformation by ethylene oxide and gamma-radiation of C3H10T1/2 cells, *Mutagen.,* **4,** 58–61 (1989).

W. R. Lee, Molecular dosimetry of chemical mutagens. Determination of molecular dose to the germ line, *Mutat. Res.,* **38,** 311–316 (1976).

S. Osterman-Golkar, L. Ehrenberg, D. Segerbäck, and I. Hallstrom. Evaluation of genetic risks of alkylating agents. II. Hemoglobin as a dose monitor. *Mutat. Res.,* **34,** 1–10 (1976).

K. Randerath, M. V. Reddy, and R. C. Gupta, ^{32}P-labeling test for DNA damage, *Proc. Natl. Acad. Sci. USA,* **78,** 6126–6129 (1981).

K. Randerath, R. H. Miller, D. Mittal, and E. Randerath, Monitoring human exposure to carcinogens by ultrasensitive postlabeling assays: Application to unidentified genotoxicants. In: *Methods for Detecting DNA Damaging Agents in Humans: Applications in Cancer Epidemiology and Prevention,* IARC Scientific Publication No. 89, eds. H. Bartsch, K. Hemminki, and I. K. O'Neill, IARC, Lyon, France, pp. 361–367 (1988).

E. K. Silbergeld, L. G. Ehrenberg, K. Hemminki, M. Hutton, R. J. Laib, R. R. Lauwerys, H-G. Neuman, G. F. Nordberg, J. Piotrowski, W. G. Thilly, and A. S. Wright. Exposures: uptake, tissue and target dose. In: *Mechanisms of Cell Injury: Implications for Human Health,* ed. B. A. Fowler, Wiley, London, pp. 405–429 (1987).

R. W. Tennant, B. H. Margolin, M. D. Shelby, E. Zeiger, J. K. Hazeman, J. Spalding, W. Caspary, M. Resnick, S. Stasiewitz, B. Anderson, and R. Minor, Predictions of chemical carcinogenicity in rodents from *in vitro* genetic toxicity assays, *Science,* **236,** 933–941 (1987).

W. P. Watson, R. J. Smith, K. R. Huckle, and A. S. Wright, Human organ culture techniques for the detection and evaluation of genotoxic agents. In: *Methods for Detecting DNA Damaging Agents in Humans: Applications in Cancer Epidemiology and Prevention,* IARC Scientific Publication No. 89, eds. H. Bartsch, K. Hemminki, and I. K. O'Neill, IARC, Lyon, France, pp. 384–388 (1988).

A. S. Wright, New strategies in biochemical studies for pesticide toxicity. In: *The Pesticide Chemist and Modern Toxicology,* ACS Symposium Series 160, American Chemical Society, Washington, D.C., eds. S. K. Bandall, G. J. Marco, L. Golberg, and M. L. Leng, pp. 285–304 (1981).

A. S. Wright, Molecular dosimetry techniques in human risk assessment: An industrial perspective. In: *Developments in the Science and Practice of Toxicology,* eds. A. W. Hayes, R. C. Schnell, and R. S. Miya, Oxford: Elsevier Science Publishers, Amsterdam, pp. 311–318 (1983).

A. S. Wright, T. K. Bradshaw, and W. P. Watson, Prospective detection and assessment of genotoxic hazards: A critical appreciation of the contribution of L. G. Ehrenberg. In: *Methods for Detecting DNA Damaging Agents in Humans: Applications in Cancer Epidemiology and Prevention,* IARC Scientific Publication No. 89, eds. H. Bartsch, K. Hemminki, and I. K. O'Neill, IARC, Lyon, France, pp. 237–247 (1988).

Chapter

2

Molecular Epidemiology: Overview of Biochemical and Molecular Basis

Curtis C. Harris
Laboratory of Human Carcinogenesis
National Cancer Institute
Bethesda, Maryland

The primary goal of molecular epidemiology is to identify individuals at high cancer risk by obtaining evidence of high exposure to carcinogens, and/or increased oncogenic susceptibility due to either inherited or acquired host factors. This emerging and multidisciplinary area of cancer research combines epidemiological and laboratory approaches. Clinical and epidemiological studies have identified populations at high cancer risk and also etiological agents, such as tobacco smoke as the major cause of bronchogenic carcinoma. Laboratory studies have extended these epidemiological findings by isolating and identifying individual carcinogenic and cocarcinogenic agents found in complex mixtures, e.g., polycyclic aromatic hydrocarbons (PAH) in coal tar, diesel exhaust, and tobacco smoke, and by discovering naturally-occurring and human-made carcinogenic and cocarcinogenic agents in our environment. Laboratory studies, especially those using animal models, have also made major contributions to our current understanding of the multistage carcinogenic process that has been suggested by clinical observations, human histopathologic studies, and investigations of animal models. Extrapolation of carcinogenesis data to humans and the interindividual variation in oncogenic susceptibility in the human population are discussed in detail in other chapters of this volume.

MULTISTAGE CARCINOGENESIS

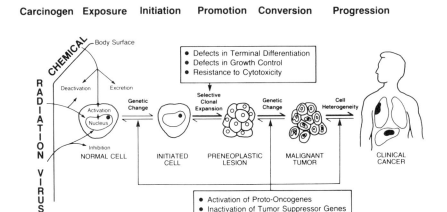

Figure 1. Schematic representation of the multistage process of carcinogenesis.

This overview will discuss opportunities and some of the experimental approaches that show promise in identifying individuals at the highest cancer risk in a population. A more comprehensive analysis can be gleaned from the selected review articles cited in the reference list and from the individual chapters in this multifaceted volume.

The concepts of tumor initiation, promotion, malignant conversion, and progression developed from studies in chemical carcinogenesis form the framework for this discussion. A schematic representation of this multistage process is depicted in Fig. 1. Humans are repeatedly exposed to complex and variable mixtures. These environmental mixtures contain both initiators and promoters, e.g., tobacco smoke, and complete carcinogens having both initiating and promoting activities. Other factors, including host factors, are also known to influence cancer risk. This concept of multistage carcinogenesis provides a link that connects laboratory animal studies, *in vitro* human cell carcinogenesis, and human carcinogenesis.

The earliest events in the multistage process of chemical carcinogenesis are considered to include exposure to carcinogens, transport of the carcinogen to the target cell, activation to ultimate carcinogenic metabolites, if the agents are procarcinogens, and DNA damage, leading to changes that result in the initiated cell (Fig. 1). Each of these events will only be briefly discussed here.

1. CARCINOGEN EXPOSURE AND METABOLISM

Evidence of human exposure to carcinogens can be indirectly obtained by detection of carcinogens in the environment. Direct evidence comes from measurement of carcinogens and/or their metabolites in body fluids, such as urine, breast milk, seminal fluid, and serum, and in samples of tissue that may accumulate the carcinogen, such as fat for lipophilic agents. Mutagenic activity found in body fluids or extracts of tissues has also been used as an indirect measurement of the putative presence of procarcinogens.

Once the carcinogen is transported into the target cell, it may be enzymatically activated, generally to an electrophilic metabolite, and may be detoxified, generally to more polar, water-soluble metabolites. Enzymatic activation may also take place at a site remote from the target organ, especially the liver. Many carcinogens in our environment are procarcinogens that require enzymatic activation. It has been recognized during the last decade that many carcinogens require multiple enzymatic steps to become activated to their ultimate carcinogenic form, and the activity of the rate-limiting enzymatic reaction responsible for activation of a carcinogen may differ among people and influence their cancer risk. Therefore the metabolic balance between activation and deactivation of a procarcinogen and the covalent binding of its ultimate carcinogenic metabolite to DNA are considered to be important events in stages of carcinogenesis, termed tumor initiation and malignant conversion (Fig. 2).

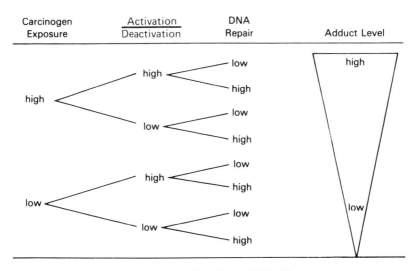

DETERMINANTS OF CARCINOGEN-DNA ADDUCT LEVELS

Figure 2. Major determinants of amounts of carcinogen-DNA adducts.

Carcinogens can also be formed by endogenous reactions, e.g., nitrosation of amines to form *N*-nitrosamines. Assays to assess an individual's capability of producing endogenous carcinogens are being developed using safe substrates. For example, nitrosation of proline to form *N*-nitrosoproline, which is excreted in urine, can be readily measured.

Most chemical carcinogens are activated by oxidative metabolism. The increasing number of types of cytochrome *P*-450 that have been identified in laboratory animals and in human tissues may in part explain interindividual differences in xenobiotic metabolism; interindividual variations in the activities of enzymes responsible for carcinogen metabolism in various human tissues and cells can vary more than 100-fold. Substrate affinities and/or capacities of these multiple forms differing from person to person are another possible important variable factor.

Certain carcinogens and drugs may be metabolically activated by the same enzymes, and indicator drugs, such as debrisoquine, antipyrine, and dapsone, may prove to be useful in probing genetic polymorphisms of carcinogen metabolism. Using the debrisoquine metabolic ratio (debrisoquine: 4-hydroxydebrisoquine in the urine), an increased rate of hydroxylation of debrisoquine has been observed in patients with bronchial cancer or advanced bladder cancer compared to that of noncancer patients. In bronchial cancer patients, this increase occurs irrespective of the number of cigarettes consumed and tumor stage. Interestingly, only some histological types, e.g., squamous-cell carcinoma and small-cell carcinoma, but not adenocarcinoma, are associated with the debrisoquine-related extensive metabolizer phenotype. Occupational exposures to either asbestos or PAHs increase the relative risk to 18- and 34-fold, in the extensive metabolizer group.

N-acetylation polymorphism is the best-known genetic condition affecting the pharmacokinetics of several drugs, e.g., isoniazid, hydralazine, procainamide, and dapsone. The acetylation status is controlled by two autosomal alleles at a single locus, with slow acetylation being manifest as a recessive trait. There is some evidence that an increased urinary bladder cancer risk in those occupationally exposed to carcinogenic aromatic amines might be due to their slow *N*-acetylation phenotype. In contrast, the fast *N*-acetylation phenotype is associated with increased risk of colon carcinoma.

2. CARCINOGEN-MACROMOLECULAR ADDUCTS

As noted in the previous section, metabolism of carcinogens from several chemical classes, including *N*-nitrosamines, PAHs, hydrazines, mycotoxins, and aromatic amines, has been studied in human tissues and cells. The enzymes responsible for the activation and deactivation of procarcinogens, the metabolites produced, and the carcinogen–DNA adducts formed by cultured human tissues and cells are generally qualitatively similar among donors and tissue types. The DNA adducts and carcinogen metabolites are also very similar to those found in most laboratory animals, an observation that supports the qualitative extrapolation of carcinogenesis data from the laboratory animal to the human situation. Because studies using experimental animals generally indicate that their cancer risk is influenced by the capacity for metabolic activation of procarcinogens to genotoxic lesions, it is likely that a similar relationship exists for humans.

In vitro models of human tissues and cells have been extensively utilized to investigate the metabolism of chemical carcinogens and the covalent binding of their metabolites to DNA. Although the major DNA adducts are qualitatively similar, quantitative differences have been found among individuals and their various tissues, and in outbred animals. These differences in enzymatic activities and number of DNA adducts generally range from 10- to 150-fold among humans and are of the same order of magnitude found in pharmacogenetic studies of drug metabolism.

Results from *in vivo* and *in vitro* studies discussed above serve as a basis for investigations in biochemical and molecular epidemiology. For example, the observation that the carcinogen–DNA adducts formed in cultured human tissues are generally the same as those found in experimental animals in which these chemicals induce cancer has encouraged investigators to search for DNA adducts in biological specimens obtained from people exposed to either specific carcinogens, such as benzo[a]pyrene, or chemotherapeutic agents. The recent development of highly sensitive methods for detecting carcinogen–DNA adducts has made this search possible (Table 1). Methods currently used include ^{32}P-nucleotide postlabeling and chromatography, synchronous scanning fluorescent spectrophotometry, mass spectrometry, electrochemical conductance detection, and immunoassays. These methods are being used to measure carcinogen–macromolecular adducts in cells from people exposed to carcinogenic chemicals and cancer chemotherapeutic agents. Although these techniques measure DNA lesions considered to be important in carcinogenesis, cancer is a multistage process, so it is unlikely that DNA adducts will be precise quantitative predictors of cancer risk.

Although hemoglobin and albumin are not considered targets for the pathobiological effects of carcinogens, these macromolecules have certain advantages for the molecular dosimetry of exposure to carcinogens. For example, red blood cells have a lifetime of about 120 days in humans, so the levels of carcinogen–hemoglobin adducts may be an integrative measure of exposure during a period of nearly four months. In addition, these protein adducts, unlike DNA adducts, are not considered to be repaired.

Current dogma has directed our efforts to measuring adducts formed by the direct action of the activated carcinogen metabolite(s). However, carcinogens may also exert their oncogenic effects via indirect damage to macro-

TABLE 2–1. Examples of Assays to Measure Carcinogen–Macromolecular Adducts

Cellular DNA adducts
  Nonhydrolyzed
    Immunoassays
    Fluorimetry
  Hydrolyzed (concentration, e.g., high-pressure liquid chromatography or immunocolumn followed by):
    ^{32}P-Postlabeling
    Immunoassay
    Electrochemical conductance
Urinary DNA base adducts
Protein adducts
  Hemoglobin
  Albumin

molecules such as carcinogen-induced formation of oxy radicals, which cause DNA damage including thymine glycol and other altered nucleic acid structures. This induction of the pro-oxidant state, i.e., increased concentrations of active oxygen, organic peroxides, and oxygen radicals, may also be of importance in tumor promotion. Both immunoassays and ^{32}P-postlabeling assays are being developed to measure radiation- and oxy-radical-induced DNA damage.

3. DNA REPAIR

DNA-repair enzymes modify DNA damage caused by carcinogens, including removal of DNA adducts. Studies of cells from donors with xeroderma pigmentosum have been particularly important in expanding our understanding of DNA excision repair and its possible relationship to risk of cancer. The rate but not the fidelity of DNA repair can be determined by measuring unscheduled DNA synthesis and removal of DNA adducts, and substantial interindividual variations in DNA-repair rates have been observed. In addition to finding excision repair rates severely depressed in xeroderma pigmentosum cells (e.g., complimentation group A), an approximately 5-fold variation among individuals in unscheduled DNA synthesis induced by UV exposure of lymphocytes *in vitro* has been found in the general population. A significant reduction in unscheduled DNA synthesis induced *in vitro* by N-acetoxy-2-acetylaminofluorene has been observed in mononuclear leukocytes from individuals with a history of cancer in first-degree relatives when compared to those without a family history.
Interindividual variation has been noted in the activity of O^6-alkylgua-

nine-DNA-alkyltransferase; this enzyme repairs alkylation damage to O^6-deoxyguanine. In addition to these person-to-person differences, wide variations in DNA-repair activities have been observed in different types of tissues, and fetal tissues exhibit 2- to 5-fold less activity than the corresponding adult tissues, and cells may have lower repair rates after terminal differentiation. The activity of this DNA-repair enzyme is inhibited by certain aldehydes and alkylating cancer chemotherapeutic agents. Decreased activity of this DNA-repair activity has been observed in fibroblasts from patients with lung cancer when compared with donors with either melanoma or noncancer controls. Therefore acquired and/or inherited deficiency in O^6-alkylguanine-DNA-alkyltransferase may be a cancer risk factor in tobacco smokers.

An unimodal distribution of repair rates of benzo[a]pyrene diol epoxide–DNA adducts has been observed using human lymphocytes *in vitro*. The interindividual variation was substantially greater than the intraindividual variation, which suggests the influence of inherited factors. The influence of these variations in DNA-repair rates in determining tissue site and risk of cancer in the general population remains to be determined.

4. ONCOGENES AND TUMOR SUPPRESSOR GENES

Carcinogenesis is a multistage process involving both genetic and epigenetic aberrations. Activation of proto-oncogenes and/or inactivation of tumor suppressor genes may occur in an early stage of carcinogenesis, i.e., tumor initiation, as well as in the later stages, e.g., tumor conversion and progression (Fig. 1). For example, activation of Ki-*ras* by base-substitution mutations is usually an early event in human colon carcinogenesis, whereas amplification of N-*myc* is associated with the tumor progression stage of human neuroblastoma. In the inherited form of retinoblastoma, a defective tumor suppressor gene, RB-1, is found in the germ line DNA, and the second RB-1 allele is inactivated most frequently by deletion or recombinational mechanisms during the initial stages of carcinogenesis. In contrast, the dual inactivation of the RB-1 gene by somatic mutation appears to occur during tumor progression of human breast and lung cancer.

Whereas the role of oncogenes in carcinogenesis is now well known to the biomedical scientist, less is known about mechanisms of tumor suppression and the genes involved. Putative tumor suppressor genes are likely to be as numerous as oncogenes, i.e., more than 40, and belong to several functional classes (Table 2). In addition, somatic cell genetic studies indicate that tumor suppressor genes are dominant when compared to oncogenes in determining the tumorigenic potential of mammalian cells.

Considering the large number of progenitor cells, clinically evident cancer is a pathobiological event of exceedingly low probability. Although systemic host factors, e.g., the immune system, may largely account for its rarity, the lack of convincing reports of "spontaneous" transformation of human cells *in vitro* and the difficulty in inducing their *in vitro* neoplastic transformation with chemical, physical, and viral oncogenic agents attest to the presence of inherent suppressing factors at the biological level of the progenitor cells. Evidence for these presumed dominant-acting "tumor suppressor" genes has arisen primarily in epidemiological studies; molecular analysis of genetic loci exhibiting DNA restriction fragment-length polymorphisms (DNA-RFLPs; Fig. 3) show reduction to homozygosity of chromosome 13 found in retinoblastoma and osteosarcoma and of chromosome 11 in Wilms's tumor and carcinomas of the lung, breast, or bladder. Furthermore, the latter data have been corroborated by somatic cell genetic studies with human cell hybrids. The tumor suppressor gene RB-1 involved in retinoblastoma and a variety of other tumors including small-cell carcinoma of the lung and breast carcinoma has been recently cloned and shown to suppress the tumorigenicity of a cell line with a defective RB-1 gene.

5. GENETIC PREDISPOSITION TO CANCER

Because of the biologically diverse, genetically heterogenous human population and the multistage nature of the carcinogenic process, there are differences in susceptibility among individuals to the oncogenic effects of environmental carcinogens.

Investigation of individuals with multiple tumors has led to the discovery of several hereditary conditions of increased cancer incidence (Table 3). These diseases may be transmitted by single-gene, polygenic, or chromosomal mechanisms. Some hereditary conditions predispose only one type of tissue of neoplasia (e.g., actinic keratosis predisposes to basal-cell carcinomas of the skin and familial polyposis coli predisposes to tumors in the

TABLE 2–2. Examples of Functions of Putative Tumor Suppressor Genes

Induce terminal differentiation
Trigger senescence
Regulate growth
Inhibit proteases
Modulate histocompatibility antigens
Regulate angiogenesis
Facilitate cell–cell communication
Maintain chromosomal stability

DNA POLYMORPHISM ANALYZED BY RESTRICTION ENZYME DIGESTION, DNA ELECTROPHORESIS, AND DNA HYBRIDIZATION

1. DNA digestion by restriction enzymes (R_1 and R_2)

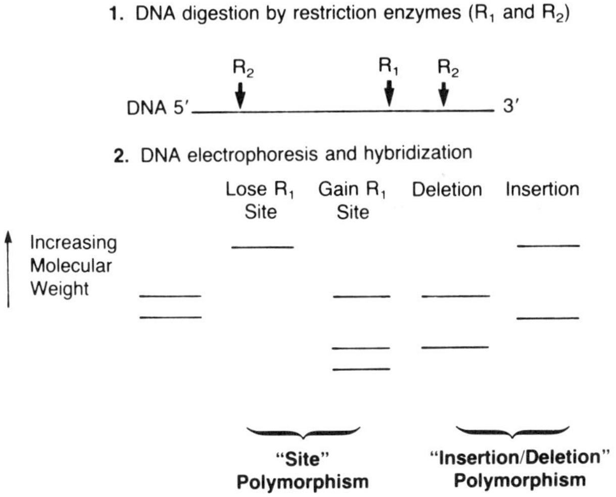

Figure 3. Schematic representation of DNA polymorphism analyzed by restriction enzyme digestion, DNA electrophoresis, and DNA hybridization.

gastrointestinal tract), while other hereditary conditions predispose to multiple tumors in different tissues (e.g., hereditary retinoblastoma, Bloom's syndrome, and ataxia telangiectasia).

Most known hereditary causes of multiple tumors were discovered because they manifested as dramtic clinical syndromes with a Mendelian pattern of inheritance. In some of these diseases, host susceptibility to specific environmental agents has been determined, e.g., ultraviolet radiation as a cause of skin cancer in patients with xeroderma pigmentosum. In contrast to these rather rare conditions, most common types of hereditary multiple cancer occur in familial patterns that probably have a polygenic basis.

Analysis of certain epidemiological data has suggested a two-hit mutation model of carcinogenesis. A genetic predisposition may be due to inheritance of the first autosomal mutation so that only one additional genetic event is required. Therefore these individuals may already have initiated cells due to germ-line genetic lesions. Retinoblastoma is an example of a human tumor in which there is a pair of tumor suppressor genes (RB-1) located at the chromosomal region 13q14, and loss (inactivation) of these diploid genes leads to tumor formation. In this context, diploidy can be viewed as a protective genetic mechanism. Possible mechanisms producing homozygosity or hemizygosity at the retinoblastoma locus in tumor cells include chromosomal nondisjunction, nondisjunction and reduplication, mitotic recombination, deletion, gene inactivation, and base-substitution mu-

TABLE 2–3. Examples of Inherited Disorders Associated with an Increased Risk of Cancer

Constitutional chromosomal abnormalities
　Down's syndrome
　Klinefelter's syndrome
Mendelian traits
　Inherited cancer syndromes
　　Retinoblastoma
　　Familial polyposis coli
　　Multiple endocrine neoplasia syndromes
　Inherited preneoplastic states
　　DNA-repair defects and chromosomal insta-
　　bility syndromes: xeroderma pigmento-
　　sum, Fanconi, Bloom's, ataxia
　　telangiectasia
　Disturbances of tissue organization
　　Hamartomatous syndromes: Peutz-Jegher's,
　　Cowden, neurofibromatosis
　　Other conditions in which disturbance of tis-
　　sue proliferation, differentiation or organi-
　　zation is associated with increased cancer
　　risk: disorders of skin keratinization (?),
　　α_1-antitrypsin deficiency (?)
　Debrisoquine metabolic polymorphism
　N-acetylation polymorphism
　Ha-*ras* variable tandem repeat polymorphism
Immune deficiency syndromes
Multifactoral predisposition
Ethnic cancer differences
Familial cancer aggregations

tation. One of these mechanisms, mitotic recombination, may also be important in the genetic predisposition to cancer found in Bloom's syndrome.

New experimental approaches to identify people genetically predisposed to cancer are being evaluated, e.g., RFLP analysis of genetic polymorphism. Until recently, genetic predisposition could be assessed only by measuring gene products, e.g., histocompatibility antigens on cell surfaces, isoenzymes, and, as discussed earlier, the pharmacogenetic phenotype. Advances in molecular biology have now made it possible to measure genetic polymorphism at the DNA level, which greatly expands the potential of assays of polymorphisms in biomedical research. Genetic polymorphism can be measured by restriction-enzyme analysis and DNA hybridization (Fig. 3). RFLPs are of two types: site or insertion-deletion polymorphism. In the case of site polymorphism, the recognition site for a given restriction enzyme in a particular region of DNA either appears or disappears as the result of point mutation. Accordingly (Fig. 3), in the site-absent phenotype, one large fragment instead of two smaller fragments is observed. Conversely, in the site-present phenotype, three fragments (one of expected size and two smaller) instead of two fragments are seen.

In the case of insertion-deletion polymorphism, variation in fragment length is the result of insertion or deletion of DNA sequences and will be detected by any restriction endonuclease that possesses recognition sites tightly fitting the region of sequence alteration. Because the insertions or deletions can assume a continuum of lengths, more than two alleles are possible. This RFLP approach has already proved beneficial in identifying individuals with a genetic predisposition to a variety of diseases, including retinoblastoma, non-insulin-dependent diabetes mellitus, Huntington's disease, and hemoglobinopathies. The same molecular approach using specific DNA probes, e.g., for oncogenes and ectopic hormones, is now being studied to determine its potential in identifying persons with a genetic predisposition to cancer. The human Ha-*ras* proto-oncogene may prove useful for these studies because it contains a hypervariable insertion-deletion polymorphism. The molecular basis for this insertion-deletion polymorphism is a region of 30 to 100 tandem nucleotide repeats with a 28-base-pair consensus sequence aligned head-to-tail that is approximately 1 kilobase downstream from the structural gene of Ha-*ras*. The function of this variable tandem repeat is unknown, but it may have gene enhancer activity and appears to be inherited in a Mendelian fashion. An increased frequency of rare alleles at this locus has been found in certain cancer patients, including those with carcinomas of the lung, breast, and bladder.

Although RFLP analysis should help to provide the general location of genes involved in common malignancies, in most cases more powerful tools will be required to precisely identify them. The recent development of pulsed-field electrophoresis, combined with innovative cloning techniques, should prove valuable in gene isolation.

Polymerase chain reaction amplification of DNA coupled with advanced DNA sequencing methods is revolutionizing molecular diagnosis of disease and inherited predisposition. This innovative approach requires only minute amounts of nucleic acids, is rapid, and is proving to be invaluable in the detection of mutations, e.g., Ki-*ras* and p53 gene mutations, in human cells.

6. OVERVIEW SUMMARY

The developing field of molecular epidemiology is intimately tied to advances in our understanding of the mechanisms of multistage carcinogenesis. The ability to measure carcinogen exposures, carcinogen metabolites, the macromolecular adducts, and the mutational spectra of these agents are currently being applied to human studies. Advances in the tools used in molecular biology are also being applied to biological samples from exposed individuals. In total, these data will provide the scientific foundation for the epidemiological investigations into the causes of human cancer at organ sites where etiological agents have yet to be described.

Interindividual Variation in Biotransformation of Carcinogens: Basis and Relevance

F. Peter Guengerich
Department of Biochemistry and Center in Molecular Toxicology
Vanderbilt University
Nashville, Tennessee

1. INTRODUCTION

Cancer incidence is not random. Both the overall occurrence and the frequency in specific tissues can show wide variations. In situations where accidental gross exposures to chemical carcinogens have occurred, only some of the individuals develop cancers (Cartwright, 1984a). Genetic influences also predispose individuals to cancers (Knudson, 1985; Hansen and Cavenee, 1987). Shifts in predominant sites of tumors in Japanese people who move to the United States are well known (Dunn, 1975), and groups who abstain from certain foods and alcohol show decreased tumor incidence (Correa, 1981; Weisburger 1987). Isolated geographical regions in both developing and industrialized countries can exhibit unusually high incidences of certain tumors (Yang, 1980).

What are some of the possible reasons for the observed differences? Many can be considered, including diet, exposure to carcinogens, hormonal balance, differences in metabolism, and defective proto-oncogenes and DNA repair. In this chapter, interindividual differences in the metabolism of chemical carcinogens will be considered as a possible basis for variation in tumor incidence.

Although many readers may feel that this hypothesis is already proven, there are really few definitive experiments in humans (Guengerich, 1988a). What are some of the reasons for considering the hypothesis? First, there are animal studies in which manipulation of levels of enzymes involved in the processing of carcinogens has been found to alter tumor incidence (see Guengerich, 1988a, and references therein). These manipulations were usually done either using genetics or enzyme induction or inhibition, and, while the experiments cannot be completely controlled, the work does collectively indicate the importance of rates of carcinogen activation. Another piece of evidence comes from studies on drug metabolism in humans. The same enzymes are involved in the processing of both carcinogens and pharmaceutical agents, and variations are well-documented in the processing of drugs in humans. Furthermore, profound physiological effects can be observed *in vivo* when an individual is metabolizing a drug at an unusually rapid or slow rate (Küpfer and Preisig, 1983). Finally, the point should be made that DNA-adduct levels vary considerably among individuals, as evidenced by available information from studies in China and elsewhere (Umbenhauer *et al.*, 1985; Groopman *et al.*, 1985). In these field studies, control for variations in exposure and variations in DNA repair is difficult, but altered rates of adduct formation need to be considered.

As studies with experimental animals have developed, at least four bases of alteration of enzyme activity are recognized: (1) cofactor supply, (2) polymorphic defects in structural genes, (3) induction (including variations in regulatory elements and the presence of inducing agents), and (4) direct enzyme inhibition and stimulation. Point (1) will not be discussed in detail here (see Thurman *et al.*, 1987); point (4) is considered briefly under some of the examples. Points (2) and (3) are the more likely reasons for seeing variations in the enzymes under consideration, although it should be emphasized that our understanding of the basis of variation in most of the enzymes to be discussed is rather rudimentary at this time.

With this introduction, the major enzymes involved in the biotransformation of chemical carcinogens will be considered, with the emphasis being on humans and the relationship to metabolism of known procarcinogens.

2. EXAMPLES OF ENZYME VARIATION AND POSSIBLE IMPLICATIONS

2.1. Cytochrome P-450

2.1.1. P-450$_{NF}$ (III A4)

P-450$_{NF}$ was isolated in this laboratory on the basis of its catalytic activity for oxidation of nifedipine, giving rise to the designation ''NF'' (Guengerich *et al.*, 1986a). Similar preparations in the literature include P-450HLFa (Kitada *et al.*, 1987) and P-450HLp (Watkins *et al.*, 1985). cDNA cloning experiments indicate a multigene family and some caveats must be given concerning certain results because of the potential existence of very closely related proteins (see Bork *et al.*, 1989).

In this section, individual P-450 enzymes are denoted by designations in the original literature or else equated to preparations used in the author's laboratory when possible, for purposes of simplification; for reference see Distlerath and Guengerich, 1987 and Guengerich, 1987b, 1989. Recently a tentative nomenclature for the P-450 genes has been published (Nebert *et al.*, 1987) and revised (Nebert *et al.*, 1989). The term "isozymes" is avoided here. Although this term has been used extensively in the past, it technically refers to situations in which a single reaction is catalyzed by several proteins. Much of the attention is now focused on more selective reactions of individual proteins (P-450 and others), and the term is often not very appropriate.

Levels of hepatic P-450$_{NF}$ vary at least 20-fold (Guengerich *et al.*, 1986b; Guengerich, 1988a), and considerable *in vivo* variation in nifedipine oxidation is also seen (Kleinbloesem *et al.*, 1984; Schellens *et al.*, 1988). Recently this enzyme or a close relative has been detected in the human small intestine (Watkins *et al.*, 1987). Several lines of both *in vitro* and *in vivo* evidence suggest that the hepatic enzyme is probably inducible by compounds such as barbiturates, macrolide antibiotics, and glucocorticoids (see Watkins *et al.*, 1985). P-450$_{NF}$ appears to oxidize a number of substrates, including drugs—nifedipine and other dihydropyridine calcium channel blockers (Guengerich *et al.*, 1986a; Böcker and Guengerich, 1986), quinidine (Guengerich *et al.*, 1986b), erythromycin (Watkins *et al.*, 1985), and cyclosporin A (Daujat *et al.*, 1988)—and steroids—testosterone (Guengerich *et al.*, 1986a), androsterone, progesterone (Waxman *et al.*, 1988), cortisol (Ged *et al.*, 1989), and 17β-estradiol (Guengerich *et al.*, 1986a). Elevated levels of P-450$_{NF}$ can enhance the 2-hydroxylation of the oral contraceptive 17α-ethynylestradiol and decrease its therapeutic effectiveness to cause unintended pregnancy (Bolt *et al.*, 1975; Guengerich, 1988a).

Of particular interest is the recent discovery that the bioactivation of aflatoxins in human liver microsomes is catalyzed primarily by P-450$_{NF}$, as indicated by *in vitro* experiments involving enzyme reconstitution, immunoinhibition, and inhibition and stimulation with chemicals. A high degree of correlation was observed (among various human liver samples) for nifedipine oxidation (or P-450$_{NF}$ protein levels) and rates of bioactivation of aflatoxin B$_1$ (AFB$_1$), aflatoxin G$_1$ (AFG$_1$), and sterigmatocystin (STG) (Table 1) (Shimada and Guengerich, 1989). P-450$_{NF}$ also appears to be the main human enzyme involved in the bioactivation of 6-aminochrysene (6-AC) and tris-(2,3-dibromopropyl)phosphate (Tris-BP) (Table 1), as judged by results obtained using the induction of the *umu* response in the *Salmonella typhimurium* TA1535/

TABLE 3–1. Pro-Carcinogens Activated by Human P-450 Enzymes

P-450$_{PA}$*	P-450$_{NF}$*, †, ‡	P-450j§
2-AA	6-AC	*N*-Benzyl-*N*-methylnitrosamine
2-AAF	AFB$_1$	*N*-Butyl-*N*-methylnitrosamine
4-ABP	AFG$_1$	*N*,*N*-Diethylnitrosamine
2-AF	BP-7,8-diol	DMN
Glu-P-1	BFA-9,10-diol	
Glu-P-2	DMBA-3,4-diol	
IQ	STG	
MeIQ	Tris-BP	
MeIQx		
Trp-P-2		

*Shimada *et al.* (1989a).
†Shimada and Guengerich (1989).
‡Shimada *et al.* (1989b).
§Results of Yoo *et al.* (1988).

pSK1002 system (with a chimeric *umu*C' '*lac*Z plasmid construct) (Shimada *et al.*, 1989a). Of particular interest is the finding that P-450$_{NF}$ also appears to be the major human liver enzyme involved in the oxidative bioactivation of the dihydrodiol derivatives of at least three carcinogenic polycyclic aromatic hydrocarbons [*trans*-7,8-dihydroxy-7,8-dihydrobenzo(*a*)pyrene (BP-7, 8-diol), *trans*-9,10-dihydroxy-9,10-dihydrobenzo(*b*)fluoranthene (BFA-9,10-diol), and *trans*-3,4-dihydroxy-3,4-dihydro-7,12-dimethylbenz(*a*)anthracene (DMBA-3,4-diol)] (Shimada *et al.*, 1989b). These results are of particular interest in that the orthologous rodent proteins are not involved in these responses. In the case of P-450$_{NF}$, none of the observed catalytic activities towards carcinogens (Table 1) would have been expected on the basis of studies in experimental animals. Flavonoids such as 7,8-benzoflavone and others have been known to stimulate some human P-450 enzymes, and P-450$_{NF}$ appears to be a principal target of this action (Shimada and Guengerich, 1989). The enzyme also becomes inhibited during the oxidation of some antibiotics (Watkins *et al.*, 1985). P-450$_{NF}$ function can be monitored noninvasively in humans by measuring urinary excretion of nifedipine products (Schellens *et al.*, 1988) or levels of endogenously derived 6β-hydroxycortisol (Ohnhaus and Park, 1979; Ged *et al.*, 1989).

2.1.2. P-450$_{MP}$ (II C)

At least four human liver P-450 enzymes are coded by this complex multigene family (Ged *et al.*, 1988), and at least two of the proteins show catalytic activity for the 4'-hydroxylation of the *S*-enantiomer of the antiepileptic drug mephenytoin. About 3% of the white population show the "poor-metabolizer" *in vivo* phenotype, with little 4'-hydroxylation and a large "hydroxylation index;" in the Japanese population the poor-metabolizer incidence is >20% (Nakamura *et al.*, 1985). To date there is little evidence of association of carcinogen metabolism with the proteins in this family. Some reports suggest that P-450$_{MP-1}$ may catalyze benzo(a)pyrene hydroxylation, although there is little evidence to indicate a major role in this reaction in human liver at this time (Shimada *et al.*, 1986; Kawano *et al.*, 1987). Kaisiary *et al.* (1987) did not find any relationship between mephenytoin hydroxylation phenotype and bladder cancer incidence. Nevertheless, the differential distribution of the poor-metabolizer phenotype in some racial groups makes the enzymes attractive candidates for further study. The molecular basis of the poor-metabolizer phenotype is unknown, although several pieces of evidence favor mutations in the structural genes as opposed to variation in mRNA levels (Ged *et al.*, 1988). Although the polymorphism is genetic, mephenytoin hydroxylation is also inducible by rifampicin (Zhou *et al.*, 1990) and perhaps other compounds.

2.1.3. P-450j (II E1)

Rat P-450j II E1 is induced by administration of ethanol, acetone, isoniazid, pyrazole, or certain other compounds. The human orthologue (also termed P-450j) is coded by a single-gene family (Umeno *et al.*, 1988), and some evidence suggests that ethanol may also induce P-450j in humans (Wrighton *et al.*, 1987). Levels of the enzyme have been found to vary at least 50-fold in individual livers (Yoo *et al.*, 1988). The enzyme oxidizes ethanol and other primary alcohols to the corresponding aldehydes (Wrighton *et al.*, 1987), but of perhaps more concern are the one-electron reduction of CCl_4 to the trichloromethyl radical and the N-demethylation of N,N-dimethylnitrosamine (DMN), which generate a methylating species (alkyldiazohydroxide) that presumably leads to tumor initiation. At low, relevant N,N-dimethylnitrosamine substrate concentrations P-450j clearly appears to be the major enzyme involved (Yoo *et al.*, 1988; Wrighton *et al.*, 1987). The specificity of P-450 enzymes towards all carcinogenic N-nitrosamines is not known, although the results of Yoo *et al.* (1988) suggest that all of the dealkylation reactions involving N,N-diethylnitrosamine, N-nitrosobenzylmethylamine, and N-nitrosobutylmethylamine are catalyzed in part by P-450j, as judged by correlations among individual liver samples (Table 1).

N-Nitrosamines occur widely in the environment and are also formed endogenously. Of particular interest are the synergistic relationships shown between ethanol consumption and tobacco use in the development of certain tumors, and a hypothesis to consider is that ethanol induces P-450j, which then activates nitrosamines more rapidly. Further experiments will be necessary to explore this possibility.

2.1.4. P-450$_{DB}$ (II D6)

P-450$_{DB}$ catalyzes the 4-hydroxylation of the antihypertensive medication debrisoquine, the first drug to be extensively studied in terms of genetic polymorphism for mixed-function oxidation (Mahgoub *et al.*, 1977; Eichelbaum *et al.*, 1979). The protein (Distlerath *et al.*, 1985; Gut *et al.*, 1986; Birgersson *et al.*, 1986) and gene (Gonzalez *et al.*, 1988) have been characterized, and the basis of defective metabolism appears to be mutations in the gene which give rise to abnormal mRNA that cannot undergo the usual splicing and translation (Gonzalez *et al.*, 1988), at least in the cases studied to date.

Several groups have reported that phenotypic poor metabolizers of debrisoquine appear to be at decreased risk from several cancers, including lung (Ayesh *et al.*, 1984; Weston *et al.*, 1988), bladder (Kaisary *et al.*, 1987), liver (Idle *et al.*, 1981), and head and neck (Roots *et al.*, 1988). Further analysis can be used to show better correlations with specific types of cancers within each category (see Weston *et al.*, 1988 and Kaisary *et al.*, 1987). One of the difficulties in advancing this approach is that the etiology of the tumors under consideration is unknown. Even in the lung tumors (Ayesh, 1984) where smoking was identified as a contributing factor, the question of which chemicals are most carcinogenic in tobacco smoke arises. It is possible that the association with tumor incidences may be due to genetic linkage with oncogenes, tumor suppressor genes, or other factors and not even have a metabolic basis—further experiments are necessary to test the general hypothesis.

A number of drugs have been identified as substrates for P-450$_{DB}$ (Distlerath *et al.*, 1985; Guengerich, 1989; Gut *et al.*, 1986; Küpfer and Preisig, 1983), but none of the carcinogens examined to date appear to be substrates (Table 1) (Wolff *et al.*, 1985; Shimada *et al.*, 1989a). The association of poor metabolism of debrisoquine with liver cancer was postulated to be related to aflatoxin consumption (Idle *et al.*, 1981), but evidence now links the bioactivation of aflatoxin B$_1$ and some related compounds to P-450$_{NF}$ instead of P-450$_{DB}$ (Wolff *et al.*, 1985; Plummer *et al.*, 1986; Shimada and Guengerich, 1989). P-450$_{DB}$ is very effectively inhibited by quinidine and other related alkaloids (Otton *et al.*, 1984), although quinidine itself is a substrate for P-450$_{NF}$ and not P-450$_{DB}$ (Guengerich *et al.*, 1986b).

2.1.5. P-450$_{PA}$ (I A2)

P-450$_{PA}$ is the high-affinity phenacetin *O*-deethylase (Distlerath *et al.*, 1985). A number of pieces of indirect evidence (Guengerich, 1989; Wrighton *et al.*, 1986) suggest that this is the product of the human P450IA2 gene, which has been characterized in several laboratories (Quattrochi *et al.*, 1986; Jaiswal *et al.*, 1987). The rat liver orthologue is known to be involved in the N-oxidation of a number of carcinogenic aromatic amines (Yamazoe *et al.*, 1984; Kadlubar and Hammons, 1987); correlation studies suggest that P-450$_{PA}$ is involved in the activation of 4-aminobiphenyl to genotoxic derivatives, but not 4,4'-methylene-bis-(2-chloroaniline) (Butler *et al.*, 1989a, b). Studies with the *umu* gene activation assay indicate that the following are all activated by P-450$_{PA}$ in the human liver (Table 1) (Shimada *et al.*, 1989a): 2-aminoanthracene (2-AA), 2-acetylaminofluorene (2-AAF), 4-aminobiphenyl (4-ABP), 2-aminofluorine (2-AF), 2-amino-6-methyldipyrido[1,2-*a*:3',2'-*d*] imidazole (Glu-P-1), aminodipyrido[1,2-*a*:3',2'-*d*]imidazole (Glu-P-2), 2-amino-3-methylimidazo[4,5-*f*]quinoline (IQ), 2-amino-3,5-dimethylimidazo[4,5-*f*]quinoline (MeIQ), 2-amino-3,8-dimethylimidazo[4,5-*f*]quinoxaline (MeIQx), and 3-amino-1-methyl-5*H*-pyrido[4,3-*b*]indole (Trp-P-2). Surprisingly, 3-amino-1,4-dimethyl-5*H*-pyrido[4,3-*b*]indole (Trp-P-1) appears to be activated by another enzyme (Shimada *et al.*, 1989a), even though it differs from Trp-P-2 by only one methyl group, and the rat orthologue P-450$_{ISF-G}$ activates both compounds (Yamazoe *et al.*, 1984).

This enzyme appears to be induced in human liver by cigarette smoking, as judged by both *in vivo* (Pantuck *et al.*, 1974) and *in vitro* studies (Sesardic *et al.*, 1988). Pantuck *et al.* (1976) also demonstrated that (*in vivo*) phenacetin clearance can be accelerated by ingestion of char-broiled food. This P-450 enzyme also appears to be involved in the oxidative demethylation of the methyl xanthines caffeine and theophylline, and these compounds provide noninvasive assays (Campbell *et al.*, 1987; Grant *et al.*, 1987; Butler *et al.*, 1989b). It should be pointed out that, in *in vitro* studies, the enzyme is extremely sensitive to inhibition by 7,8-benzoflavone, while P-450$_{NF}$ is stimulated by the same compound (*vide supra*). Another P-450 enzyme, P$_1$-450, is also sensitive to inhibition by 7,8-benzoflavone but current evidence indicates that this enzyme is primarily extrahepatic in humans while P-450$_{PA}$ is essentially only hepatic (*vide infra*).

2.1.6. P$_1$-450 (I A1)

Historically the interest in this particular P-450 enzyme has been great since the discovery of aryl hydrocarbon hydroxylase activity in animal cells. Kellermann *et al.* (1973a, 1973b) reported that cigarette smokers with lung cancer showed greater polycyclic hydrocarbon inducibility of aryl hydrocarbon hydroxylase activity in their (cultured) lymphocytes. This report stimulated considerable effort to use peripheral cells as test systems for determining the susceptibility of individuals to lung cancer. However, it has been difficult to establish whether the previously reported trimodal variation in inducible activity really exists, for many factors affect the assay (Paigen *et al.*, 1981). A correlation between the incidence of lung cancer in smokers and the level of *basal* lymphocyte aryl hydrocarbon hydroxylase activity has been reported (Kouri *et al.*, 1982), although it cannot be determined if one trend is the result of the other. The level of P$_1$-450 mRNA is correlated with aryl hydrocarbon hydroxylase activity in lymphocytes of individual humans (Jaiswal *et al.*, 1985b), and basal lymphocyte and lung aryl hydrocarbon hydroxylase levels are also correlated with each other (Kärki *et al.*, 1987). Conclusions concerning genetic regulation are controversial (Paigen *et al.*, 1977).

Although many studies have been done on aryl hydrocarbon hydroxylase in humans, the enzyme itself has never been purified. Jaiswal *et al.* (1985a) used a human breast carcinoma cell library to isolate a cDNA and establish its sequence, which is 80% identical to the coding for the orthologous mouse enzyme. The amino acid sequence is most similar to that of P-450$_{PA}$ (*vide supra*); antibodies raised to one of the two proteins (or the rat or rabbit orthologues) often recognize both proteins (Guengerich *et al.*, 1982, Guengerich, 1987a, 1987b), and the genes are both placed in the IA family (Nebert *et al.*, 1987, 1989). However, as in rats, the evidence suggests that P-450$_{PA}$ is essentially only a hepatic enzyme. P$_1$-450, however, appears to be primarily an extrahepatic enzyme. For instance, antibodies directed against the rat orthologue inhibit aryl hydrocarbon hydroxylase activity in extrahepatic tissue extracts but not in liver (Fujino *et al.*, 1982). For many years it has been known that cigarette smoking induces aryl hydrocarbon hydroxylase activity in extrahepatic sites such as placenta, lymphocytes, skin, and lung but not in liver (Pelkonen *et al.*, 1986), although smoking does enhance hepatic phenacetin *O*-deethylation (by inducing P-450$_{PA}$). Immunoblotting studies with human liver show only a polypeptide band corresponding in mobility to rat P-450$_{ISF-G}$ (orthologue of P-450$_{PA}$) but not rat P-450$_{\beta NF-B}$ (orthologue of P$_1$-450) in most samples. Finally, oligonucleotide probes specific for P$_1$-450 did not hybridize to substantial amounts of hepatic mRNA in human liver samples, even those obtained from cigarette smokers (Cresteil *et al.*, 1988).

Thus, if P$_1$-450 has a major role in influencing risk from carcinogens, it will probably be in cases involving extrahepatic carcinogenesis and where hepatic metabolism does not have an overwhelming influence. The induction paradigm of aryl hydrocarbon hydroxylase in livers of animals may not be applicable to humans except in situations where gross accidental exposures occur (e.g., dioxins, polyhalogenated biphenyls). The literature tends to show enhanced extrahepatic tumors due to polycyclic hydrocarbons in genetically responsive mice (Levitt *et al.*, 1979; Nebert, 1981); however, the induction of enzymes by polycyclic hydrocarbons exerts a protective effect for hepatocarcinogenesis and the formation of extrahepatic tumors by polycyclic hydrocarbons and aromatic amines (Richardson *et al.*, 1952; Miller *et al.*, 1958; Tawfic, 1965; Anderson and Seetharam, 1985). The relationship between aryl hydrocarbon hydroxylase and lung cancer in humans has also been questioned (Paigen *et al.*, 1977). Work by De Flora *et al.* (1987) has suggested that smoking actually shifts the enzymatic balance in favor of pulmonary detoxication of procarcinogens.

The catalytic specificity of human P_1-450 has not been carefully defined. Conversion of benzo(a)pyrene to fluorescent phenols (3-,9-) appears to occur (Fujino *et al.*, 1982; Shimada *et al.*, 1989b). Although the protein sequence is similar to the animal orthologues, it should be pointed out that the orthologous rat and mouse enzymes (98% identical) differ markedly in the oxidation of warfarin (Kaminsky *et al.*, 1984).

2.1.7. Other P-450 Enzymes

In addition to the P-450 enzymes discussed here, others exist in humans and have been characterized in liver microsomes [P-450$_{AA}$, P-450$_9$, P-450$_{10}$, pHP450(1)] and the microsomes (P-450arom, P-450$_{c-21}$, P-450$_{17\alpha}$) and mitochondria (P-450scc, P-450$_{11\beta}$) of steroidogenic tissues (for a review see Guengerich, 1989). However, there is little information available concerning the metabolism of carcinogens by these enzymes and they will not be discussed here. Recently Philpot and his associates (Philpot, 1988) have isolated and characterized cDNA clones from a human lung library that appear to be most related to (the sequences of) rabbit P-450 2 and P-450 5, two enzymes that do have roles in the metabolism of carcinogens in rabbit lung. These enzymes appear to be expressed primarily in lung and not liver.

2.2. Flavin-Containing Mono-oxygenase

This enzyme is a flavoprotein that catalyzes the mono-oxygenation of amines and sulfur compounds. Most characterization studies have been done with the hog liver enzyme by Ziegler and his associates (Ziegler, 1988). The evidence now suggests that multiple forms of this enzyme, with distinct physical properties and catalytic specificities, can exist within a single species (rabbit) and show distinct tissue distribution. It is not clear whether or not multiple forms of the enzyme exist in humans, although preliminary fractionation work in the author's laboratory would suggest that distinct proteins can be separated (Guengerich *et al.*, 1988).

The level of the enzyme in different human liver samples is known to vary considerably (Gold and Ziegler, 1973), as judged by rates of *N,N*-dimethylaniline *N*-oxidation. In addition, the levels of a protein detected on immunoblots with rabbit antihog flavin-containing mono-oxygenase vary considerably among humans (Guengerich *et al.*, 1988). Rates among individuals can vary considerably in *N*-oxidation of trimethylamine, probably as a reflection of levels of the flavin-containing mono-oxygenase (Higgins *et al.*, 1972; Al-Waiz *et al.*, 1987). A small group (1%–2%) of people have very little ability to oxidize endogenously produced (or exogenous) trimethylamine and are plagued by a characteristic "fish oil" smell. To date no attempts have been made to link this apparent genetic polymorphism to cancer incidence, although the flavin-containing monooxygenase is known to be involved in the oxidation of a number of carcinogens (in some cases other enzymes—P-450 enzymes, peroxidases—may catalyze the same reactions). Flavin-containing monooxygenase catalyzes the oxygenation of several procarcinogenic *N*-methyl arylamines [which may be formed *in vivo* by *N*-methylation (Ziegler *et al.*, 1988)], 2-aminofluorene (Frederick *et al.*, 1982), 2-naphthylamine (Hammons *et al.*, 1985), hydrazines (Ziegler, 1988), and thioacetamide (Chieli and Malvaldi, 1985). In addition, a number of carcinogens are also oxidized to products with reduced capability for tumor initiation, including pyrrolizidine alkaloids (Williams *et al.*, 1989) and nicotine (Damani *et al.*, 1988).

2.3. Peroxidases

Mammalian peroxidases are capable of activating some procarcinogens, particularly amines. The best-studied system is prostaglandin synthase, which has been demonstrated to catalyze the arachidonate-dependent oxidation of benzo(*a*)pyrene-7,8-dihydrodiol and aflatoxin B$_1$ to epoxides, apparently the ultimate carcinogens (Battista and Marnett, 1985), and of several aromatic amines to mutagenic products (Krauss and Eling, 1984; Marnett, 1985). The chemical and stereochemical nature of certain DNA adducts specifically formed by this pathway (as opposed to P-450) has permitted estimation of the extent of its participation in DNA-adduct formation: in mouse skin a significant amount of some benzo(*a*)pyrene-7,8-dihydrodiol-derived adducts are formed via the prostaglandin synthase reaction (Marnett, 1985); in dog bladder benzidine adducts are formed via this route (Kadlubar *et al.*, 1987). It is possible that such peroxidase reactions may have substantial contributions in extrahepatic tissues; the extent of variation of prostaglandin synthase in humans has not been examined.

2.4. Acetyltransferase

The acetylation polymorphism, which is genetic in its nature, was known before the oxidative polymorphisms involving P-450 enzymes. *N*-Acetylation is involved in the metabolism of many drugs and carcinogenic aromatic amines. Epidemiological studies suggest that, with respect to aromatic amines, slow acetylators are more likely to contract bladder cancer but are at decreased risk for colorectal cancer (Cartwright, 1984b; Ilett *et al.*, 1987; Hanssen *et al.*, 1985; Mommsen *et al.*, 1985; Evans *et al.*, 1983; Karakaya *et al.*, 1986). However, the difference in cancer incidence for the two phenotypes in either tissue is not overwhelming. The point should be made that the etiology is uncertain in these studies. Philip *et al.* (1988) reported that the *N*-acetylation phenotype did not appear to contribute to lung cancer incidence. Levels of acetyltransferase varied over a 60-fold range among

35 liver cytosol samples examined, and sulfamethazine *N*-acetyltransferase, 2-aminofluorene *N*-acetyltransferase, and acetyl CoA-dependent *O*-deacetylation of *N*-hydroxy-2-aminofluorene (to form DNA adducts) activities were all highly correlated with each other (Flammang *et al.*, 1987). All three activities, plus *N,O*-acetyltransferase activity, seem to be catalyzed by the same enzyme (King and Glowinski, 1983).

It appears that phenotypic rapid acetylators are likely to be at increased risk to aromatic amine-induced cancers in those tissues containing appreciable levels of *N*-hydroxy arylamine *O*-acetyltransferase, such as colon. The carcinogens under consideration here include a wide variety of primary aromatic amines. Finally, it should be emphasized that the *N,O*-acyl transfer, which is thought to generate the ultimate carcinogen, must follow an oxidation step, and sulfation (*vide infra*) can also be a competing reaction.

2.5. Glutathione *S*-Transferases

The glutathione *S*-transferases can be grouped into four gene families, which are expressed in experimental animals and humans. The α–ε group consists of basic proteins, μ is a neutral class, and the π enzyme is acidic. These are all cytosolic enzymes—another enzyme, the microsomal glutathione *S*-transferase, is quite distinct. The α–ε and the μ groups appear to be coded by multigene families, and the complexity in humans is not yet known. The α–ε, μ, and microsomal glutathione *S*-transferases are found in liver. In addition, the α–ε class has been detected in skin (Del Boccio *et al.*, 1987), and μ is expressed in leukocytes (Seidegård *et al.*, 1987). The π enzyme is normally found in placenta and in several extrahepatic tissues; in humans, its presence in liver is associated with tumor development. Its presence in neoplastic tissue can help impart resistance to alkylating chemotherapeutic agents (Batist *et al.*, 1986). Fetuses have α–ε and π enzymes but not μ enzymes (Guthenberg *et al.*, 1986).

The significance of the glutathione *S*-transferases lies in the reactions they catalyze. Usually they provide a protective role in that they catalyze the conjugation of potentially deleterious electrophiles with glutathione as a first step in elimination. In this regard, the catalytic specificity of the rat liver enzymes has been extensively studied (see Mantle *et al.*, 1987, and chapters therein). Human μ and π enzymes are most efficient in conjugating benzo(*a*)pyrene-7,8-dihydrodiol-9,10-oxide (Robertson *et al.*, 1986), and the μ enzyme also conjugates benzo(*a*)pyrene-4,5-oxide and styrene-7,8-oxide (Dostal *et al.*, 1988). However, some procarcinogens can also be *activated* by conjugation with glutathione. The α and ε enzymes catalyze the conjugation of ethylene dibromide, which leads to the attack of a resulting episulfonium ion on DNA (Koga *et al.*, 1986; Cmarik *et al.*, 1990). Glutathione conjugation is also involved in the genotoxicity of CH_2Cl_2—while the nature of the human glutathione transferases involved in the reaction is unknown, human liver activity has been assayed (and found to vary), and the results led to a reassessment of the risks of low doses of CH_2Cl_2 to humans (Reitz *et al.*, 1989).

While levels of individual glutathione *S*-transferases are known to vary among people (Warholm *et al.*, 1983), and hypotheses linking enzyme activity to cancer risk have not been as common as with the P-450 enzymes and acetyltransferases. However, levels of individual proteins can be determined in liver samples using techniques of electrophoresis and immunoelectrophoresis (Hussey *et al.*, 1986) and chromatography (Ostlund-Farrants *et al.*, 1987). In the latter reference, high-performance liquid chromatography was used to resolve two human α proteins. Levels of the major enzymes have been measured in samples of human liver, and no correlation has been found among levels of the various glutathione *S*-transferases among themselves or with any of the human liver P-450 enzymes examined (Ketterer and Guengerich, 1988), arguing against coordinate regulation. Human leukocytes express transferase μ (Seideegård *et al.*, 1987) and it may be possible to do phenotyping using this approach if the expression of the enzyme can be shown to be similar in leukocytes and in major tissues such as liver. A similar approach could be used with skin punches (Del Boccio *et al.*, 1987). The development of drugs that can be conjugated by specific glutathione *S*-transferases to provide noninvasive assays would be useful to phenotyping efforts and hypotheses regarding association with risk.

2.6. *N*-Methyl Transferases

N-Methylation *per se* probably does not activate any proximate carcinogens; however, Ziegler *et al.* (1988) have shown that a rabbit *N*-methyl transferase can convert primary arylamines such as benzidine and 4-aminobiphenyl to the corresponding secondary amines. These secondary amines can serve as substrates for flavin-containing mono-oxygenase, while the primary amines do not (Kadlubar *et al.*, 1976). The resulting hydroxylamines can be esterified, and departure of the leaving group then leaves a formal nitrenium ion to react with DNA.

Several *N*-methyl transferases have been characterized in experimental animal systems (Ansher and Jakoby, 1986). In humans an *N*-methyltransferase that catalyzes the $^\tau N$-methylation of histamine has been characterized in red blood cells. Scott *et al.* (1988) have reported that inheritance plays a role in the variation in this activity seen among humans, although the distribution was concluded to be unimodel. Most levels of the activities were within a factor of 5.

2.7. Sulfotransferases

Sulfotransferases act to improve the water solubility and excretion of alcohols, but the addition of a sulfate also has the effect of introducing a good leaving group onto unstable compounds such as hydroxylamines. In a classic series of experiments, the Millers and their associates were able to show that mice genetically deficient in sulfation were less prone to liver tumors from 1'-hydroxyestragole and that the sulfation inhibitor pentachlorophenol could also lower the tumor incidence (Boberg *et al.*, 1987). They were also able to identify a major role for sulfation in the activation of 2-acetylaminofluorene to an ultimate carcinogen (Lai *et al.*, 1985, 1987).

A number of these enzymes are found in each animal species. Human tissues contain at least two forms of phenol sulfotransferase, with characteristic substrate selectivities (Weinshilboum, 1986). The platelet activity appears to reflect the enzymatic capacity in other tissues, as judged by correlations of platelet activity with urinary excretion of sulfate products of drugs. The levels of both forms of sulfotransferase appear to vary by more than an order of magnitude among individuals (Weinshilboum, 1986). Familial variations among the ''P'' or ''TS'' form of phenol sulfotransferase have been reported (Van Loon and Weinshilboum, 1984), but further studies on genetic polymorphism and its possible relevance to carcinogenic risk remain to be done.

2.8. UDP Glucuronosyl Transferases

In general the UDP glucuronosyl transferases are involved in the detoxication of drugs and carcinogens, although in some cases (e.g., aromatic amines) glucuronidation can introduce a good leaving group and favor nitrenium ion formation, especially in acidic environments such as the bladder (Kadlubar *et al.*, 1976). Several UDP glucuronsyl transferase enzymes have been characterized in rat and rabbit liver, and the number of genes coding proteins is probably at least 10 (Jackson *et al.*, 1987). Post-translational modification is also involved. Recently Tephley and his associates have been able to purify UDP glucuronosyl transferases from human liver (Irshaid and Tephley, 1987). However, the extent to which the contribution of variation is a risk factor in carcinogenesis is unknown.

2.9. Epoxide Hydrolase

Three major forms of epoxide hydrolase (or at least three categories) appear to exist: (1) the microsomal epoxide hydrolase with relatively broad specificity, (2) cytosolic epoxide hydrolase, and (3) the microsomal cholesterol-5,6-oxide hydrolase. The first two enzymes have been purified from several species, including humans (Guengerich *et al.*, 1979; Wang *et al.*, 1982). The third enzyme, which hydrolyzes the mutagenic cholesterol-5,6-oxide, a product of lipid peroxidation, has not been purified from any species. Epoxide hydrolases can be considered to have protective roles in that they remove electrophilic epoxides. A situation in which epoxide hydrolase can be considered to be an activating enzyme involves the hydrolysis of benzo(a)pyrene-7,8-oxide to its dihydrodiol, which can then serve as a substrate for P-450-catalyzed oxidation to the ultimate carcinogen, the 7,8-dihydrodiol-9,10-oxide (Wood *et al.*, 1976).

Oesch and his colleagues have measured epoxide hydrolase activities in several human tissues. The specific activity of hepatic microsomal epoxide hydrolase activity [measured using benzo(a)pyrene-4,5-oxide as a substrate] varied 63-fold (among 166 individual biopsy samples) (Mertes *et al.*, 1985). In the same report the specific activity of cytosolic epoxide hydrolase (measured using *trans*-stilbene oxide as a substrate) varied 539-fold. No correlation between the two activities was found. Considerably less interindividual variation was seen in the lung (5-fold) (Oesch *et al.*, 1980) and lymphocyte (4-fold) (Glatt *et al.*, 1980) microsomal epoxide hydrolase activities. Major differences due to smoking, sex, alcohol intake, and presence of tumors were not observed. Any relationship between levels of any of the epoxide hydrolases and tumor susceptibility remains to be established.

3. SUMMARY AND CONCLUSIONS

The view that differences in the enzymatic capacity to bioactivate and detoxicate carcinogens contribute to interindividual variation in cancer risk is attractive and might explain many observations made in the field. Many enzymes have been demonstrated to be capable of activating carcinogens in experimental animals. In humans, the orthologous enzymes have been shown to vary considerably among individuals in terms of *in vivo* disposition of pharmaceutical agents. However, more direct studies in humans are still lacking.

In reviewing the human enzymes involved in carcinogen activation, the candidates most likely to play major roles are three cytochrome P-450 enzymes: P-450$_{PA}$, P-450$_{NF}$, and P-450j. The enzymes play key roles in the activation of known carcinogenic arylamines, aflatoxins, polycyclic aromatic hydrocarbon dihydrodiols, and short-chain alkyl nitrosamines (Table 1). Furthermore, all three of these enzymes are known to be highly inducible, with induction resulting from materials as common as cigarette smoke, ethanol, and barbiturates. Nevertheless, epidemiological implication of roles in cancer risk is still lacking. P$_1$-450 is also suspected of playing an important role in carcinogenesis. The enzyme is highly inducible and its animal orthologues are potent activators of polycyclic hydrocarbons and aromatic amines; however, a number of pieces of direct evidence implicating P$_1$-450 in the metabolism in humans are still missing. Acetyltransferase variation shows a link with tumor susceptibility, although the differences are not so marked to make interpretation of the results simple. The inverse relationship between debriso-quine 4-hydroxylation and incidence of certain cancers remains without a metabolic explanation. Many other enzymes with roles in carcinogen metabo-lism are known to exhibit considerable variation in humans. However, the catalytic specificity has not been defined with the human enzymes in terms of procarcinogenic substrates.

In summary, then, several points should be made. With several situa-tions, opportunities clearly exist for noninvasive measurement of enzyme activities in individuals exposed to known carcinogens. Such information would be of great assistance in considerations concerning risks. Enzyme induction may actually play a more significant role than strictly genetic determination of an enzymatic activity, as judged by the examples available to date. The level of a particular enzyme present may not always be able to be considered in "good" or "bad" terms—that is, the situation is dictated by the chemicals (procarcinogens) that the host encounters. In the setting of exposure to one chemical a high level of a given enzyme might be desirable, while in another setting a lower level would be advantageous. In this regard the literature regarding P-450$_{DB}$, P-450$_{PA}$, and P-450$_{NF}$ should be considered. Furthermore, "poor metabolizers" with regard to acetyltransferase activity are more prone to develop bladder cancer but less prone to develop colon cancer (than extensive metabolizers) (*vide supra*). Finally, in the considera-tion of chemopreventive agents the point should be made that even a single compound can have opposing effects on two parts of a pathway involving multiple steps of bioactivation—7,8-benzoflavone inhibits oxidation of benzo(*a*)pyrene to its 7,8-epoxide but enhances oxidation of the resulting 7,8-dihydrodiol to the ultimate carcinogen, benzo(*a*)pyrene-7,8-dihydrodiol-9,10-oxide (Shimada *et al.*, 1989b). All of these considerations suggest that the enzymatic activation of chemical carcinogens is a complex situation in humans; however, understanding the paradigm can lead to a full assessment of the contributions of individual chemicals to human cancer and may lead to productive approaches of intervention.

ACKNOWLEDGMENTS

The research in the author's laboratory has been supported in part by U.S. Public Health Service Grants No. CA 44353 and No. ES 00267.

REFERENCES

M. Al-Waiz, R. Ayesh, S. C. Mitchell, J. R. Idle, and R. L. Smith, A genetic polymorphism of the *N*-oxidation of trimethylamine in humans, *Clin. Pharmacol. Ther.* **42**, 588–594 (1987).

L. M. Anderson and S. Seetharam, Protection against tumorigenesis by 3-methylcholanthrene in mice by β-naphthoflavone as a function of inducibility of methylcholanthrene metabolism, *Cancer Res.* **45**, 6384–6389 (1985).

S. S. Ansher and W. B. Jakoby, Amine N-methyltransferases from rabbit liver, *J. Biol. Chem.* **261**, 3996–4001 (1986).

R. Ayesh, J. R. Idle, J. C. Ritchie, M. J. Crothers, and M. R. Hetzel, Metabolic oxidation phenotypes as markers for susceptibility to lung cancer. *Nature* **312**, 169–170 (1984).

G. Batist, A. Tulpule, B. K. Sinha, A. G. Katki, C. E. Myers, and K. H. Cowan, Overexpression of a novel anionic glutathione transferase in multidrug-resistant human breast cancer cells, *J. Biol. Chem.* **261**, 15544–15549 (1986).

J. R. Battista and L. J. Marnett, Prostaglandin H synthase-dependent epoxidation of aflatoxin B₁. *Carcinog.* **6**, 1227–1229 (1985).

C. Birgersson, E. T. Morgan, H. Jörnvall, and C. von Bahr, Purification of a desmethylimipramine and debrisoquine hydroxylating cytochrome P-450 from human liver, *Biochem. Pharmacol.* **35**, 3165–3166 (1986).

E. W. Boberg, A. Liem, E. C. Miller, and J. A. Miller, Inhibition by pentachlorophenol of the initiating and promoting activities of 1'-hydroxysafrole for the formation of enzyme-altered foci and tumors in rat liver, *Carcinog.* **8**, 531–539 (1987).

H. M. Bolt, H. Kappas, and M. Bolt, Effect of rifampicin treatment on the metabolism of oestradiol and 17α-oestradiol by human liver microsomes, *Eur. J. Clin. Pharmacol.* **8**, 301–307 (1975).

R. W. Bork, T. Muto, P. H. Beaune, P. K. Srivastava, R. S. Lloyd, and F. P. Guengerich, Characterization of mRNA species related to human liver cytochrome P-450 nifedipine oxidase and the basis of regulation of catalytic activity, *J. Biol. Chem.,* **264**, 910–919 (1989).

R. H. Böcker and F. P. Guengerich, Oxidation of 4-aryl- and 4-alkyl-substituted 2,6-dimethyl-3,5-bis-(alkoxycarbonyl)-1,4-dihydropyridines by human liver microsomes and immunochemical evidence for the involvement of a form of cytochrome P-450, *J. Med. Chem.* **29**, 1596–1603 (1986).

M. A. Butler, F. P. Guengerich, and F. F. Kadlubar, Metabolic oxidation of the carcinogens 4-aminobiphenyl and 4,4'-methylene-*bis*-(2-chloroaniline) by human hepatic microsomes and by purified rat hepatic cytochrome P-450 monooxygenases, *Cancer Res.,* **49**, 25–31 (1989a).

M. A. Butler, M. Iwasaki, F. P. Guengerich, and F. F. Kadlubar, Human cytochrome P-450ₚₐ (I A2), the phenacetin O-deethylase, is primarily responsible for the hepatic 3-demethylation of caffeine and N-oxidation of carcinogenic arylamines, *Proc. Natl. Acad. Sci. USA* **86**, 7696–7700 (1989b).

M. E. Campbell, D. M. Grant, T. Inaba, and W. Kalow, Biotransformation of caffeine, paraxanthine, theophylline, and theobromine by polycyclic aromatic hydrocarbon-inducible cytochrome(s) P-450 in human liver microsomes, *Drug Metab. Disp.* **15**, 237–249 (1987).

R. A. Cartwright, Cancer epidemiology. In: *Chemical Carcinogens,* 2nd Ed., ed. C. E. Searle, Vol. 1, pp. 1–39. American Chemical Society, Washington, D.C., pp. 1–39 (1984a).

R. A. Cartwright, Epidemiological studies on *N*-acetylation and C-center oxidation in neoplasia. In: *Genetic Variability in Responses to Chemical Exposure,* eds.

G. S. Omenn and H. V. Gelboin, Banbury Reports, Vol. 16, Cold Spring Harbor Laboratory Press, New York, pp. 359–368 (1984b).

E. Chieli and G. Malvaldi, Role of the P-450 dependent and FAD-containing monooxygenases in the bioactivation of thioacetamide, thiobenzamide and their sulphoxides, *Biochem. Pharmacol.* **34**, 395–396 (1985).

J. C. Cmarik, P. B. Inskeep, M. J. Meredith, D. J. Meyer, B. Ketterer, and F. P. Guengerich, Selectivity of rat and human glutathione *S*-transferases in activation of ethylene dibromide by glutathione conjugation and DNA binding and induction of unscheduled DNA synthesis in human hepatocytes, *Cancer Res.*, **50**, 2747–2752 (1990).

P. Correa, Epidemiological correlations between diet and cancer frequency, *Cancer Res.* **41**, 3685–3690 (1981).

T. Cresteil, A. K. Jaiswal, and H. C. Eisen, Regulation of human P_1-450 gene. In: *Liver Cells and Drugs,* ed. A. Guillouzo, Colloque INSERM, Vol. 164, John Libbey Eurotext, London, pp. 51–58 (1988).

L. A. Damani, W. F. Pool, P. A. Crooks, R. K. Kaderlik, and D. M. Ziegler, Stereoselectivity in the N^1-oxidation of nicotine isomers by flavin-containing monooxygenase, *Mol. Pharmacol.* **33**, 702–705 (1988).

M. Daujat, P. Clair, T. Pineau, L. Pichard, C. Bonfils, C. Dalet, and P. Maurel, Regulation of cytochrome P-4503c. In: *Liver Cells and Drugs,* ed. A. Guillouzo, Colloque INSERM, Vol. 169, John Libbey Eurotext, London, pp. 41–50 (1988).

S. De Flora, S. Petruzzelli, A. Camoirano, C. Bennicelli, M. Romano, M. Rindi, L. Ghelarducci, and C. Giuntini, Pulmonary metabolism of mutagens and its relationship with lung cancer and smoking habits, *Cancer Res.* **47**, 4740–4745 (1987).

G. Del Boccio, C. Di Ilio, P. Ålin, H. Jörnvall, and B. Mannervik, Identification of a novel glutathione transferase in human skin homologous with Class Alpha glutathione transferase 2-2 in the rat, *Biochem. J.* **244** 21–25 (1987).

L. M. Distlerath and F. P. Guengerich, Enzymology of human liver cytochromes P-450. In: *Mammalian Cytochromes P-450,* ed. F. P. Guengerich, CRC Press, Boca Raton, FL, pp. 133–198 (1987).

L. M. Distlerath, P. E. B. Reilly, M. V. Martin, G. G. Davis, G. R. Wilkinson, and F. P. Guengerich, Purification and characterization of the human liver cytochromes P-450 involved in debrisoquine 4-hydroxylation and phenacetin O-deethylation, two prototypes for genetic polymorphism in oxidative drug metabolism, *J. Biol. Chem.* **260**, 9057–9067 (1985).

L. A. Dostal, C. Guthenberg, B. Mannervik, and J. R. Bend, Selectivity and regioselectivity of purified human glutathione transferases π, α–ε, and μ with alkene and polycyclic arene oxide substrate, *Drug Metab. Disp.* **16**, 420–424 (1988).

J. E. Dunn, Jr., Cancer epidemiology in populations of the United States—with emphasis on Hawaii and California—and Japan, *Cancer Res.* **35**, 3240–3245 (1975).

M. Eichelbaum, N. Spannbrucker, B. Steincke, and H. J. Dengler, Defective N-oxidation of sparteine in man: a new pharmacogenetic defect, *Eur. J. Clin. Pharmacol.* **16** 183–187 (1979).

D. A. P. Evans, L. C. Eze, and E. J. Whibley, The association of the slow acetylator phenotype with bladder cancer, *J. Med. Genet.* **20,** 330–333 (1983).

T. J. Flammang, Y. Yamazoe, F. P. Guengerich, and F. F. Kadlubar, The S-acetyl coenzyme A-dependent metabolic activation of the carcinogen N-hydroxy-2-aminofluorene by human liver cytosol and its relationship to the aromatic amine N-acetyltransferase phenotype, *Carcinog.* **8,** 1967–1970 (1987).

C. B. Frederick, J. B. Mays, D. M. Ziegler, F. P. Guengerich, and F. F. Kadlubar, Cytochrome P-450 and flavin-containing monooxygenase-catalyzed formation of the carcinogen N-hydroxy-2-aminofluorene, and its covalent binding to nuclear DNA, *Cancer Res.* **42,** 2671–2677 (1982).

T. Fujino, S. S. Park, D. West, and H. V. Gelboin, Phenotyping of cytochromes P-450 in human tissues with monoclonal antibodies, *Proc. Natl. Acad. Sci. USA* **79,** 3682–3686 (1982).

C. Ged, J. M. Rouillon, L. Pichard, J. Comalbert, N. Bressot, P. Buries, H. Michel, P. Beaune, and P. Maurel, The increase in urinary excretion of 6β-hydroxycortisol as a marker of hepatic cytochrome P-450 III A induction, *Brit. J. Clin. Pharmacol.* **28,** 373–387 (1989).

C. Ged, D. R. Umbenhauer, T. M. Bellew, R. W. Bork, P. K. Srivastava, N. Shinriki, R. S. Lloyd, and F. P. Guengerich, Characterization of cDNAs, mRNAs, and proteins related to human liver microsomal cytochrome P-450 (*S*)mephenytoin 4′-hydroxylase, *Biochem.* **27,** 6929–6940 (1988).

H. Glatt, E. Kaltenbach, and F. Oesch, Epoxide hydrolase activity in native and mitogen-stimulated lymphocytes of various human donors, *Cancer Res.* **40,** 2552–2556 (1980).

M. S. Gold and D. M. Ziegler, Dimethylaniline *N*-oxidase and aminopyrine *N*-demethylase activities of human liver tissue, *Xenobiotica* **3,** 179–189 (1973).

F. J. Gonzalez, R. C. Skoda, S. Kimura, M. Umeno, U. M. Zanger, D. W. Nebert, H. V. Gelboin, J. P. Hardwick, and U. A. Meyer, Characterization of the common genetic defect in humans deficient in debrisoquine metabolism, *Nature* **331,** 442–446 (1988).

D. M. Grant, M. E. Campbell, B. K. Tang, and W. Kalow, Biotransformation of caffeine by microsomes from human liver: kinetics and inhibition studies, *Biochem. Pharmacol.* **36,** 1251–1260 (1987).

J. D. Groopman, P. R. Donahue, J. Zhu, J. Chen, and G. N. Wogan, Aflatoxin metabolism in humans: detection of metabolites and nucleic acid adducts in urine by affinity chromatography, *Proc. Natl. Acad. Sci. USA* **82,** 6492–6496 (1985).

F. P. Guengerich. Enzymology of rat liver cytochromes P-450. In: *Mammalian Cytochromes P-450,* ed. F. P. Guengerich, Vol. 1, CRC Press, Boca Raton, FL, pp. 1–54 (1987a).

F. P. Guengerich, Cytochrome P-450 enzymes and drug metabolism. In: *Progress in Drug Metabolism,* eds. J. W. Bridges, L. F. Chasseaud, and G. G. Gibson, Vol. 10, Chapter 1, Taylor and Francis, London, pp. 1–54 (1987b).

F. P. Guengerich, Roles of cytochrome P-450 enzymes in chemical carcinogenesis and cancer chemotherapy, *Cancer Res.* **48,** 2946–2954 (1988a).

F. P. Guengerich, Oxidation of 17α-ethynylestradiol by human liver cytochrome P-450, *Mol. Pharmacol.* **33,** 500–508 (1988b).

F. P. Guengerich, Biochemical characterization of human cytochrome P-450 enzymes, *Annu. Rev. Pharmacol. Toxicol.,* **29,** 241–264 (1989).

F. P. Guengerich, G. A. Dannan, and M. V. Martin, unpublished results (1988).

F. P. Guengerich, G. A. Dannan, S. T. Wright, M. V. Martin, and L. S. Kaminsky, Purification and characterization of liver microsomal cytochromes P-450: electrophoretic, spectral, catalytic, and immunochemical properties and inducibility of eight isozymes isolated from rats treated with phenobarbital or β-naphthoflavone, *Biochem.* **21**, 6019–6030 (1982).

F. P. Guengerich, M. V. Martin, P. H. Beaune, P. Kremers, T. Wolff, and D. J. Waxman, Characterization of rat and human liver microsomal cytochrome P-450 forms involved in nifedipine oxidation, a prototype for genetic polymorphism in oxidative drug metabolism, *J. Biol. Chem.* **261**, 5051–5060 (1986a).

F. P. Guengerich, D. Müller-Enoch, and I. A. Blair, Oxidation of quinidine by human liver cytochrome P-450, *Mol. Pharmacol.* **30**, 287–295 (1986b).

F. P. Guengerich, P. Wang, M. B. Mitchell, and P. S. Mason, Rat and human liver microsomal epoxide hydratase: purification and evidence for the existence of multiple forms, *J. Biol Chem.* **254**, 12248–12254 (1979).

J. Gut, T. Catin, P. Dayer, T. Kronbach, U. Zanger, and U. A. Meyer, Debrisoquine/sparteine-type polymorphism of drug oxidation: purification and characterization of two functionally different human liver cytochrome P-450 isozymes involved in impaired hydroxylation of the prototype substrate bufuralol, *J. Biol. Chem.* **261**, 11734–11743 (1986).

C. Guthenberg, M. Warholm, A. Rane, and B. Mannervik, Two distinct forms of glutathione transferase from human foetal liver, *Biochem. J.* **235**, 741–745 (1986).

G. J. Hammons, F. P. Guengerich, C. C. Weis, F. A. Beland, and F. F. Kadlubar, Metabolic oxidation of carcinogenic arylamines by rat, dog, and human hepatic microsomes and by purified flavin-containing and cytochrome P-450 monooxygenases, *Cancer Res* **45**, 3578–3585 (1985).

M. F. Hansen and W. K. Cavenee, Genetics of cancer predisposition, *Cancer Res.* **47**, 5518–5527 (1987).

H-P. Hanssen, D. P. Agarwal, H. W. Goedde, H. Bucher, H. Huland, W. Brachmanu, and R. Ovenbeck, Association of N-acetyltransferase polymorphism and environmental factors with bladder carcinogenesis, *Eur. Urol.* **11**, 263–266 (1985).

T. Higgins, S. Chaykin, K. B. Hammond, and J. R. Humbert, Trimethylamine N-oxide synthesis: a human variant, *Biochem. Med.* **6**, 392–396 (1972).

A. J. Hussey, P. K. Stockman, G. J. Beckett, and J. D. Hayes, Variations in the glutathione S-transferase subunits expressed in human livers, *Biochim. Biophys. Acta* **874**, 1–12 (1986).

J. R. Idle, A. Mahgoub, T. P. Sloan, R. L. Smith, C. O. Mbanefo, and E. A. Bababunmi, Some observations on the oxidation phenotype status of Nigerian patients presenting with cancer, *Cancer Lett.* **11**, 331–338 (1981).

K. F. Ilett, B. M. David, P. Detchon, W. M. Castleden, and R. Kwa, Acetylation phenotype in colorectal carcinoma, *Cancer Res.* **47**, 1466–1469 (1987).

Y. M. Irshaid and T. R. Tephley, Isolation and purification of two human liver UDP-glucuronosyltransferases, *Mol Pharmacol.* **31** 27–34 (1987).

M. R. Jackson, L. R. McCarthy, D. Harding, S. Wilson, M. W. H. Coughtrie, and B. Burchell, Cloning of a human liver microsomal UDP-glucuronosyltransferase cDNA, *Biochem J.* **242**, 581–588 (1987).

A. K. Jaiswal, F. J. Gonzalez, and D. W. Nebert, Human dioxin-inducible cytochrome P₁-450: complementary DNA and amino acid sequence, *Science* **228**, 80–83 (1985a).

A. K. Jaiswal, F. J. Gonzalez, and D. W. Nebert, Human P₁-450 gene sequence and correlation of mRNA with genetic differences in benzo(a)pyrene metabolism, *Nucl. Acids Res.* **13**, 4503–4520 (1985b).

A. K. Jaiswal, D. W. Nebert, and F. J. Gonzalez, Human P₃-450: cDNA and complete amino acid sequence, *Nucl. Acids Res.* **14**, 6773–6774 (1987).

F. F. Kadlubar and G. J. Hammons, The role of cytochrome P-450 in the metabolism of chemical carcinogens. In: *Mammalian Cytochromes P-450,* ed. F. P. Guengerich, Vol. 2, CRC Press, Boca Raton, FL, pp. 81–130 (1987).

F. F. Kadlubar, M. A. Butler, B. E. Hayes, F. A. Beland, and F. P. Guengerich, Role of microsomal cytochromes P-450 and prostaglandin H synthase in 4-aminobiphenyl-DNA adduct formation. In: *Microsomes and Drug Oxidations,* Proceedings, Seventh International Symposium, eds. J. O. Miners, D. J. Birkett, R. Drew, and M. E. McManus, Taylor and Francis, London, pp. 370–379 (1988).

F. F. Kadlubar, K. L. Dooley, R. W. Benson, D. W. Roberts, M. A. Butler, C. H. Teitel, J. R. Bailey, and J. F. Young, Pharmacokinetic model of aromatic amine-induced urinary bladder carcinogenesis in Beagle dogs administered 4-aminobiphenyl. In: *Carcinogenic and Mutagenic Responses to Aromatic Amines and Nitroarenes,* eds. C. M. King, L. J. Romano, and D. Schuetzle, Elsevier, New York, pp. 173–180 (1987).

F. F. Kadlubar, J. A. Miller, and E. C. Miller, Microsomal N-oxidation of the hepatocarcinogen N-methyl-4-aminoazobenzene and the reactivity of N-hydroxy-N-methyl-4-aminoazobenzene, *Cancer Res.* **36**, 1196–1206 (1976).

F. F. Kadlubar, J. A. Miller, and E. C. Miller, Hepatic microsomal N-glucuronidation and nucleic acid binding of N-hydroxyarylamines in relation to urinary bladder carcinogenesis, *Cancer Res.* **37**, 805–814 (1977).

A. Kaisary, P. Smith, E. Jaczq, C. B. McAllister, G. R. Wilkinson, W. A. Ray, and R. A. Branch, Genetic predisposition to bladder cancer: ability to hydroxylate debrisoquine and mephenytoin as risk factors, *Cancer Res.* **47**, 5488–5493 (1987).

L. S. Kaminsky, G. A. Dannan, and F. P. Guengerich, Composition of cytochrome P-450 isozymes from hepatic microsomes of C57BL/6 and DBA/2 mice assessed by warfarin metabolism, immunoinhibition, and immunoelectrophoresis with anti-(rat cytochrome P-450), *Eur. J. Biochem.* **141**, 141–148 (1984).

A. E. Karakaya, I. Cok, S. Sardas, P. Gögüs, and O. S. Sardas, N-Acetyltransferase phenotype of patients with bladder cancer, *Human Toxicol.* **5**, 333–335 (1986).

N. T. Kärki, R. Pokela, L. Nuutinen, and O. Pelkonen, Aryl hydrocarbon hydroxylase in lymphocytes and lung tissue from lung cancer patients and controls, *Int. J. Cancer* **39**, 565–570 (1987).

S. Kawano, T. Kamataki, T. Yasumori, Y. Yamazoe, and R. Kato, Purification of human liver cytochrome P-450 catalyzing testosterone 6β-hydroxylation, *J. Biochem. (Tokyo)* **102**, 493–501 (1987).

G. Kellermann, M. Luyten-Kellermann, and C. R. Shaw, Genetic variation of aryl hydrocarbon hydroxylase in human lymphocytes, *Am. J. Hum. Genetics* **25**, 327–331 (1973a).

G. Kellermann, C. R. Shaw, and M. Luyten-Kellermann, Aryl hydrocarbon hydroxylase inducibility and bronchogenic carcinoma, *N. Engl. J. Med.* **298**, 934–937 (1973b).

B. Ketterer and F. P. Guengerich, unpublished results (1988).

C. B. King and I. B. Glowinski, Acetylation, deacetylation and acyl transfer. *Environ. Health Perspect.* **49**, 43–50 (1983).

M. Kitada and T. Kamataki, Partial purification and properties of cytochrome P-450 from homogenates of human fetal livers, *Biochem. Pharmacol.* **28**, 793–797 (1979).

M. Kitada, T. Kamataki, K. Itahashi, T. Rikihisa, and Y. Kanakubo, P-450HLFa, a form of cytochrome P-450 purified from human fetal livers, is the 16α-hydroxylase of dehydroepiandrosterone 3-sulfate, *J. Biol. Chem.* **262**, 13534–13537 (1987).

C. H. Kleinbloesem, P. van Brummelen, H. Faber, M. Danhof, N. P. E. Vermuelen, and D. D. Breimer, Variability in nifedipine pharmacokinetics and dynamics: a new oxidation polymorphism in man, *Biochem. Pharmacol.* **33**, 3721–3724 (1984).

A. G. Knudson, Jr., Hereditary cancer, oncogenes, and antioncogenes, *Cancer Res.* **45**, 1437–1443 (1985).

N. Koga, P. B. Inskeep, T. M. Harris, and F. P. Guengerich, S-[2-(N[7]-Guanyl)ethyl]glutathione, the major DNA adduct formed from 1,2-dibromoethane, *Biochem.* **25**, 2192–2198 (1986).

R. E. Kouri, C. E. McKinney, S. J. Slomiany, D. R. Snodgrass, N. P. Wray, and T. L. McLemore, Positive correlation between high aryl hydrocarbon hydroxylase activity and primary lung cancer as analyzed in cryopreserved lymphocytes, *Cancer Res.* **42**, 5030–5037 (1982).

R. S. Krauss and T. E. Eling, Arachidonic acid-dependent cooxidation: a potential pathway for the activation of chemical carcinogens *in vivo, Biochem. Pharmacol.* **33**, 3319–3324 (1984).

A. Küpfer and R. Preisig, Inherited defects of drug metabolism, *Seminars Liver Disease* **3**, 341–354 (1983).

C-C. Lai, J. A. Miller, E. C. Miller, and A. Liem, N-Sulfooxy-2-aminofluorene is the major ultimate electrophilic and carcinogenic metabolite of N-hydroxy-2-acetylaminofluorene in the livers of infant male C57BL/6J × C3H/HeJF₁(B6C3F₁) mice, *Carcinogen.* **6**, 1037–1045 (1985).

C-C. Lai, E. C. Miller, J. A. Miller, and A. Liem, Initiation of hepatocarcinogenesis in infant male B6C3F₁ mice by N-hydroxy-2-aminofluorene or N-hydroxy-2-acetylaminofluorene depends primarily on metabolism to N-sulfooxy-2-aminofluorene and formation of DNA-(deoxyguanosin-8-yl)-2-aminofluorene adducts, *Carcinog.* **8**, 471–478 (1987).

N. P. Lang, D. Z. J. Chu, C. F. Hunter, D. C. Kendall, T. J. Flammang, and F. F. Kadlubar, Role of aromatic amine acetyltransferase in human colorectal cancer, *Arch. Surg.* **121**, 1259–1261 (1986).

R. C. Levitt, J. M. Fysh, N. M. Jensen, and D. W. Nebert, The A*h* locus: biochemical basis for genetic differences in brain tumor formation in mice, *Genetics* **92**, 1205–1210 (1979).

A. Mahgoub, L. G. Dring, J. R. Idle, R. Lancaster, and R. L. Smith, Polymorphic hydroxylation of debrisoquine in man, *Lancet* **2**, 584–585 (1977).

T. J. Mantle, C. B. Pickett, and J. D. Hayes (eds.), *Glutathione S-Transferase and Carcinogenesis,* Taylor and Francis, London, (1987).

L. J. Marnett, Arachidonic acid metabolism and tumor initiation. In: *Arachidonic Acid Metabolism and Tumor Initiation,* ed. L. J. Marnett, Martinus Nighoff Publishing, Boston, pp. 39–82 (1985).

I. Mertes, R. Fleischmann, H. R. Glatt, and F. Oesch, Interindividual variations in the activities of cytosolic and microsomal epoxide hydrolase in human liver, *Carcinog.* **6**, 219–233 (1985).

E. C. Miller, J. A. Miller, R. R. Brown, and J. C. Macdonald, On the protective action of certain polycyclic aromatic hydrocarbons against carcinogenesis by aminoazodyes and 2-acetylaminofluorene, *Cancer Res.* **18**, 469–477 (1958).

S. Mommsen, N. M. Barford, and J. Aagaard, N-Acetyltransferase phenotypes in the urinary bladder carcinogenesis of a low-risk population, *Carcinog.* **6**, 199–201 (1985).

K. Nakamura, F. Goto, W. A. Ray, C. B. McAllister, E. Jacqz, G. R. Wilkinson, and R. A. Branch, Interethnic differences in genetic polymorphism of debrisoquine and mephenytoin hydroxylation between Japanese and Caucasian populations, *Clin. Pharmacol. Ther.* **38**, 402–408 (1985).

D. W. Nebert, Genetic differences in susceptibility to chemically induced myelotoxicity and leukemia, *Environ. Health Perspect.* **39**, 11–22 (1981).

D. W. Nebert, M. Adesnik, M. J. Coon, R. W. Estabrook, F. J. Gonzalez, F. P. Guengerich, I. C. Gunsalus, E. F. Johnson, B. Kemper, W. Levin, I. R. Phillips, R. Sato, and M. R. Waterman, The P450 gene superfamily. Recommended nomenclature, *DNA* **6**, 1–11 (1987).

D. W. Nebert, D. R. Nelson, M. Adesnik, M. J. Coon, R. W. Estabrook, F. J. Gonzalez, F. P. Guengerich, I. C. Gunsalus, E. F. Johnson, B. Kemper, W. Levin, I. R. Phillips, R. Sato, and M. R. Waterman, The P-450 superfamily: update on listing of all genes and recommended nomenclature of the chromosomal loci, *DNA,* **8**, 1–13 (1989).

F. Oesch, H. Schmassmann, E. Ohnhaus, U. Althaus, and J. Lorenz, Monooxygenase, epoxide hydrolase, and glutathione-S-transferase activities in human lung. Variation between groups of bronchiogenic carcinoma and non-cancer patients and interindividual differences, *Carcinog.* **1**, 827–835 (1980).

E. E. Ohnhaus and B. K. Park, Measurement of urinary 6-β-hydroxycortisol excretion in an *in vivo* parameter in the clinical assessment of the microsomal enzyme-inducing capacity of antipyrine, phenobarbitone and rifampicin, *Eur. J. Clin. Pharmacol.* **15**, 139–145 (1979).

A-K. Ostlund-Farrants, D. J. Meyer, B. Coles, C. Southan, A. Aitken, P. J. Johnson, and B. Ketterer, The separation of glutathione transferase subunits by using reverse-phase high-pressure liquid chromatography, *Biochem. J.* **245**, 423–428 (1987).

S. V. Otton, T. Inaba, and W. Kalow, Competitive inhibition of sparteine oxidation in human liver by β-adrenoceptor antagonists and other cardiovascular drugs, *Life Sci.* **34**, 73–80 (1984).

B. Paigen, H. L Gurtoo, J. Minowada, L. Houten, R. Vincent, K. Paigen, N. B. Parker, E. Ward, and N. T. Hayner, Questionable relation of aryl hydrocarbon hydroxylase to lung-cancer risk, *N. Engl. J. Med.* **297**, 346–350 (1977).

B. Paigen, E. Ward, A. Reilly, H. L. Gurtoo, J. Minowada, K. Steenland, M. B. Havens, and P. Sartori, Seasonal variation of aryl hydrocarbon hydroxylase activity in human lymphocytes, *Cancer Res.* **41**, 2757–2761 (1981).

E. J. Pantuck, K-C. Hsiao, A. H. Conney, W. A. Garland, A. Kappas, K. E. Anderson, and A. P. Alvares, Effect of charcoal-broiled beef on phenacetin metabolism in man, *Science* **194**, 1055–1957 (1976).

E. J. Pantuck, K-C. Hsiao, A. Maggio, K. Nakamura, R. Kuntzman, and A. H. Conney, Effect of cigarette smoking on phenacetin metabolism, *Clin. Pharmacol. Ther.* **15**, 9–17 (1974).

P. Pelkonen, M. Pasanen, H. Kuha, B. Gachalyi, M. Kairaluoma, E. A. Sotaniemi, S. S. Park, F. K. Friedman, and H. V. Gelboin, The effect of cigarette smoking on 7-ethoxyresorufin O-deethylase and other monooxygenase activities in human liver: analyses with monoclonal antibodies, *Br. J. Clin. Pharmacol.* **22**, 125–134 (1986).

P. A. Philip, D. L. Fitzgerald, R. A. Cartwright, M. D. Peake, and H. J. Rogers, Polymorphic *N*-acetylation capacity in lung cancer, *Carcinog.* **9**, 491–493 (1988).

R. M. Philpot, personal communication (1989).

A. Plummer, A. R. Boobis, and D. S. Davies, Is the activation of aflatoxin B_1 catalyzed by the same form of cytochrome P-450 as that 4-hydroxylating debrisoquine in rat and/or man? *Arch. Toxicol.* **58**, 165–170 (1986).

L. C. Quattrochi, U. R. Pendurthi, S. T. Okino, C. Potenza, and R. H. Tukey, Human cytochrome P-450 4 mRNA and gene: part of a multigene family that contains *Alu* sequences in its mRNA, *Proc. Natl. Acad. Sci. USA* **83**, 6731–6735 (1986).

R. H. Reitz, A. L. Mendrala, and F. P. Guengerich, *In vitro* metabolism of methylene chloride in human and animal tissues: use in physiologically-based pharmacokinetic models, *Toxicol. Appl. Pharmacol.,* **97**, 230–246 (1989).

H. L. Richardson, A. R. Stier, and E. Borsos-Nachtnebel, Liver tumor inhibition and adrenal histologic response in rats to which 3′-methyl-4-dimethylaminoazobenzene and 20-methylcholanthrene were simultaneously administered, *Cancer Res.* **12**, 356–361 (1952).

I. G. C. Robertson, C. Guthenberg, B. Mannervik, and B. Jernstrom, Differences in stereoselectivity and catalytic efficiency of three human glutathione transferases in the conjugation of glutathione with 7β,8α-dihydroxy-9α,10α-oxy-7,8,9,10-tetrahydrobenzo(*a*)pyrene, *Cancer Res.* **46**, 2220–2224 (1986).

I. Roots, N. Drakoulis, J. Brockmöller, J. Ritter, G. Heinemeyer, and T. Janik, Prevalence of poor metabolizers of debrisoquine among patients with cancer of stomach, pharynx, larynx, lung or bladder. In: *Abstracts, Second International Meeting, International Society for the Study of Xenobiotics,* Kobe, 16–20 May, p. 110 (1988).

J. H. M. Schellens, P. A. Soons, and D. D. Breimer, Lack of bimodality in nifedipine plasma kinetics in a large population of healthy subjects, *Biochem. Pharmacol.,* **37**, 2507–2510 (1988).

M. C. Scott, J. A. Van Loon, and R. M. Weinshilboum, Pharmacogenetics of N-methylation: heritability of human erythrocyte histamine N-methyltransferase activity, *Clin. Pharmacol. Ther.* **43**, 256–262 (1988).

J. Seidegård, C. Guthenberg, R. W. Pero, and B. Mannervik, The *trans*-stilbene oxide-active glutathione transferase in human mononuclear leucocytes is identical with the hepatic glutathione transferase μ, *Biochem. J.* **246**, 783–785 (1987).

D. Sesardic, A. R. Boobis, R. J. Edwards, and D. S. Davies, A form of cytochrome P-450 in man, orthologous to form d in the rat, catalyses the O-deethylation of phenacetin and is inducible by cigarette smoking. *Brit. J. Clin. Pharmacol.* **26**, 363–372 (1988).

T. Shimada and F. P. Guengerich, Evidence for P-450$_{NF}$, the nifedipine oxidase, being the principal enzyme involved in the bioactivation of aflatoxins in human liver, *Proc. Natl. Acad. Sci. USA,* **86**, 462–465 (1989).

T. Shimada, M. Iwasaki, M. V. Martin, and F. P. Guengerich, Human liver microsomal cytochrome P-450 enzymes involved in the bioactivation of pro-carcinogens detected by *umu* gene response in *Salmonella typhimurium* TA1535/pSK1002, *Cancer Res.,* **49**, 3218–3228 (1989a).

T. Shimada, M. V. Martin, D. Pruess-Schwartz, L. J. Marnett, and F. P. Guengerich, Roles of individual human cytochrome P-450 enzymes in the bioactivation of benzo(a)pyrene, 7,8-dihydroxy-7,8-dihydrobenzo(a)pyrene, and other dihydrodiol derivatives of polycyclic aromatic hydrocarbons, *Cancer Res.,* **49**, 6304–6312 (1989b).

T. Shimada, K. S. Misono, and F. P. Guengerich, Human liver microsomal cytochrome P-450 mephenytoin 4-hydroxylase, a prototype of genetic polymorphism in oxidative drug metabolism: purification and characterization of two similar forms involved in the reaction, *J. Biol. Chem.,* **261**, 909–921 (1986).

H. N. Tawfic, Studies on ear duct tumors in rats. Part II: Inhibitory effects of methylcholanthrene and 1,2-benzanthracene on tumor formation by 4-dimethylaminostilbene, *Acta Pathol. Jpn.* **15**, 255–260 (1965).

R. G. Thurman, S. A. Belinsky, and F. C. Kauffman, Regulation of monooxygenation in intact cells. In: *Mammalian Cytochrome P-450,* ed. F. P. Guengerich, Vol. 2, CRC Press, Boca Raton, FL, pp. 131–152 (1987).

D. R. Umbenhauer, C. P. Wild, R. Monterano, R. Saffhill, J. M. Boyle, N. Huh, U. Kirstein, R. Thomale, M. F. Rajewsky, and S. H. Lu, O^6-Methyldeoxyguanosine in oesophageal DNA among individuals at high risk of oesophageal cancer, *Int. J. Cancer.* **36**, 661–665 (1985).

M. Umeno, O. W. McBride, C. S. Yang, H. V. Gelboin, and F. J. Gonzalez, Human ethanol-inducible P-450 II E1: complete gene sequence, promoter characterization, chromosome mapping, and cDNA-directed expression, *Biochemistry* **27**, 9006–9013 (1988).

J. Van Loon and R. M. Weinshilboum, Human platelet phenol sulfotransferase: familial variatiosn in the thermal stability of the TS form, *Biochem. Genet.* **22**, 997–1013 (1984).

P. Wang, J. Meijer, and F. P. Guengerich, Purification of human liver cytosolic epoxide hydrolase and comparison to the microsomal enzyme, *Biochem.* **21**, 5769–5776 (1982).

M. Warholm, C. Guthenberg, and B. Mannervik, Molecular and catalytic properties of glutathione transferase μ from human liver: an enzyme efficiently conjugating epoxides, *Biochem.* **22**, 3610–3617 (1983).

P. B. Watkins, S. A. Wrighton, P. Maurel, E. G. Schuetz, G. Mendez-Picon, G. A. Parker, and P. S. Guzelian, Identification of an inducible form of cytochrome P-450 in human liver, *Proc. Natl. Acad. Sci. USA* **82**, 6310–6314 (1985).

P. B. Watkins, S. A. Wrighton, E. G. Schuetz, D. T. Molowa, and P. S. Guzelian,

Identification of glucocorticoid-inducible cytochromes P-450 in the intestinal mucosa of rats and man, *J. Clin. Invest.* **80**, 1029–1036 (1987).

D. J. Waxman, C. Attisano, F. P. Guengerich, and D. P. Lapenson, Cytochrome P-450 steroid hormone metabolism catalyzed by human liver microsomes, *Arch. Biochem. Biophys.* **263**, 424–436 (1988).

W. W. Weber and D. W. Hein, *N*-Acetylation pharmacogenetics, *Pharmacol. Res.* **37**, 25–79 (1985).

R. M. Weinshilboum, Phenol sulfotransferase in humans: properties, regulation, and function, *Fed. Proc.* **45**, 2223–2228 (1986).

J. H. Weisburger, Mechanisms of nutritional carcinogenesis associated with specific human cancers. In: *ISI Atlas of Science: Pharmacology,* Information Sciences Institute, Philadelphia, pp. 162–167 (1987).

A. Weston, J. C. Willey, M. J. Newman, G. E. Haugen, D. K. Manchester, J. S. Choi, T. Krontis, B. Light, D. L. Mann, and C. C. Harris, Application of biochemical and molecular techniques to the epidemiology of human lung cancer. In: *Microsomes and Drug Oxidations,* Proceedings of the Seventh International Symposium, eds. J. Miners, D. J. Birkett, R. Drew, and M. McManus, Taylor and Francis, London, pp. 380–391 (1988).

D. E. Williams, R. L. Reed, B. Kedzierski, D. M. Ziegler, and D. R. Buhler, The role of flavin-containing monooxygenase in the N-oxidation of the pyrrolizidine alkaloid senecionine, *Drug Metab. Disp.,* **17**, 387–392 (1989).

T. Wolff, L. M. Distlerath, M. T. Worthington, J. D. Groopman, G. J. Hammons, F. F. Kadlubar, R. A. Prough, M. V. Martin, and F. P. Guengerich, Substrate specificity of human liver cytochrome P-450 debrisoquine 4-hydroxylase probed using immunochemical inhibition and chemical modeling, *Cancer Res.* **45**, 2116–2122 (1985).

S. A. Wrighton, C. Campanile, P. E. Thomas, S. L. Maines, P. B. Watkins, G. Parker, G. Mendez-Picon, M. Haniu, J. E. Shively, W. Levin, and P. S. Guzelian, Identification of a human liver cytochrome P-450 homologous to the major iso-safrole-inducible cytochrome P-450 in the rat, *Mol. Pharmacol.* **29**, 405–410 (1986).

S. A. Wrighton, P. E. Thomas, D. E. Ryan, and W. Levin, Purification and characterization of ethanol-inducible human hepatic cytochrome P-450HLj, *Arch. Biochem. Biophys.* **258**, 292–297 (1987).

Y. Yamazoe, M. Shimada, K. Maeda, T. Kamataki, and R. Kato, Specificity of four forms of cytochrome P-450 in the metabolic activation of several aromatic amines and benzo(*a*)pyrene, *Xenobiotica* **14**, 549–552 (1984).

C. S. Yang, Research on esophageal cancer in China: a review, *Cancer Res.* **40**, 2633–2644 (1980).

J-S. H. Yoo, F. P. Guengerich, and C. S. Yang, Metabolism of *N*-nitrosodialkylamines in human liver microsomes, *Cancer Res.* **88**, 1499–1504 (1988).

H. H. Zhou, L. B. Anthony, A. J. J. Wood, and G. R. Wilkinson, Induction of polymorphic 4′-hydroxylation of *S*-Mephenytoin by rifampicin, *Brit. J. Clin. Pharmacol.* **30**, 471–475 (1988).

D. M. Ziegler, Flavin-containing monooxygenases: catalytic mechanism and substrate specificities, *Drug Metab. Rev.* **19**, 1–32 (1988).

D. M. Ziegler, S. S. Ansher, T. Nagata, F. F. Kadlubar, and W. B. Jakoby, N-Methylation: a potential mechanism for metabolic activation of carcinogenic primary alkylamines, *Proc. Natl. Acad. Sci. USA* **85**, 2514–2517 (1988).

Validation of Molecular Epidemiologic Methods

Frederica P. Perera
Columbia University School of Public Health
Division of Environmental Sciences
New York, New York

1. INTRODUCTION

Providing the impetus for validation of molecular epidemiological methods is the tremendous annual toll of cancer: more than 460,000 deaths each year in the United States alone. A small percentage of the cancer burden is attributable to genetic factors alone: thus most cancers can be prevented by controlling environmental exposures to carcinogens. The conventional tools to identify and assess potential cancer risks are—individually—sadly deficient. Thus, by epidemiologic methods, only a few exposures (cigarette smoking, certain occupational exposures, and drugs) have been conclusively shown to cause cancer in humans. Yet we know that a large variety of mutagens and animal carcinogens are present in the workplace and indoor and ambient air, as well as in the drinking water and food supply. While some are of natural origin, contamination by industrial or human-made chemicals is significant (See Perera and Boffetta, 1988 for review). Compounding the challenge to researchers is the complex etiology of human cancer, exemplified by the demonstrated synergism between cigarette smoking and occupational exposure to asbestos, radiation, or nickel.

How might molecular epidemiology contribute to cancer prevention? Simply put, the expectation by risk assessors is that biologic markers will not only permit early identification of potential risks (hazards) to humans but will provide comparative dosimetry and response data, facilitating quantitative extrapolation from experimental animals to humans. For the epidemiologist, the hope is that biologic markers can be incorporated into analytic studies (cohort and case control) in such a way as to increase the power of such studies to identify a causal relationship between exposure to a specific agent and increased risk of cancer. For example, plasma cotinine and aflatoxin B_1 (AFB_1) in urine can currently be considered reliable indicators of both exposure and internal dose of tobacco smoke and dietary AFB_1, respectively. As such they can serve to assign subjects to "exposed" and "unexposed" groups on a more objective basis than self-reported exposure, job history, or even environmental monitoring, thereby increasing the power of a study by reducing random exposure misclassification. Moreover, certain biologic markers of preclinical response (occurring before progression to cancer was inevitable) might ultimately replace clinical disease as an end point. By providing an earlier and a more commonly occurring outcome, such markers would mitigate the latency problem in cancer epidemiology, increase study power, and allow effective intervention (Gann *et al.*, 1985). Implicit in the above discussion is the anticipation that biomarkers will significantly expand our knowledge of the mechanisms or stage at which individual carcinogens exert their effect. Obviously, such information would permit more effective intervention and prevention strategies.

Seen against this backdrop of high expectations, the present state of the art may, at first glance, seem disappointing. We can point to numerous biomonitoring studies where persons with known (usually high) exposures to genotoxic compounds have been sampled; but there are few examples of studies in which a hypothesis regarding cancer causation has been tested within a "molecular epidemiologic" study (Högstedt, 1988). Moreover, there is only one case of a well-developed application of molecular epidemiology to quantitative risk assessment (Calleman *et al.*, 1978; Ehrenberg *et al.*, 1986; Högstedt, 1986). In addition, as can be clearly seen from Table 1, there is a dearth of methods for carcinogens that do not interact efficiently with the genetic material.

On the other hand, the feasibility of biomonitoring and molecular epidemiology has been demonstrated. The diversity of methods now being validated and the increasing interest in seriously addressing problems posed by the marriage of experimental technology and epidemiology are both highly encouraging.

For in-depth discussion of these issues, the reader is referred to Bridges, 1980; Farmer *et al.*, 1986; Harris *et al.*, 1987; Hattis, 1987; Hulka and Wilcosky, 1988; International Agency for Research on Cancer (IARC), 1984; 1988; National Research Council (NCR), 1987; Perera and Weinstein, 1982; Perera, 1987; Schulte, 1987, 1988; and Wogan, 1988. The purpose of this chapter is to provide a snapshot of the current state of the art and to set out a strategy for validation of this approach.

2. OVERVIEW OF METHODS AND APPLICATIONS: SIGNIFICANCE OF FINDINGS

Table 1 summarizes the database according to the category of biologic marker. The three classes discussed differ in terms of the point at which they occur on the continuum between initial exposure to an environmental agent and overt clinical disease. While *internal dose* is a measurement of the amount of carcinogen or metabolites present in cells, tissues, or body fluids, *biologically effective dose* reflects the amount that has interacted with cellular macromolecules at a target site or an established surrogate. As such, this class of markers is more mechanistically relevant to carcinogenesis than internal dose but poses more challenging analytical problems. The third category comprises markers that indicate an irreversible *biologic effect* resulting from a toxic interaction, either at the target or an analogous site, which is known or believed to be pathogenically linked to cancer. Potential markers of *susceptibility* such as genetic and/or acquired preexposure traits affecting carcinogen metabolism or DNA repair are outside the scope of this chapter. There is, however, considerable interest in the role of factors such as AHH inducibility (Khouri *et al.*, 1985; Karki *et al.*, 1987), debrisoquine hydroxylation (Harris *et al.*, 1987; Conney, 1982; Ayesh *et al.*, 1984), *N*-acetylation (Cartwright *et al.*, 1982), and glutathione-*S*-transferase (Seidegard and Pero, 1985) in determining individual response to carcinogenic exposures. Moreover, using molecular biological techniques, it is now possible to analyze genetic polymorphism at the DNA level to identify individuals with a genetic predisposition to particular types of cancer (Francomano and Kazazian, 1986). Many investigators are now studying the relationship between genetic polymorphism and lung cancer (Brauch *et al.*, 1987; Yokota *et al.*, 1987; Willey *et al.*, 1988). Many of the same methodologic issues discussed below apply to markers of susceptibility, and they engender some unique ethical concerns as well (Ashford, 1986).

As shown in Table 1, biologic markers have been studied in three types of populations: those with defined environmental (including occupational) exposures to carcinogens (cigarette smokers, workers, persons with certain dietary exposures), clinical populations with exposure to carcinogenic drugs, and cancer patients. Each one of the methods listed has shown elevated levels of the particular marker in at least one of the exposed populations. Two consistent findings have been significant interindividual variation in levels of markers among persons with comparable exposure and an apparently important contribution of ''background'' exposures manifest in so-called ''unexposed'' controls. These points will be discussed in greater detail below in the context of criteria for validation of biologic markers.

Although small-scale and generally of limited design, this body of research has demonstrated that most of the methods are adequately sensitive

TABLE 4–1. A. Studies of Chemical-Specific Markers in Humans*

Compound Analyzed	Exposure Source	Biologic Sample	Population	Reference
Internal Dose				
Nitrosaminoacids	*N*-nitroso compounds in diet	Urine	Chinese residing in areas of low and high cancer risk	Lu et al., 1986
N-nitrosoproline	Cigarette smoke	Urine	Smokers, nonsmokers; unexposed	Hoffmann et al., 1986; Garland et al., 1986
1-Hydroxybenzo(a) pyrene	Coal tar products	Urine	Workers, smokers, coal tar treated patients	Bos and Jongeneelen, 1988
Aflatoxin B$_1$ (AFB$_1$)	Diet	Urine	Chinese residing in a high exposure area	Groopman et al., 1985
Biologically Effective Dose				
N-3-(2-Hydroxy-ethyl) histidine; *N*-(2-Hydroxy-ethyl) valine	Ethylene oxide	RBC	Workers, smokers, unexposed	Calleman et al., 1978; van Sittert et al., 1985; Farmer et al., 1986; Tornqvist et al., 1986
Alkylated Hb	Propylene oxide	RBC	Workers	Osterman-Golkar et al., 1984
4-Aminobiphenyl-HB	Cigarette smoke	RBC	Smokers, nonsmokers	Bryant et al., 1987
AFB$_1$guanine	Diet	Urine	Chinese and Kenyans residing in a high exposure and high and low risk area respectively	Groopman et al., 1985; Autrup et al., 1983
3-Methyladenine	Methylating agents	Urine	Unexposed	Shuker and Farmer, 1988

Adduct	Carcinogen/exposure	Tissue	Population	References
PAH–DNA	PAH in cigarette smoke, in workplace	WBC, lung tissue, placenta	Lung cancer patients, smokers, workers	Perera et al., 1982; Shamsuddin et al., 1985; Vahakangas et al., 1985; Harris et al., 1985; Haugen et al., 1986; Everson et al., 1986, 1988; Perera et al., 1987, 1988a, 1988b; Garner et al., 1988; Weston et al., 1988; Haugen et al., 1986
Antibodies to PAH–DNA O^6-Methyldeoxy-guanosine	PAH in workplace Nitrosamines in diet	Serum Esophageal and stomach mucosa	Workers Chinese and European cancer patients	Umbenhauer et al., 1985; Wild et al., 1986
Cisplatinum-DNA, Cisplatinum-protein	Cisplatin	WBC, Hb, plasma proteins	Chemotherapy patients	Reed et al., 1986; Fichtinger-Schepman et al., 1987;
8-Methoxypsoralen–DNA Styrene oxide–DNA	8-Methoxypsoralen Styrene	Skin WBC	Psoriasis patients Workers	Mustonen et al., 1988 Santella et al., 1988a Liu et al., 1988a, 1988b

B. Studies Using Non-Chemical-Specific Markers in Humans*

Compound or End Point Source	Exposure Sample	Biologic Sample	Population	Reference
		Internal Dose		
Mutagenicity of urine	Cigarette smoke, various industrial exposures	Urine	Smokers, workers	Vainio et al., 1984, for review
Thymine glycol	Agents that cause oxidative damage to DNA	Urine	Volunteers	Cathgart et al., 1984
		Biologically Effective Dose		
Spectrum of DNA adducts	Betel and tobacco chewing, smoking, industrial exposures, wood smoke	Placenta, lung tissue, oral mucosa, WBC, bone marrow, colonic mucosa	Smokers, workers, volunteers	Everson et al., 1986; Dunn and Stich, 1986; Chacko and Gupta, 1987; Phillips et al., 1986, 1988a, 1988b; Reddy et al., 1987
		Early Biologic Effect or Response		
Single strand breaks	Styrene	WBC	Workers	Walles et al., 1988
Unscheduled DNA synthesis	Propylene oxide	WBC	Workers	Pero et al., 1982
Sister chromatid exchange	Various industrial exposures, radiation	WBC	Workers	Carrano and Moore, 1982, for review
Micronuclei	Organic solvents, heavy metals, cigarette smoke, betel quid	WBC, oral mucosa	Workers	Högstedt et al., 1983; Stich and Dunn, 1988

Chromosomal aberrations	Various industrial exposures, radiation	WBC	Workers	Evans, 1982, for review
DNA hyperploidy	Aromatic amines	Bladder and lung cells	Workers	Hemstreet, 1988
HPRT mutation	Chemotherapeutic agents, radiation	WBC	Patients, workers	O'Neill *et al.*, 1987; Messing *et al.*, 1986
GPA mutation	Chemotherapeutic agents, radiation	RBC	Patients, Japanese atom bomb survivors	Langlois *et al.*, 1987; Jensen *et al.*, 1986
Oncogene activation	PAH, cigarette smoke	Serum	Workers, cancer patients	Brandt-Rauf, 1988; Perera *et al.*, 1988c

* Abbreviations: 4-ABP, 4-aminobiphenyl; PAH, polycyclic aromatic hydrocarbons; RBC, red blood cells; WBC, white blood cells; ^{32}P, ^{32}P-postlabeling; HPRT, hypoxanthine-guanine phosphoribosyltransferase; GPA, glycophorin A.

Source: Perera, 1987, 1988; Wogan, 1988.

for human studies. However, the results are often limited by technical variability in the assays, small sample sizes, lack of appropriate controls, failure to account for confounding variables, and a paucity of data on exposure.

For these reasons, it is often difficult to compare results from different studies—even where the exposure source is similar. Tables 2–5 summarize recent studies that have evaluated DNA or protein adducts in the same "model" populations using a variety of methods. To facilitate comparisons between them, quantitative results have been converted to the same units (mole per mole). Apparent discrepant results such as those shown in Tables 2–5 are attributable to factors such as different laboratory procedures, the small numbers of subjects studied, the different methods used to ascertain exposure, and the varying intervals between exposure and collection of the sample. Interestingly, only one study has demonstrated significant smoker-nonsmoker differences in placental adduct levels (Everson *et al.*, 1988). Adduct concentrations in lymphocytes and placental tissue from smokers have otherwise been only marginally higher or similar to those in nonsmokers (Tables 2 and 3).

3. CRITERIA AND STRATEGIES FOR VALIDATION

Validation is the process by which certain basic requirements or criteria for biological markers are met through a combination of experimental and human pilot studies. By way of illustration of the extent to which a commonly used biologic marker meets these criteria, let us evaluate the extent to which aromatic DNA and protein adducts in peripheral blood cells meet these standards, bearing in mind that the discussion is generalizable to other markers.

3.1. The Fundamental Characteristics of the Assay must be Established

The *low dose sensitivity and reproducibility* are key concerns. For example, the sensitivity of the DNA methods is in the range of one adduct per 10^6–10^{10} nucleotides (or 1–10,000 adducts per cell) with the postlabeling method being several orders of magnitude more sensitive than the immunoassay. Methods for carcinogen-protein adducts have estimated levels of detection well below human exposure levels (Santella, 1988; IARC, 1988; Lohman, 1988; Tannenbaum and Skipper, 1984).

Within-laboratory variability must also be established so that differences in marker concentration are not erroneously attributed to intra- or interindividual variation. Recently, a collaborative interlaboratory comparison of immunoassays to polycylnic aromatic hydrocarbon-DNA (PAH-DNA) adducts successfully resolved a number of discrepancies between quantitative results

TABLE 4–2. DNA Adducts in White Blood Cells of Smokers and Nonsmokers*

Assay	Population (*n*)		Range (μmol/mol)	% Positive	Mean (Pos.)[†] (μmol/mol)	Reference
PAH–DNA	S	(22)	0.054–0.17	41	0.096	Perera *et al.*, 1987
(ELISA)	NS	(24)	0.035–0.14	33	0.08	
PAH–DNA	S	(32)	<0.032–0.29	9	0.192	Weston *et al.*, 1988
(USERIA)	NS	(49)	<0.032–0.13	4	0.096	
DNA adducts	S	(4)	<0.002–0.014	100	0.0065	Phillips *et al.*, 1986
(³²P)	NS	(6)	0.004–.01	100	0.0065	

* Abbreviations: μmol/mol, micromole of adduct per mole of DNA; ELISA, enzyme-linked immunosorbent assay; USERIA, ultrasensitive enzymatic radioimmunoassay; ³²P, ³²P-postlabeling; S, smoker; NS, nonsmoker; see also Table 1.
[†] The value given is the mean for positive samples; i.e., those with detectable levels of adducts.

TABLE 4–3. DNA Adducts in Placental Tissue*

Assay	Population (n)	Range (μmol/mol)	% Positive	Mean (Pos.)[†] (μmol/mol)	Reference
PAH–DNA (ELISA)	S (24)	ND–0.67	87	0.36	Everson et al., 1988
	NS (24)	ND–0.28	87	0.24	
PAH–DNA (IAC, SFS, GC–MS)	S (15)	...	36	0.08	Weston et al., 1988
	NS (13)	...	36	(Pooled pos. samples)	
Adducts (^{32}P)	S (30)	ND–16[‡]	93	4.8[‡]	Everson et al., 1988
	NS (23)	NS–1.5[‡]	17	0.15[‡]	
O^6-medGua	S (10)	<0.12–0.4	20	...	Foiles et al., 1988
	NS (10)	<0.12–0.4	30	...	

* Abbreviations: IAC, immunoaffinity column; NS, nonsmokers; ND, nondetectable; O^6-medGua, O^6-methyldeoxyguanosine; see also Table 1 and 2.

[†] The value given is the mean for positive samples; i.e., those with detectable levels of adducts.

[‡] Intensity score.

TABLE 4–4. PAH–DNA Adducts in White Blood Cells of Workers*

Assay	Population (*n*)	Range (µmol/mol)	% Positive	Mean (Pos.)[†] (µmol/mol)	Reference
PAH–DNA (ELISA)	Coke oven (20)	<0.032–0.13	35	0.064 (0.022)	Shamsuddin *et al.*, 1985
PAH–DNA (ELISA)	Coke oven (27)	<0.032–10.9	66	2.27 (1.51)	Harris *et al.*, 1985
PAH–DNA (ELISA)	Coke oven (38)	<0.032–4.38	34	0.51 (0.18)	Haugen *et al.*, 1986
PAH–DNA (ELISA)	Foundry (35)	<0.01–0.90	86	0.20 (0.17)	Perera *et al.*, 1988
DNA adducts (^{32}P)	Foundry[‡] (25)	<0.002–0.1	100	0.018	Phillips *et al.*, 1988

* Abbreviations: see also Tables 1 and 2.
[†] The value given is the mean for positive samples; i.e., those with detectable levels of adducts. Values in parentheses are means for all samples assayed.
[‡] Medium and high exposures.

TABLE 4-5. Adducts in Cisplatinum-Treated Patients*

Assay	Tissue	N	Range	Reference
DNA Adducts (ELISA)	PBL	45	0–>0.032 fmol/mol	Reed et al., 1986
DNA adducts (LC; ELISA)	PBL	6	0–3.2 fmol/mol	Fichtinger-Schepman et al., 1987
DNA adducts (Immunocytochem.)	Buccal mucosa	9	0–>150 nuclear density	den Engelse et al., 1988
Protein adducts (AAS)	Hb	32	0.2–4 ng/mg	Hemminki et al., 1988b

* Abbreviations: fmol, femtomole; ng, nanogram; mg, milligram; LC, liquid chromatography; AAS, atomic absorption spectrophotometry; N, number; PBL, peripheral blood leukocytes; Hb, hemoglobin; see also Tables 1 and 2.

in the enzyme-linked immunosorbent assay (ELISA) and ultrasensitive enzymatic radioimmunoassay (USERIA) (Santella *et al.*, 1988b). More such collaborative ventures, involving exchange of standards and samples, are badly needed.

Has there been *confirmation of assay results* by other methods? Another tool in understanding the characteristics of an individual assay is the use of corroborative methods on the same sample. Recently, a discrepancy between the enzyme-linked immunosorbent assay (which showed significant levels of PAH-DNA adducts in placental tissue of smokers and nonsmokers) and the postlabeling method (which did not) was at least partially resolved. As shown in Table 5, PAH-DNA adducts were definitively identified in placental tissue by both synchronous fluorescence spectrophotometry (SFS) and gas chromatography-mass spectrometry (GC-MS) (Weston *et al.*, 1988). Thus the postlabeling method was not detecting low but measurable quantities of PAH adducts present. It is recommended that each biomarker be confirmed by a different, complementary method.

Is the marker *chemical or exposure specific;* i.e., selective for a particular chemical, exposure source, or period of exposure? The extent to which the method is selective determines the applicability and the interpretation of results. In contrast to methods that quantitate cisplatinum-DNA and AFB_1-guanine adducts, the widely used antibodies to PAH-DNA adducts detect multiple diol-epoxide adducts formed by this class of compounds (Santella *et al.*, 1988b). Even less chemical-specific is the postlabeling method, which provides an idiosyncratic fingerprint of a spectrum of adducts. While characterization of these adducts is somewhat problematic, this method can be useful in evaluating complex mixtures and unknown exposures (Randerath *et al.*, 1988).

Biologic markers can integrate exposure via multiple routes (inhalation, oral, dermal), multiple sources (ambient and indoor air, workplace air, cigarette smoke, diet, drinking water), and across all patterns of exposure (past, current, intermittent, continuous). This is an advantage since risks can be assumed to be additive. However, it is also a disadvantage in that many environmental chemical such as PAHs are ubiquitous in the environment and it is difficult to distinguish the effect of any particular exposure source. Markers vary greatly with respect to source specificity: for example, 4-aminobiphenyl-hemoglobin (4-ABP-Hb) has far fewer non-cigarette-smoking-related ''background'' sources than ethylene oxide (EtO), PAH, and *N*-nitroso compounds.

The extent to which the marker will document specific time periods of exposure will depend upon the pharmacokinetics of the chemical and the persistence of the marker in the biologic sample assayed (itself a function of the turnover rate of the sample and repair processes). For example, given that the erythrocyte has a 120-day lifetime and lacks repair systems, it is a good short-term cumulating dosimeter. By contrast, most adduct measurements are carried out on the total amount of DNA extracted from white blood cells (WBCs) having widely varying lifetimes. These measurements therefore require a more complex temporal interpretation than hemoglobin since they can reflect past as well as current or recent exposure. T cells, which constitute about 13–25% of peripheral WBCs, have a half-life of about 3 years. Thus in cases of discontinued past exposure, these cells will be the sole contributors to adduct measurements. In such cases even the most long-lived markers in lymphocytes (which have a relatively low repair activity) will be diluted out by cell turnover. In situations involving current exposure, the short-lived granulocytes, B cells, and monocytes, of course, will dominate.

There is evidence that dose rate or pattern of exposure may influence biologic effects (Generosos *et al.,* 1986). Yet no human studies have compared the extent of adduct formation resulting from intermittent, high versus low-level constant exposures.

Characterizing the selectivity of a method is one of the greater challenges in validation. For example, in the case of an immunoassay for carcinogen-DNA adducts, complex studies are necessary to establish the cross-reactivity of the antibody with other chemical complexes. Thus one must identify the various environmental sources of these chemicals in order to account for "background" exposure in human studies. Finally, pharmacokinetic modeling and even social sampling in humans are required to define the persistence of adducts in various cell types.

The same criteria of *adequate sensitivity, specificity, and predictive value* that apply to the validation of screening methods should be met by biomarkers. Is the marker able to distinguish truly exposed (or unexposed) and truly "at risk" (or not at risk) individuals with a high degree of accuracy? To answer this question, calculation of false positive and false negative rates and positive and negative predictive values should be based on a series of independent samples from the population [Griffith *et al.,* in press; National Academy of Sciences, in preparation]. While considerable progress has been made in evaluating the relationship between exposure and adducts, the ability of adducts to predict risk of cancer has not been assessed in prospective studies—either in laboratory animals or in humans. This is a glaring research need that can best be met by prospective animal bioassays and by nested case-control studies that take advantage of banks of stored samples from human subjects followed longitudinally to determine cancer risk.

Finally, how *feasible* is the marker? That is, how acceptable is it to the public, how cost-effective is it, and how permanent is it in stored samples?

3.2. The Dose-Response Relationship and the Extent of Interindividual and Intraindividual Variability in Humans must be Characterized

It is often said that to be valid a marker must be highly correlated with an individual's estimated exposure. This assumption ignores the fact that the search for biologic markers was triggered in large part by the awareness of large differences in individual processing of xenobiotics. Moreover, if this assumption were true, there would be little point in collecting biologic information. Indeed, all evidence to date suggests that there is a high degree of variability in DNA- and protein-adduct levels among persons smoking the same amount, among individuals working in the same plant, and even among chemotherapy patients treated with standardized doses of cisplatinum (Fichtinger-Schapmann *et al.*, 1987). For example, a proportional relationship has not been observed between the reported amount of smoking and adduct levels (4-ABP-Hb, PAH-DNA) in peripheral blood cells of volunteers. This is probably because of variability in metabolic processing of xenobiotics and/or in repair as well as errors in estimating individual exposure. There should be less interindividual variability in response to direct-acting carcinogens not requiring metabolic activation. However, while an initial study showed a good correlation between estimated levels of EtO in a sterilization plant and Hb adduct levels in an initial study (Calleman *et al.*, 1978), a subsequent study involving decreased air concentrations did not (van Sittert *et al.*, 1985). Another striking finding has been the widely variable but detectable level of DNA and hemoglobin adducts in persons without the exposure of concern (e.g., nonsmokers and individuals without occupational exposure to PAHs, styrene, and EtO) (see Perera, 1988, for review). Indeed in most studies, including those in Tables 2–4, the range of adducts in controls has overlapped with that observed in the exposed group. These two general observations could result from a combination of variation in "background" sources and heightened sensitivity or resistance on the part of some individuals. They could also be a reflection of intraindividual variability. Therefore it is important to elucidate both interindividual and intraindividual variability in the biological marker of interest by means of large-scale surveys with repeated sampling.

Although one would not expect or require a significant correlation between estimated individual exposure and adduct levels, there should be a difference in the mean or medium values of exposed and unexposed groups when potential confounding variables are accounted for. For example, in a study of foundry workers, there was significant variation in PAH-DNA adduct levels by immunoassay within the exposed group (ranging from nondetectable to 2.8 fmol of adduct per μg of DNA). However, a good dose-response relationship was seen between estimated PAH exposure (high, medium, low) and adducts by both the immunoassay and the postlabeling methods (Perera *et al.*, 1988a; Hemminki *et al.*, 1988a; Phillips *et al.*, 1988).

3.3. The Mechanistic or Biologic Relevance to Cancer must be Established

The biological basis for measuring DNA adducts derives from extensive experimental data supporting their role in the initiation and possibly in the progression of cancer (Miller and Miller, 1981; Lutz, 1979; Harris *et al.*, 1987; Yuspa and Poirier, 1988; Weinstein *et al.*, 1984; Perera, 1988; Wogan and Gorelick, 1985). Despite this impressive body of information, however, much remains to be uncovered about the quantitative relationship between adduct formation (as a necessary but not sufficient event) and cancer risk. Factors that are likely to play a role in cancer risk include the specific type and biological effectiveness of adducts formed (e.g., location in tissue, cell type, and site of binding on the genome), the rate and accuracy of repair prior to cell replication, and the presence of endogenous or exogenous promoting agents or cocarcinogens. Ideally, therefore, one would want this type of detailed characterization for the chemical-DNA adduct being measured. By sharp contrast, most of the methods in Table 1 provide information on total or multiple adducts and are incapable of pinpointing the ''critical'' adducts on DNA. Considerable basic research is needed to identify cell and site-specific adducts that are most biologically effective and to develop adequately sensitive methods for human monitoring. The first steps have, however, been taken in methods development (Baan *et al.*, 1988; Olivero *et al.*, 1988).

As evidenced by experimental studies, protein such as hemoglobin can, in many cases of acute or continuous exposure to carcinogens, serve as a valid surrogate for DNA (Neumann, 1984). However, as a step in validation, it is necessary to establish that relationship for each carcinogen of interest. Interpretation of results should also account for the varying kinetics of DNA and protein adducts even in ''steady-state'' situations.

The larger question—which still remains largely unanswered—is whether tissue accessible for biomonitoring such as peripheral blood cells are a reasonable surrogate for the target tissue such as the lung. Experimental studies with benzo[a]pyrene (BP) suggest that comparable levels of adducts are formed in WBCs and other tissues (Stowers and Anderson, 1985) but comparisons in humans are only now underway in lung cancer patients and controls. Illustrating the potential problem of organ, tissue, and even cell specificity of DNA adducts is the observation that bronchial cells of smokers showed clear cell-type specific induction of BP-DNA adducts (Baan *et al.*, 1988). Therefore studies are needed to understand the biological relevance of adducts in surrogate tissue such as peripheral blood cells.

3.4. Potential Confounding Factors must be Identified

Paraphrasing the classic definition used by epidemiologists (Kelsey *et al.,* 1986), confounders in molecular epidemiology can be defined as variables that are causally related to the end point (here biologic marker) under study and are associated with the exposure under study but are not a consequence of this exposure. Confounders could cause an investigator to believe that an association exists when it does not and vice versa.

In addition to factors associated with the exposure itself, there is potential confounding by other exposures or sources that are detected by the laboratory method used and which are either not known or inadequately characterized. By their very nature, biologic markers are sensitive to confounding (Riboli and Saracci, 1984). For example, potential confounding variables in studies of DNA adducts include age, sex, race, lifestyle habits (cigarette smoking, alcohol consumption, and diet), drugs, other environmental exposures (including those that alter metabolism of the carcinogen of interest), genetic factors, and preexisting health impairment. This fact imposes the requirement that epidemiologists be involved in the design of studies to allow for either matching or control of these variables in data analysis.

4. CONCLUSION

The burgeoning new field of molecular epidemiology is now at a critical point in its development. The progress to date has been rapid and exciting. Difficult questions must now be addressed jointly by laboratory experts and epidemiologists before definitive large-scale studies of cancer causation can be attempted. Although considerable groundwork remains to be done, the great potential of this approach to enhance risk assessment and ultimately to prevent cancer should justify this effort.

REFERENCES

N. A. Ashford, Policy considerations for human monitoring in the workplace, *J. Occup. Med.* **28,** 563–568 (1986).

H. Autrup, K. A. Bradley, A. K. M. Shamsuddin, J. Wakhisi, and A. Wasunna, Detection of putative adduct with fluorescence characteristics identical to 2,3-dihydro-2-(7-guanyl)-3-hydroxyaflatoxin B1 in human urine collected in Murang'a District, Kenya, *Carcinog.* **4,** 1193–1195 (1983).

R. Ayesh, J. Idle, J. Richie, M. Crothers, and M. Hetzel, Metabolic oxidation phenotypes as markers for susceptibility to lung cancer, *Nature* **312,** 169–170 (1984).

R. A. Baan, P. T. M. van den Berg, M.-J. S. T. Steenwinkel, and C. J. M. van der Wulp, Detection of benzo(a)pyrene diol-epoxide by quantitative immunofluorescence microscopy and ^{32}P-postlabeling; immunofluorescence analysis of benzo(a)pyrene-DNA adducts in bronchial cells from smoking individuals. In IARC, 1988, *op. cit.,* pp. 146–154 (1988).

R. P. Bos and F. J. Jongeneelen, Nonselective and selective methods for biological monitoring of exposure to coal tar products. In IARC, *op. cit.*, pp. 389–395 (1988).

P. W. Brandt-Rauf, New markers for monitoring occupational cancer: The example of oncogene proteins. *J. Occup. Med.* **30**, 399–404 (1988).

H. Brauch, B. Johnson, J. Hovis, T. Yano, A. Gazdar, O. S. Pettengill, S. Graziano, G. D. Sorenson, B. J. Poiesz, J. Minna, M. Linehan and B. Zbar, Molecular analysis of the short arm of chromosome 3 in small-cell and non-small-cell carcinoma of the lung, *N. Engl. J. Med.* **317**(18), 1109–1113 (1987).

B. A. Bridges, An approach to the assessment of the risk to man from DNA damaging agents, *Arch. Toxicol. Suppl.* **3**, 271–281 (1980).

M. S. Bryant, P. L. Skipper, S. R. Tannenbaum, and M. MacLure, Hemoglobin adducts of 4-aminobiphenyl in smokers and nonsmokers, *Cancer Res.* **47**, 602–608 (1987).

C. J. Calleman, L. Ehrenberg, B. Jansson, S. Osterman-Golkar, D. Segerbäck, K. Svensson, and C. A. Wachtmeister, Monitoring and risk assessment by means of alkyl groups in hemoglobin in persons occupationally exposed to ethylene oxide, *J. Environ. Pathol. Toxicol.* **2**, 427–442 (1978).

A. V. Carrano and D. H. Moore, The rationale and methodology for quantifying sister chromatic exchange frequency in humans. In: *Mutagenicity: New Horizons in Genetic Toxicology,* ed. J. A. Heddle, Academic Press, New York, pp. 267–304 (1982).

R. A. Cartwright, R. W. Glashhan, H. J. Rogers, D. Barham-Hale, R. A. Ahmad, E. Higgins, M. A. Kahn, Role of *N*-acetyltransferase phenotypes in bladder carcinogenesis: A pharmacokinetic epidemiological approach to bladder cancer, *Lancet* **2**, 842–846 (1982).

R. Cathcart, E. Schweirs, R. L. Saul, B. N. Ames, Thymine glycol and thymidine glycol in human and rat urine: A possible assay for oxidative DNA damage. *Procl. Natl. Acad. Sci. USA* **81**, 5633–5637 (1984).

M. Chacko and R. C. Gupta, Evaluation of DNA damage in the oral mucosa of cigarette smokers and nonsmokers by ^{32}P-adduct assay, *Proc. Am. Assoc. Cancer Res.* **28**, 101 (1987).

A. H. Conney, Induction of microsomal enzymes by foreign chemicals and carcinogenesis by polycyclic aromatic hydrocarbons, *Cancer Res.* **42**, 4875–4917 (1982).

B. P. Dunn and H. F. Stich, ^{32}P-postlabeling analysis of aromatic DNA adducts in human oral mucosal cells. *Carcinogenesis* **7**, 1115–1120 (1986).

L. den Engelse, P. M. A. B. Terhaggen, and A. C. Begg, Cisplatin-DNA interaction products in sensitive and resistant cell lines and in buccal cells from cisplatin-treated cancer patients. In: *Proceedings of the American Association for Cancer Research* **29**, 339 (abstract no. 1348) (1988).

L. Ehrenberg, S. Osterman-Golkar, and M. Törnqvist, Macromolecule adducts, target dose, and risk assessment. In: *Genetic Toxicology of Environmental Chemicals, Part B: Genetic Effects and Applied Mutagenesis,* eds. C. Ramel, B. Lambert, and J. Magnusson, Liss, New York pp. 253–260 (1986).

H. J. Evans, Cytogenetic studies on industrial populations exposed to mutagens. In: *Indicators of Genotoxic Exposure,* eds. B. A. Bridges, B. E. Butterworth, and I. B. Weinstein, Banbury Report No. 19, Cold Spring Harbor Laboratory, Cold Spring Harbor, NY (1982).

R. E. Everson, E. Randerath, R. M. Santella, R. C. Cefalo, T. A. Avitts, and K. Randerath, Detection of smoking-related covalent DNA adducts in human placenta, *Science* **231**, 54–57 (1986).

R. E. Everson, E. Randerath, R. M. Santella, T. A. Avitts, I. B. Weinstein, and K. Randerath, Quantitative association between DNA damage in human placenta and maternal smoking and birth weight, *J. Natl. Cancer Inst.* **80**, 567–576 (1988).

P. B. Farmer, E. Bailey, S. M. Gorf, M. Törnqvist, S. Osterman-Golkar, A. Kautiiainen, and D. P. Lewis-Enright, Monitoring human exposure to ethylene oxide by the determination of haemoglobin adducts using gas chromatography-mass spectrometry, *Carcinog.* **7**, 637–640 (1986).

P. B. Farmer, H-G. Neumann, and D. Henschler, Estimation of exposure of man to substances reacting covalently with macromolecules, *Arch. Toxicol.* **60**, 251–260 (1987).

A. M. J. Fichtinger-Schepman, A. T. van Oosterom, P. H. M. Lohman, and F. Berends, Interindividual human variation in cisplatinum sensitivity, predictable in an in vitro assay? *Mutat. Res.* **190**, 59–62 (1987).

P. G. Foiles, L. M. Miglietta, S. A. Akerkar, and S. S. Hecht, Detection of O-methyldeoxyguanosine in human placental DAN, *Cancer Res.* **48**, 4184–4188 (1988).

C. Francomano and H. H. Kazazian, Jr., DNA analysis in genetic disorders, *Annu. Rev. Med.* **37**, 377–395 (1986).

P. H. Gann, D. L. Davis, and F. Perera, Biologic markers in environmental epidemiology: Constraints and opportunities, Fifth Workshop of the Scientific Group on Methodologies for the Safety Evaluation of Chemicals (SGOMSEC), Mexico City, August 12–16 (1985).

W. A. Garland, E. Kuenzig, F. Rubio, H. Kornychuk, E. P. Norkus, A. H. Conney, Urinary excretion of nitrosodimethylamine and nitrosopoline in humans: Interindividual differences and the effect of administered ascorbic acid and -tocopherol. *Cancer Res.* **46**, 5392–5400 (1986).

R. C. Garner, B. Tierney, and D. H. Phillips, A comparison of ^{32}P-postlabeling and immunological methods to examine human lung DNA for benzo(a)pyrene adducts. In IARC, 1988, *op. cit.,* pp. 196–200 (1988).

W. M. Generoso, K. T. Cain, L. A. Highes, *et al.,* Ethylene oxide dose and dose-rate effects in the mouse dominant-lethal tes, *Environ. Mutagen.* **8**, 1–7 (1986).

J. Griffith, R. C. Duncan, J. R. Goldsmith, and B. S. Hulka, Biochemical and biological markers: Implications for epidemiologic studies (in press).

J. D. Groopman, P. R. Donahue, J. Zhu, J. Chen, and G. N. Wogan, Aflatoxin metabolism and nucleic acid adducts in urine by affinity chromatography, *Proc. Natl. Acad. Sci. USA* **82**, 6492–6496 (1985).

C. C. Harris, A. Weston, J. Willey, G. Trivers, and D. Mann, Biochemical and molecular epidemiology of human cancer: Indicators of carcinogen exposure, DNA damage and genetic predisposition, *Env. Health Pers.* **75**, 109–119 (1987).

C. C. Harris, K. Vahakangas, N. J. Newman, G. E. Trivers, A. Shamsuddin, N. Sinopoli, D. L. Mann, and W. Wright, Detection of benzo(a)pyrene diol epoxide-DNA adducts in peripheral blood lymphocytes and antibodies to the adducts in serum from coke oven workers, *Proc. Natl. Acad. Sci. USA* **82**, 6672–6676 (1985).

D. Hattis, The value of molecular epidemiology in quantitative health risk assessment. In: *Environmental Impacts on Human Health: The Agenda for Long-Term Research and Development*, eds. S. Draggen, J. J. Cohrssen, and R. E. Morrison, Praeger, New York, (1987).

A. Haugen, G. Becher, C. Benestad, K. Vahakangas, G. E. Trivers, M. J. Newman, and C. C. Harris, Determination of polycyclic aromatic hydrocarbons in the urine, benzo[a]pyrene diol epoxide-DNA adducts in lymphocyte DNA, and antibodies to the adducts in sera from coke oven workers exposed to measured amounts of polycyclic aromatic hydrocarbons in the work atmosphere, *Cancer Res.* **46,** 4178–4183 (1986).

K. Hemminki, F. P. Perera, D. H. Phillips, K. Randerath, M. V. Reddy, and R. M. Santella, Aromatic DNA adducts in white blood cells of foundry workers. In IARC, 1988, *op. cit.,* pp. 190–195 (1988a).

K. Hemminki, P. Hietanen, R. Mustonen, and A. Alhonen, Determination of cis-diamminedichloroplatinum (II) in white blood cell DNA, plasma proteins and hemoglobin of cancer patients. In: *Proceedings of the American Association for Cancer Research, 29,* 260 (abstract no. 1033) (1988b).

G. P. Hemstreet, P. A. Schulte, K. Ringen, W. Stringer, and E. B. Altekruse, DNA hyperploidy as a marker for biological response to bladder carcinogen exposure, *Int. J. Cancer,* (in press).

Hoffman *et al.,* 1986 — we have in Procite: 1983 in Cancer Res., 1989 in Springer Handbook of Pharmacology, and 1987 in IARC.

B. Högstedt, B. Akesson, K. Axell, B. Gullberg, F. Mitelman, R. W. Pero, S. Skerving, and H. Welinder, Increased frequency of lymphocyte micronuclei in workers producing reinforced polyester resin with low exposure to styrene, *Scand. J. Work Environ. Health* **49,** 271–276 (1983).

L. C. Högstedt, Future perspectives, needs and expectations of biological monitoring of exposure to genotoxicants in prevention of occupational disease. In: *Monitoring of Occupational Genotoxicants. Progress in Clinical and Biological Research,* eds. M. Sorsa and H. Norppa, Liss, New York, pp. 231–243 (1986).

L. C. Högstedt, Summary: Epidemiological Applications. In IARC, 1988, *op. cit.,* pp. 21–22 (1988).

B. S. Hulka and T. Wilcosky, Biological markers in epidemiologic research, *Arch. Environ. Health* **43,** 83–89 (1988).

IARC, Monitoring of Humans with Exposures to Carcinogens, Mutagens and Epidemiological Applications. In: *Cancer Occurrence, Causes and Control,* IARC Scientific Publication Series, ed. L. Tomatis, IARC, Lyon, France, Sec. 6.3 (1986).

IARC, *Methods of Monitoring Human Exposure to Carcinogenic and Mutagenic Agents,* IARC Scientific Publication No. 59, eds. A. Berlin, M. Draper, K. Hemminki, and H. Vainio, IARC, Lyon, France, (1984).

IARC, *Methods for Detecting DNA Damaging Agents in Humans: Applications in Cancer Epidemiology and Prevention,* IARC Scientific Publication No. 89, eds. H. Bartsch, K. Hemminki, and I. K. O'Neill, IARC, Lyon, France (1988).

R. H. Jensen, R. G. Langlois, and W. L. Bigbee, Determination of somatic mutations in human erythrocytes by flow cytometry. In: *Genetic Toxicology of Environmental Chemicals, Part B: Genetic Effects and Applied Mutagenesis,* eds. C. Ramel,

B. Lambert, and J. Magnusson, Alan R. Liss, New York, pp. 177–184 (1986).

N. K. Karki, R. Pakela, L. Nuutinen, and O. Pelkonen, Aryl hydrocarbon hydroxylase in lymphocytes and lung tissue from lung cancer patients and controls, *Int. J. Cancer* **39**, 565–570 (1987).

J. L. Kelsey, W. D. Thompson, and A. S. Evans, *Methods in Observational Epidemiology*, Oxford Univ. Press, New York (1986).

M. J. Khouri, C. A. Newill, and G. A. Chase, Epidemiologic evaluation of screening, *Am. J. Pub. Health* **75**, 1204–1208 (1985).

R. G. Langlois, W. L. Bigbee, S. Kgoizumi, N. Nakamura, M. A. Bean, M. Akiyama, and R. H. Jensen, Evidence for increased somatic cell mutations at the glycophorin A locus in atom bomb survivors, *Science* **236**, 445–448 (1987).

S. F. Liu, S. M. Rappaport, K. Pongracz, and W. J. Bodell, Detection of styrene oxide-DNA adducts in lymphocytes of a worker exposed to styrene, in IARC, 1988 methods for detecting DNA damaging agents in humans: Applications in cancer epidemiology and prevention. Eds. H. Bartsch, K. Hemminki, I. K. O'Neill. Int. Agency for the Research of Cancer, *Sci. Publ.* No. 89, Lyon pp. 217–222 (1988a).

S. F. Liu, S. M. Rappaport, and W. J. Bodell, Detection of styrene oxide-DNA adducts in lymphocytes of workers exposed to styrene, *Proceedings of the American Association for Cancer Research* **29**, 259 (abstract no. 1029) (1988b).

P. H. M. Lohman, Summary: Adducts, in IARC, 1988, methods for detecting DNA damaging agents in humans: Applications in cancer epidemiology and prevention. Eds. H. Bartsch, K. Hemminki, I. K. O'Neill. Int. Agency for the Research of Cancer, *Sci. Publ.* No. 89, Lyon pp. 13–20 (1988).

S.-H. Lu, H. Ohshima, H.-M. Fu, Y. Tian, F.-M. Li, M. Blettner, J. Wahrendorf and H. Bartsch, Urinary excretion of N-nitrosamino acids and nitrate by inhabitants of high- and low-risk areas for esophageal cancer in northern China: Endogenous formation of nitrosopoline and its inhibition by vitamin C. *Cancer Res.* **46**, 1485–1491 (1986).

W. K. Lutz, In vivo covalent binding of organic chemicals to DNA as a quantitative indicator in the process of chemical carcinogenesis, *Mutat. Res.* **65**, 289–356 (1979).

K. Messing, A. M. Seifert, and W. E. C. Bradley, In vivo mutant frequency of technicians professionally exposed to ionizing radiation. In: *Monitoring of Occupational Genotoxicants*, eds. M. Sorsa and H. Norppa, Alan R. Liss, New York, pp. 87–97 (1986).

E. C. Miller and J. A. Miller, Mechanisms of chemical carcinogenesis, *Cancer* **47**, 1055–1064 (1981).

R. Mustonen, K. Hemminki, A. Alhonen, P. Hietanen, and X. Kiilunen, Determination of cisplatin in blood compartments of cancer patients. In IARC, methods for detecting DNA damaging agents in humans: Applications in cancer epidemiology and prevention. Eds. H. Bartsch, K. Hemminki, I. K. O'Neill. Int. Agency for the Research of Cancer, *Sci. Publ.* No. 89, Lyon pp. 329–332 (1988).

National Research Council, Committee on Biological Markers, Biological markers in environmental health research, *Environ. Health Perspect.* **64**, 3–9 (1987).

H. G. Neumann, Dosimetry and dose-response relationships. In IARC, 1984, *op. cit.*, pp. 115–126 (1984).

J. P. O'Neill, M. J. McGinniss, J. K. Berman, L. M. Sullivan, J. A. Nicklas, and R. J. Albertini, Refinement of a T-lymphocyte cloning assay to quantify the in vivo thioguanine-resistant mutant frequency in humans, *Mutagen.* **2,** 87–94 (1987).

O. Olivero, H. S. Huitfelt, and M. C. Poirier, Chromosome site-specific immuno-histochemical detection of DNA adducts in N-acetoxy-acetylaminofluorene-exposed CHO cells. In *Proceedings of the American Association for Research on Cancer* **29,** 94 (abstract no. 373) (1988).

S. Osterman-Golkar, E. Bailey, P. B. Farmer, S. M. Gorf, and J. H. Lamb, Monitoring exposure to propylene oxide through the determination of hemoglobin alkylation, *Scand. J. Work Environ. Health* **10,** 99–102 (1984).

F. P. Perera, Molecular cancer epidemiology: A new tool in cancer prevention, *J. Natl. Cancer Inst.* **78,** 887–898 (1987).

F. P. Perera, The significance of DNA and protein adducts in human biomonitoring studies, *Mutat. Res.* **205,** 255–269 (1988).

F. P. Perera, K. Hemminki, T.L. Young, D. Brenner, G. Kelly, and R. M. Santella, Detection of polycyclic aromatic hydrocarbon-DNA adducts in white blood cells of foundry workers, *Cancer Res.* **48,** 2288–2291 (1988b).

F. P. Perera, R. M. Santella, D. Brenner, T. L. Young, and I. B. Weinstein, Application of biological markers to the study of lung cancer causation and prevention. In IARC, 1988, *op. cit.,* pp. 451–459b (1988c).

F. P. Perera, R. M. Santella, D. Brenner, M. C. Poirier, A. A. Munshi, H. K. Fischman, and J. Van Ryzin, DNA adducts, protein adducts and sister chromatid exchange in cigarette smokers and nonsmokers, *J. Natl. Cancer Inst.* **79,** 449–456 (1987).

F. P. Perera and I. B. Weinstein, Molecular epidemiology and carcinogen-DNA adduct detection: New approaches to studies of human cancer causation, *J. Chron. Dis.* **35,** 581–600 (1982).

F. P. Perera, M. C. Poirier, S. H. Yuspa, J. Nakayama, A. Jaretzki, M. C. Curnen, D. M. Knowles, and I. B. Weinstein, A pilot project in molecular cancer epidemiology: Determination of benzo[a]pyrene-DNA adducts in animal and human tissues by immunoassay, *Carcinog.* **3,** 1405–1410 (1982).

F. Perera and P. Boffetta, Perspectives on comparing risks of environmental carcinogens, *J. Natl. Cancer Inst.* **80,** 1281–1293 (1988).

R. W. Pero, T. Bryngelsson, B. Widegren, B. Högstedt, and H. Welinder, A reduced capacity for unscheduled DNA synthesis in lymphocytes from individuals exposed to propylene oxide and ethylene oxide, *Mutat. Res.* **104,** 193–200 (1982).

D. H. Phillips, A. Hewer, and P. L. Grover, Aromatic DNA adducts in human bone marrow and peripheral blood leukocytes, *Carcinog.* **7,** 2071–2075 (1986).

D. H. Phillips, K. Hemminki, A. Alhonen, A. Hewer, and P. L. Grover, Monitoring occupational exposure to carcinogens: detection by ^{32}P-postlabeling of aromatic DNA adducts in white blood cells from iron foundry workers, *Mutat. Res.* **204,** 531–541 (1988).

K. Randerath, R. H. Miller, D. Mittal, and E. Randerath, Monitoring human exposure to carcinogens by ultrasensitive postlabeling assays: Application to unidentified gentoxicants. In IARC, 1988, *op. cit.,* pp. 361–367 (1988).

M. V. Reddy and K. Randerath, ^{32}P-postlabeling assay for carcinogen-DNA adducts: Nuclease P1-mediated enhancement of its sensitivity and applications. *Env. Health Persp.* **76,** 41–47 (1987).

E. Reed, S. Yuspa, L. A. Zwelling, R. F. Ozols, and M. C. Poirier, Quantitation of cis-diamminedichloroplatinum (II) (cisplatin)-DNA-intrastrand adducts in testicular and ovarian cancer patients receiving cisplatin chemotherapy, *J. Clin. Invest.* **77**, 545–550 (1986).

E. Riboli and R. Saracci, Marqeurs d'exposition et de lesions precoces en l'epidemiologie du cancer, *Rev. Epidemiol. Sante Publique* **33**, 304–311 (1984).

R. M. Santella, Application of new techniques for detection of carcinogen adducts to human population monitoring, *Mutat. Res.* **205**, 271–282 (1988).

R. M. Santella, X. Y. Yang, V. De Leo, and F. P. Gasparro, Detection and quantification of 8-methoxypsoralen-DNA adducts. In IARC, 1988, *op. cit.*, pp. 333–340 (1988a).

R. M. Santella, A. Weston, F. P. Perera, G. T. Trivers, C. C. Harris, T. L. Young, D. Nguyen, B. M. Lee, and M. C. Poirier, Interlaboratory comparison of antisera and immunoassays for benzo(a)pyrene-diol-epoxide-I-modified DNA, *Carcinog.* **9**, 1265–1269 (1988b).

P. A. Schulte, A conceptual framework for the validation and use of biologic markers, *Environ. Res.* **48**, 129–144 (1989).

P. A. Schulte, Methodologic issues in the use of biologic markers in epidemiologic research, *Am. J. Epidem.* **126**, 1006–1016 (1987).

J. Seidegard and R. W. Pero, The hereditary transmission of high glutathione transferase activity towards trans-stilbene oxide in human mononuclear leukocytes, *Hum. Genet.* **69**, 66–68 (1985).

A. K. M. Shamsuddin, N. T. Sinopoli, K. Hemminki, R. R. Boesch, and C. C. Harris, Detection of benzo[a]pyrene-DNA adducts in human white blood cells, *Cancer Res.* **45**, 66–68 (1985).

D. E. G. Shuker and P. B. Farmer, Urinary excretion of 3-methyladenine in humans as a marker of nucleic acid methylation. In IARC, 1988, *op. cit.*, pp. 92–96 (1988).

H. F. Stich and B. P. Dunn, DNA adducts, micronuclei and leukoplakias as intermediate endpoints in intervention trials. In IARC, 1988, *op. cit.*, pp. 137–145 (1988).

S. J. Stowers and M. W. Anderson, Formation and persistence of benzo[a]pyrene metabolite-DNA adducts, *Environ. Health Perspect.* **62**, 31–39 (1985).

S. R. Tannenbaum and P. L. Skipper, Biological aspects to the evaluation of risk: Dosimetry of Carcinogens in man *Fundam. Appl. Toxicol.* **4**, S367–S370 (1984).

M. Törnqvist, S. Osterman-Golkar, A. Kautiainen, A. Jensen, P. B. Farmer, and L. Ehrenberg, Tissue doses of ethylene oxide in cigarette smokers determined from adduct levels in hemoglobin, Carcinog. **7**, 1519–1521 (1986).

D. Umbenhauer, C. P. Wild, R. Montesano, R. Saffhill, J. M. Boyle, N. Huh, U. Kirstein, J. Thomale, M. F. Rajewsky, and S. H. Lu, O Methyldeoxyguanosine in oesophageal DNA among individuals at high risk of oesopageal cancer, *Int. J. Cancer* **36**, 661–665 (1985).

K. Vahakangas, G. Trivers, A. Haugen *et al.*, Detection of benzo(a)pyrene diol-epoxide-DNA adducts by synchronous fluorescence spectrophotometry and ultrasensitive enzymatic radioimmunoassay in coke oven workers. *J. Cell Bioch.* (Suppl.) (1985) Suppl. 9c: 1271.

H. Vainio, M. Sorsa, and K. Falck, Bacterial urinary assay in monitoring exposure to mutagens and carcinogens. In IARC, 1984, *op. cit.*, pp. 247–258 (1984).

N. J. van Sittert, G. de Jong, M. G. Clare, R. Davies, B. J. Dean, L. J. Wren, and A. S. Wright, Cytogenetic, immunological, and hemotological effects in workers in an ethylene oxide manufacturing plant, *Br. J. Ind. Med.* **42**, 19–26 (1985).

S. A. S. Walles, H. Norppa, S. Osterman-Golkar, and J. Maki-Paakkanen, Single-strand breaks in DNA of peripheral lymphocytes of styrene-exposed workers. In IARC, *op. cit.*, pp. 223–226 (1988).

I. B. Weinstein, S. Gatoni-Celli, P. Kirschmeier, M. Lambert, W. Hsiao, J. Backer, and A. Jeffrey, Multistage carcinogenesis involves multiple genes and multiple mechanisms, *Cancer Cells 1. The Transformed Phenotype*, Cold Spring Harbor Laboratory, Cold Spring Harobor, NY, pp. 229–237 (1984).

A. Weston, J. C. Willey, D. K. Manchester, V. L. Wilson, B. R. Brooks, J.-S. Choi, M. C. Poirier, G. E. Trivers, M. J. Newman, D. L. Mann, and C. C. Harris, Dosimeters of human carcinogen exposure: Polycyclic aromatic hydrocarbon macromolecular adducts. In IARC, 1988, *op. cit.*, pp. 181–189 (1988).

C. P. Wild, D. Umbenhauer, B. Chapot, and R. Montesano, Monitoring of individual human exposure to aflatoxins (AF) and N-nitrosamines (NNO) by immunoassays, *J. Cell. Biol.* **30**, 171–179 (1986).

J. C. Willey, A. Weston, A. Haugen, T. Krontiris, J. Resau, E. McDowell, B. Trump, and C. C. Harris, DNA restriction fragment length polymorphism analysis of human bronchogenic carcinoma. In IARC, 1988, *op. cit.*, pp. 439–450 (1988).

G. N. Wogan and N. J. Gorelick, Chemical and biochemical dosimetry of exposure to genotoxic chemicals, *Environ. Health Perspect.* **62**, 5–18 (1985).

G. N. Wogan, Detection of DNA Damage in Studies on Cancer Etiology and Prevention. In IARC, 1988, *op. cit.*, pp. 32–51 (1988).

J. Yokota, M. Wada, Y. Shimosato, M. Terada and T. Sugimura, Loss of heterozygosity on chromosomes 3, 13 and 17 in small-cell carcinoma and on chromosome 3 in adenocarcinomas of the lung, *Proc. Natl. Acad. Sci. USA* **84**, 9252–9256 (1987).

S. H. Yuspa and M. C. Poirier, Chemical carcinogenesis: From animal models in one decade, *Adv. Cancer Res.*, **50**, 25–70 (1988).

Toxic Exposures in a Community Setting: The Epidemiologic Approach

David M. Ozonoff
Department of Environmental Health Sciences
Boston University School of Public Health
Boston, Massachusetts

Daniel Wartenberg
Department of Environmental and Community Medicine
Robert Wood Johnson Medical School
University of Medicine and Dentistry of New Jersey
Piscataway, New Jersey

1. INTRODUCTION

When John Snow made his pioneering investigations of the cause of cholera in the mid-nineteenth century, he not only launched epidemiology but applied it to understanding an environmental hazard. Subsequently epidemiologic methods were much used in the investigation of time-space clusters of infectious disease. This was greatly aided by a new understanding of the etiology of these diseases (the germ theory) and new laboratory developments, such as Koch's pure-culture technique, which allowed etiology-based diagnoses to be made with excellent sensitivity and specificity.

Unfortunately the promise of these classical uses of environmental epidemiology has not borne comparable fruit in understanding the effects of toxic chemical exposure in a community setting. In this chapter we discuss the technical problems associated with doing such community-based studies and then suggest ways in which new understanding and new laboratory techniques may help overcome the formidable technical obstacles we face.

2. COMMUNITY-BASED TOXICS PROBLEMS

We use the phrase "community-based toxics problems" to describe three sets of situations often encountered by health officials.

1. Exposure-driven situations: These occur when it is discovered or suspected that individuals have been exposed to a chemical by virtue of their residential location in a community. This is to distinguish them from individuals exposed by virtue of some other common characteristic, such as similar occupations. Here the main concern is the consequences of this exposure for the health of these individuals or their families at some time in the future. They want to know, "What will happen to me?"
2. Outcome-driven situations: An apparent time-space cluster of disease occurrences often gives rise to a suspicion that some factor in the environment may be responsible. This may initiate a search for the putative environmental cause, as the victims ask, "Why me?"
3. Mixed situations: In these cases the combination of a known or suspected exposure with a perception of unusual disease occurrence may prompt an investigation to confirm or disconfirm the putative connection. People ask, "Are we sicker than our neighbors?"

Any of these situations may become highly politicized, compounding the difficulty of already overburdened health officials in responding adequately. To date, the conventional tools available to health departments have been unequal to the task, and worse, officials have often used these tools unimaginatively and in ways that have lost them the confidence of communities (Ozonoff and Boden, 1987).

3. EPIDEMIOLOGY AND TOXICOLOGY

The two principal disciplines used to understand the effects of toxic agents on human populations are toxicology and epidemiology. The essential distinction between them is that toxicology is based on experiment while epidemiology is based on observation, i.e., in toxicology the investigator actually manipulates the independent variable (exposure), while the epidemiologist observes an existing (natural) event.

Each has its strengths and weaknesses. Toxicological experiments can frame questions and answer them precisely but narrowly, leaving the applicability of the question and its answer to actual public health problems much in doubt. Epidemiology, on the other hand, because it observes a "natural experiment" occurring under existing conditions in the species of interest (humans) is clearly pertinent to public health. Since epidemiology depends on Nature to be its research assistant, however, the experiments so observed are often of crude design and yield ambiguous and difficult to interpret answers. It is obvious that the two techniques complement each other and must both be used to make an adequate evaluation of a community toxics problem. Toxicological methods are well described in other papers in this volume. In this chapter we will discuss the use of epidemiological tools.

4. EPIDEMIOLOGIC STUDY OF COMMUNITY-BASED TOXICS PROBLEMS

The problems involved in studying community-based toxics problems with epidemiologic methods are formidable:

1. Community toxics problems typically involve a fairly small exposed population ("neighborhood-sized"). If the health outcome is relatively rare, as is cancer, there is usually insufficient statistical power to detect even very powerful effects with any confidence. The question of statistical power is discussed in more detail below.
2. Exposures are usually poorly specified, in terms of the agents involved, the extent of exposure, and the pattern of exposure, and the exposures are often to poorly characterized mixtures rather than single agents. Moreover, people whose residences are close geographically may have widely differing exposures, depending on the amount of time spent indoors and outdoors, their movements in the neighborhood, and the amount of time spent in the neighborhood. Thus children who play outdoors in the vicinity of their homes may have substantially different exposures than their parents who work in another area for most of the day.
3. Some health effects of concern, such as cancer, have long latent periods. Substantial amounts of time must elapse after the exposure before the effect can be seen. Should effects finally be observed, even instantaneous cessation of exposure would allow additional cancers to appear for the entire length of the latent period. Thus, from the public health point of view, it is not acceptable to wait until long latency effects are observed before deciding on a course of action.

4. Some health effects, again cancer being a good example, are relatively uncommon in the population although they are of substantial importance when they occur. Such "rare" events are very difficult to detect in small populations with epidemiologic methods.

How can these difficulties be overcome?

5. EPIDEMIOLOGIC STUDY DESIGNS

"Observing a natural experiment," which is the essence of epidemiology, is, less passive and more active than the description might suggest. Indeed much of the art and craft of epidemiology consists of knowing where to look for "experiments" in the real world, and what elements of the world to observe once they have been found. The subsequent analytical process of arranging observations in a way to yield the maximum amount of information, while it may involve complex mathematical techniques, is secondary to the process of locating and making the proper kinds of observations. This prior process is called study design. Study design is well developed and sophisticated, and many complex designs are available to epidemiologists. However, for our purposes it suffices to discuss the three classical designs, as all others are variants or hybrids of these.

5.1. "What will Happen to Me?" The Cohort Design

The first, and most well-known, design, is called a cohort (or prospective) study. Cohort studies are rather good analogues of a laboratory experiment in that a group of exposed but initially disease-free individuals is followed over time and their health experience compared with a similar unexposed group. If the only differences in the two groups were exposure, as would be the case in an experiment, one could ascribe differences in health outcomes to differences in exposure. Unfortunately there may be many differences between the two groups ("confounding variables"), some of which may be possible alternative explanations for the observed difference (or lack of difference in the case of masking) in health effects. Various techniques are available for the control of confounding, both in the analysis and in the design, but the problem is a generic one in nonrandomized situations.

An obvious limitation of cohort studies is the need for very large numbers of exposed and unexposed individuals when the health outcome of interest is uncommon. For example, with an annual incidence of at most a few in every 10,000, most cancers require large cohorts indeed. On the other hand, while rare diseases are difficult to study by this method, rare exposures require only the collection of a sufficiently large exposed group for study. And while only one exposure can be studied at a time, many different health outcomes can be looked at simultaneously.

TABLE 5–1. Advantages and Disadvantages of Cohort and Case-Control Studies

	Cohort	Case-Control
Advantages	Multiple outcomes can be investigated	Multiple exposures can be investigated
	Good for rare exposures	Good for rare conditions
	Intuitive interpretation	Timely results
Disadvantages	Only single exposure can be investigated for each cohort	Only a single disease can be investigated for each case group.
	May require large cohorts	Exposure determination may be more subject to bias
	May require substantial time for long-latency diseases	Control group may be difficult to select

5.2. "Why Me?" The Case-Control Design

Case-control design tries to discover what is different about those that have contracted a disease compared with those that have not. Here one compares "cases" (those with the disease of interest) with the "controls" (those without the disease) on some set of exposure variables or personal characteristics to see if they differ. Only one disease may be studied but one can compare cases and controls on several different exposure variables simultaneously. Unlike the cohort study, this design is efficient for rare diseases—one need only accrue enough cases to allow a comparison with a control group—but is poor if the prevalence of exposure in the population is rare.

Thus in terms of their strengths and weaknesses, cohort and case-control designs are almost mirror images of each other (cf. Table 1). Some examples illustrate their use. Suppose that you were concerned that a large number of residents exposed to radiation as a result of a nuclear power plant accident might be at risk of developing cancer. More than one cancer might result, say lung and thyroid. One possible method of study would be to set up a registry of residents in the exposed community. These individuals would be followed over time and the rates of lung and thyroid cancer determined at appropriate intervals and compared with the rates in a suitable reference group such as the population of the state where the community was located. This is a typical "exposure-driven" situation.

Consider now the "outcome-driven" case. A cluster of childhood leukemia cases appears in a community that is discovered to have several of its water-supply wells contaminated. By using existing records it may be possible to tell which households in the town were likely to have received water from the contaminated wells and which were not. The sources of water for all of the town's households is thus determinable. The town's leukemia cases are then compared on their sources of water with a set of randomly picked controls of the same age and sex to see if the proportion of case households that received contaminated water is significantly higher than control households (Lagakos *et al.*, 1986).

There are also instances where neither design will work. This is true when both the disease and the exposure are rare in the population being studied. To illustrate, suppose we were interested in whether exposure to radiofrequency electromagnetic fields, say from a powerful military radar facility located in a community, were associated with a risk of cancer. A cohort study might compare those in the community whose residences were either near the main beam or its side lobes with community residents who lived outside the area of potential exposure. Since radar beams are fairly directional, the number of people exposed is likely to be small, at least relative to the number required to have a chance of detecting an increase in a disease that may strike only one person in 1000 annually. Even a many-fold increase may easily be missed if the number in the exposed cohort is only a few dozen or so. On the other hand, the case-control design fares no better. Here the problem is not the rarity of the disease, but the rarity of the exposure. In order to collect enough cases of a particular cancer, a reasonably sized base population will be required, say, large enough to give rise to 50 to 100 cases. In this population the number of people actually exposed to the radar beam is likely to be very small, thus making the comparison moot. This is a situation that is not amenable to the usual epidemiologic study methods. Here inferences from the study of other populations (e.g., military radar workers) or experimental studies on animals or *in vitro* would be necessary to make an indirect judgment about risk in this community.

5.3. "Are We Sicker than Our Neighbors?" The Cross-Sectional Design

The third of the classical designs differs from the other two primarily in that the exposure and the effect are both measured simultaneously. In a cohort study the exposure precedes the effect and both the exposed and the unexposed groups begin the observation period disease-free. In a case-control study the comparison between the case and noncase groups involves exposures that occurred before the disease developed. Thus in either case a determination of disease status is necessary for two points in time. In a cross-sectional study a comparison is made between the rate of disease in those exposed versus those not exposed, as determined at only one point in time. For example, exposed and unexposed groups might be compared for the proportion in each of individuals with detectable DNA adducts. Since the determination of exposure and effect are both made at the same time, it is not possible to know if the adducts were there prior to the exposure or not.

Thus the important causal characteristic of time priority (the cause must precede the effect) is not inherent in the design as it is with cohort and case-control studies. While this may be a serious problem for interpretation, the relative ease and lesser expense of cross-sectional studies have made them popular.

6. THE POWER OF A STUDY

The essence of the epidemiologic study design, then, is a comparison of health experience between two or more groups. Fundamental to the method is that the groups be the same except for the single independent variable being examined, which for our purposes is exposure (the model here goes back to John Stuart Mill and his "Method of Differences"). Differences in health outcome that might be a result of noncomparability of the groups is a major problem for epidemiologists and, as already noted, many techniques have been developed to handle this possibility either in the design (e.g., stratification—separate comparison groups—for some important population characteristic, such as age or sex), or in the analysis (e.g., use of multivariate analysis to "control for" differences in the two groups).

Assuming for the moment that problems of confounding have been considered, there is still another way that two groups might differ in their health experiences without the independent variable being the source. Indeed this might happen as a result of normal variation in each group producing the observed pattern by chance alone. Epidemiologists employ various statistical tests to assure themselves that the possibility that chance is the explanation for the observed pattern is small (conventionally "small" means less than 5%; this is usually expressed as, "$p<0.05$").

It is important to realize that in deciding whether an observed difference is the result of chance variation or not we can err in two entirely different ways: we can declare a chance variation to be real, or we can judge a real difference to be a result of chance. The former condition is called a type I error, the latter a type II error. (Type I and type II errors are another way to express the more familiar "false positive" and "false negative".) One would prefer not to make either type of error, but the possibility always exists. One can choose which type of error is least acceptable by erecting stringent criteria that minimize the chance it will occur, but those same criteria will also increase the likelihood of the other type of error. For instance, one could require that the chance of a difference occurring randomly be less than 1% instead of the usual 5%, but this might also increase the chance that real differences are ignored. Depending upon whether we are more concerned about missing a real effect or declaring a spurious result to be real, we can, in principle, adjust our decision criteria for deciding whether an observed difference is "real" or not.

It turns out, however, that we do not usually allow ourselves the flexibility we have. By convention we have erected the "$p < 0.05$" criterion for statistical significance. This is actually a stringent standard meant to protect against the type I error (minimizing false positives). At the same time it may carry with it the consequence of a high likelihood of a type II error. Another way to express type II error is by stating the power of a study, a measure of the likelihood that a given real effect will be correctly judged not due to chance. Thus we might have a study with 80% power to detect a 3-fold difference in cancer rates, meaning that 80% of the time we will correctly judge that a true 3-fold difference was not due to chance. With the other 20% of the time, sampling variation will lead us to conclude that the observed difference is a random variation, and we will be wrong (i.e., we will have committed a type II error). Clearly the more stringent the safeguards against a type I error, the less the power of the study for a given set of observations.

It should be emphasized that the conventional decision criterion of $p < 0.05$ is neither logically required by the scientific method nor is it always the most appropriate. Indeed, in clinical medicine the decision criteria are such that an "innocent" symptom is more likely to be considered guilty than a truly serious symptom considered benign. The ostensible reasons for this (questions of defensive medical practice aside) are that the downside risks of missing a valid warning sign are more serious than the extra time, expense, inconvenience, and possible discomforts of too thorough a work-up.

We do not use the same approach when evaluating the symptoms of a distressed community, however. Epidemiology, while an observational science, has built its methods in imitation of experimental disciplines, where the risk of adding a false association to the body of science is considered unacceptable. This is entirely reasonable when the type II error can be minimized by performing another experiment. In the case of once-only observations of events in the real world, however, this possibility does not exist. Its appropriateness for an ailing community may be no greater than that for an ailing patient.

For this reason it is important to take account of the power of a study. As we will see, there are several reasons, besides conventional statistical methods, why the power of a community study might be weak. We now discuss those reasons and the ways that new toxicological methods might help to increase study power without increasing type I error.

7. METHODS TO IMPROVE STATISTICAL POWER

Ideally we would like to increase statistical power, i.e., increase the chances of detecting a real effect, without affecting the chances of wrongly declaring an effect when none was there (type I error). One may be able to do this by enrolling more individuals in the study, as power increases as the size of the compared populations increases. Here the limitations are only time and resources. Often, however, there are additional limitations inherent in the given conditions. This is especially true in community toxic exposures where the factor limiting power is likely to be the size of the exposed population. All is not lost, however. Power may also be increased by increasing the precision and accuracy of exposure or disease determinations.

Intrinsic to the epidemiologic method is a comparison of rates of disease in an exposed versus an unexposed population. If you mix the exposed and unexposed or the diseased and nondiseased characteristics (or both) you clearly have harmed your chances of seeing differences. New laboratory techniques to measure exposure or biological effects may be able to make these classifications more accurate. Some of these techniques are well described elsewhere in this volume. We comment here on where they fit from the epidemiologic perspective.

First, it is important to note that some biomarkers may be used either as a measure of exposure or effect. For example, hemoglobin alkylation may be considered either a measure of internal exposure or a measure of biological effect, depending upon your point of view. The choice as to which aspect will be chosen may have to do with what other measures of exposure or effect are available or which seem appropriate in a given circumstance (Griffith *et al.*, 1990). In either case we may gain not only in our ability to place individuals in the correct category (''exposed'' or ''unexposed,'' ''affected'' or ''unaffected'') but also in differentiating individuals within categories by means of quantitative measures of exposure or effect rather than all-or-none categorizations. This will also increase power.

To see the implications of misclassification, we will examine one epidemiologic measure, the odds ratio. This statistic arises when subjects are cross-classified by disease and exposure status, resulting in a 2 × 2 table (Table 2). The four entries in this table express how many people are exposed and have the disease we are studying, how many are exposed but not diseased, how many are unexposed but have the disease, and how many are neither exposed nor have the disease. We then calculate the odds of getting the disease given that one was exposed relative to the odds of getting the

TABLE 5–2. 2 × 2 Table Definitions

	Disease	No Disease	Total
Exposed	a	b	N_E
Not Exposed	c	d	N_{notE}
Total	N_D	N_{notD}	N_{total}

Odds Ratio = a/c / b/d = ad/bc

TABLE 5–3. No Misclassification

	Disease	No Disease	Total
Exposed	40	20	60
Not Exposed	60	80	140
Total	100	100	200

Odds Ratio = 40/60 / 20/80 = 2.67 $P < 0.01$

disease if one were not exposed. In the case of no association between exposure and disease, the odds of getting disease will be the same whether or not one is exposed, giving an odds ratio of 1.00. If exposure increases the likelihood that one will get the disease, the odds ratio exceeds one, and it is less than one if exposure somehow protects against disease occurrence.

When we make errors in cross-classifying our subjects we add imprecision to the odds ratio. If we make these errors randomly (i.e., without respect to whether the individual is exposed or not) we push the odds ratio towards 1.0, called a bias towards the null. This both weakens our ability to detect a statistically significant effect and decreases our estimate of the size of the effect. If we misclassify subjects with a preference (bias) towards any of the four cells in the 2 × 2 table (differential misclassification), the odds ratio can change up or down, depending on the direction of the bias. The preference can occur with respect to exposure, disease, or both.

For example, in a study of contaminated drinking water, we may divide subjects into groups of exposed and unexposed individuals based on some method of estimating exposure. Invariably we make a few errors in this classification, for example, when people who are supplied with contaminated well water do not use this source but instead drink bottled water, or when those thought to be unexposed are exposed not at home but at work or school. Suppose that the *true* odds ratio is 2.67 (cf. Table 3) and that we randomly misclassify all of our subjects 10% of the time with respect to exposure only, i.e., of the 60 who were truly exposed, 6 are misclassified as unexposed, and of the 140 who were truly unexposed, 14 are misclassified as exposed. The disease classification remains unchanged. This now produces a calculated odds ratio of 2.06 (Table 4). Thus with a random misclassifi-

TABLE 5–4. 10% Non-Differential Exposure Misclassification

	Disease	No Disease	Total
Exposed	42	26	68
Not Exposed	58	74	132
Total	100	100	200

Odds Ratio = 42/58 / 26/74 = 2/06 $p < 0.05$

ODDS RATIOS WITH MISCLASSIFICATION

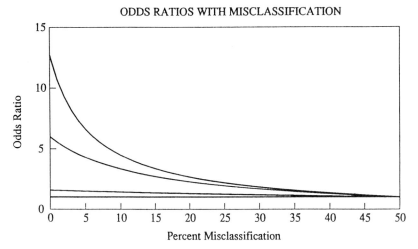

Figure 1. This figure shows the effect of non-differential exposure classification on the odds ratio. As misclassification increases (the abscissa), the odds ratio (the ordinate) tends towards 1.0. The four different curves represent situations with different true odds ratios. The greater the true odds ratio, the greater the absolute change in the odds ratio for the same amount of misclassification.

cation of 10%, the odds ratio has fallen over 35% of its excess over a no-effect level of 1.0. With greater misclassification, the odds ratio falls even more (cf. Fig. 1). A similar result follows from disease misclassification.

The lesson is that even relatively small errors in assigning exposure or disease classifications may have dramatic consequences on measures of epidemiologic effect. Given the rather crude nature of exposure determination in environmental epidemiology today it is likely that misclassification routinely exceeds 10%. Improvements obtainable by new laboratory methods are likely to result in gains in statistical power that exceed the gains in accuracy represented by the test itself. Before a method is used in the field, a determination of its sensitivity and specificity must be made for the given population under study. This important question is treated further in other papers (Griffith *et al.*, 1990, Ozonoff, 1987).

8. CONCLUSION

This rapid survey necessarily omitted some topics of epidemiologic importance, such as confounding and bias (cf. Rothman, 1986). We have instead concentrated on the aspect of epidemiologic investigation most likely to be affected by developments in toxicologic technique, better determination of exposure, and biological effect classification. The development of new laboratory techniques in this area carries with it the promise of improved statistical power in studying the effects of exposures of communities to potentially toxic chemicals. This should encourage toxicologists and epidemiologists to work together to perform the necessary tasks of validation and method development to enable us to move these new methods from the laboratory bench to the field.

REFERENCES

J. Griffith, R. C. Duncan, and B. S. Hulka, Biochemical and biological markers: Implications for epidemiologic studies. Arch. Environ. Health **44**, 375–382, 1989.

S. W. Lagakos, B. J. Wesson, and M. Zelen, An analysis of contaminated well water and health effects in Woburn Massachusetts, J. Am. Stat. Soc. **81**, 583–596 (1986).

D. Ozonoff, Using New Techniques in Epidemiology, *Comments on Toxicology (Comments on Modern Biology, Part B)*, pp. 349–362 (1987).

D. Ozonoff and L. Boden, Truth and Consequences: Health Department Responses to Environmental Problems, *Sci. Tech. Human Values* **12**, 70–77 (1987).

K. J. Rothman, *Modern Epidemiology,* Little, Brown, Boston (1986).

Monitoring of *In Vivo* Dose by Macromolecular Adducts: Usefulness in Risk Estimation

Margareta Törnqvist and Siv Osterman-Golkar
Department of Radiobiology
Arrhenius Laboratories for Natural Sciences
Stockholm University
Stockholm, Sweden

1. WHAT IS RISK ASSESSMENT?

The word *risk* is understood and defined differently by different persons and in different contexts. Since we are discussing cancer here, we restrict the definition of risk to mean the probability (P) of contracting a tumor disease (cancer morbidity risk) or of dying from cancer (cancer mortality risk). Since on average one-half of cancer cases are cured, the mortality risk is about one-half the morbidity risk. Individual risks are expressed as the probability of the event per year or during the whole lifetime; a risk of 10^{-4} per year due to a certain exposure then means that the probability of contracting cancer is one in 10,000 per year or, put another way, that one person out of 10,000 with identical exposure is expected to contract the disease. For an exposed population the collective risk may be calculated as the number of cases expected per year or during the whole lifetime. To the extent that dose-response relationships can be considered linear, the collective risk is simply the product $P(\hat{D}) \times N$, where $\hat{D}$ is the average dose and N the size of the population.

It is known that individuals vary strongly in factors that determine their sensitivity to carcinogenic agents (Willey *et al.*, 1988; Harris, 1989). These factors are both hereditary and acquired. Since we lack the means of quantifying individual's sensitivity at present, the estimated individual risks are the mean values for the populations considered.

Risk assessment of chemical carcinogens involves identification of risk factors with respect to structure, origin, etc., and estimation of the magnitude of the risk (cf. the model developed by O'Riordan, 1979).

The risk identification could be seen as a semiquantitative step, as it must include a decision whether the factor constitutes a non-negligible risk or not. Thus the clarification of dose-response relationships, e.g., with regard to the existence or nonexistence of safe thresholds, is a component of the risk identification. When the factor is a mixed exposure (such as, e.g., urban air pollution), risk identification is concerned with the mixture as such as well as efforts to identify fractions or individual components responsible for the total effect.

Risk estimation is the quantitative step in risk assessment. This activity seeks a quantification of risk under current exposure conditions. This step comprises an estimation of the risk per unit of exposure dose and an exposure assessment. In certain cases a calculation of the contributions from specific sources is included.

2. WHY IS RISK ESTIMATION OF CHEMICALS NEEDED?

By means of sensitive short-term tests for mutagenicity, a rapidly increasing number of chemicals are disclosed as being genotoxic, i.e., potential cancer initiators. Since genotoxic action is associated with chemical changes of DNA, the ability of chemicals to give rise to DNA adducts must be taken as an indication of genotoxicity. We are at present seeing a rapid development, with respect to sensitivity as well as product identification, of techniques for the determination of DNA adducts, directly or indirectly via protein adducts, in humans and animals. This development is expected to lead to an additional avalanche of detected genotoxic agents, which have to be considered carcinogenic and mutagenic.

If these developments do not involve a quantitative aspect, with careful consideration of the magnitude of the risks associated with the findings, the world and especially, regulatory agencies will soon encounter an impossible situation with the banning of chemicals that are potentially useful to human well-being and health, or with a perhaps unnecessary uneasiness about approaching disease in exposed persons or their offspring. Another danger is that limited resources are put into investigations of and preventive measures against negligible or acceptable risk factors identified by means of sensitive methods, and at the same time, factors constituting large risks remain unelucidated.

Conversely, the quantitative aspect is required for the characterization of test systems and analytical methods with respect to their ability of detecting nonacceptable risks, a property that may be described in terms of the statistical power (Ehrenberg, 1984). Particularly when a factor, e.g., because of low solubility or high toxicity, cannot be tested at high concentration, it is important to calculate the probability ("type II error") that a negative test is a false negative.

Also of importance is the relative risk quantification, e.g., the comparison of alternative solutions of technical problems such as energy generation or food preservation. Risk comparisons may also have the purpose of facilitating, to administrators or the public, a realistic perception of risks.

3. METHODS FOR CANCER RISK ESTIMATION

For some 50 chemicals or exposure situations, carcinogenicity has been demonstrated in disease-epidemiological studies of exposed human populations [International Agency for Research on Cancer (IARC), 1987]. In most of these cases any risk coefficient, calculated for the purpose of estimating risks at the low exposure levels that usually occur, is very uncertain, partly because of the statistical error of the determined incidence but, above all, because of the uncertainties in the exposure assessment. The usefulness of epidemiological methods is limited for a number of reasons: low sensitivity, long latency times for the effects, low disease specificity of the causative agent, confounding factors, and the uncertain exposure assessment. Because of these drawbacks and the fact that the risk factors already cause considerable harm when they are detected by epidemiological methods, we must in most cases refer to experimental data.

Positive results from laboratory experiments may be used for risk identification factors, although negative results cannot, without complementary information, be taken to prove absence of genotoxicity (cf. Ehrenberg, 1984). A major difficulty in estimating risks from experimental data lies in the inability of such data to predict the background promotive and cocarcinogenic conditions in human populations with which initiators interact at low doses in the origin of tumors. Long-term animal cancer tests also have the drawbacks of low sensitivity, high costs, and the uncertainty in extrapolations of risk from observations at high dose in animals to low doses in humans.

It is thus evident that methods for risk assessment are required that are sensitive, are specific to the causative agents, have a short latency time (to permit early warning), and permit risk estimation (Ehrenberg and Osterman-Golkar, 1980). The researchers at this laboratory have been working on a solution of this problem.

Although this solution, "the rad-equivalence approach," may seem specific, these studies at the same time have shown the necessity of abandoning the "overcompartmentalization of the risk assessment arena" (Barr, 1985). Instead, the subject has to be treated in a concerted research effort that combines all sources of information. For instance, metabolic and pharmacokinetic studies, using limited numbers of animals, are sources of information of great importance to risk estimation, and disease-epidemiological methods should be used, despite their limited power, as much as possible to check the reasonableness of estimated risks (cf. Barr, 1985).

4. DOSIMETRY AND THE RAD-EQUIVALENCE APPROACH

4.1. Macromolecular Adduct Levels as a Measure of Dose

Dose has by definition the dimension (intensity of exposure) $\times$ (time of exposure), for radiation, e.g., (rad $\cdot$ h^{-1}) $\cdot$ h = rad. In the case with chemicals, concentration (C) is the intensity parameter and dose (D) has thus the dimension concentration $\times$ time, e.g., millimolar hour (mM $\cdot$ h):

$$D = \int C \, dt. \tag{1}$$

A great number of mutagens and carcinogens are electrophiles reactants (R_iX), with an ability to bind covalently to nucleophiles constituents (Y_j) in cells, including certain groups in macromolecules (Miller and Miller, 1966).

The rates of formation of products (R_iY_j) of reaction of electrophilic compounds with tissue nucleophiles are determined by the respective second-order rate constants (k_{ij}) for the reaction $R_iX + Y_j \rightarrow R_iY_j + X$ and the concentration of R_iX ($[R_iX]$):

$$\frac{d[R_iY_j]}{dt} = k_{ij}[R_iX][Y_j]. \tag{2}$$

Integration of Eq. (2) gives (since $[R_iY_j] \ll [Y_j]$)

$$\frac{[R_iY_j]}{[Y_j]} = k_{ij} \int [R_iX]dt = k_{ij} D, \tag{3}$$

i.e., the degree of modification (alkylation, etc.) of Y_j is proportional to the tissue dose.

The cellular target for a genotoxic compound is DNA (cf. Brookes, 1966). The dose of concern is accordingly the dose in the environment of DNA in target tissues—usually called the target dose.

4.2. Target Dose to Blood Dose Ratio

In human dosimetry studies DNA in white blood cells, hemoglobin adducts in red blood cells, or serum albumin can be used as a surrogate for DNA in target tissues. Estimates of target dose from measured levels of adducts to the monitor molecule contain two components: first, the relationship between dose at the monitor site and in the target tissue, and second, the period of time over which the dose is monitored.

Proportionality between binding to DNA and to blood proteins has been demonstrated in experimental animals for a number of carcinogens, including ethylene oxide, methyl and ethyl methanesulfonate, urethane, and several aromatic amines (review: Farmer *et al.*, 1987; data on urethane, Svensson, 1988). For many compounds the dose appears to be approximately the same in various tissues (e.g., ethylene oxide; Segerbäck, 1983). Other compounds, particularly those that are metabolized to short-lived reactive intermediate compounds (e.g., dimethylnitrosamine, Osterman-Golkar *et al.*, 1976; vinyl chloride; Osterman-Golkar *et al.*, 1977) give a dose gradient with the highest dose at the site of bioactivation, often the liver. The size of this effect may be studied in experimental animals.

The other important issue in this context, which is related to the lifetime of the adduct measured, is the establishment of steady-state adduct levels during long-term exposure and the relation of these levels to dose. At chronic or intermittent exposure a steady-state adduct level (s.s.) is attained when the rate of disappearance of adducts is equal to the rate of their formation [cf. Eq. (2)]

$$k_{-ij}[R_iY_j]_{\text{s.s.}} = k_{ij}[R_iX][Y_j] \qquad (4)$$

where k_{-ij} is the first-order rate constant for elimination of the adduct. The steady-state adduct level may then be written

$$\frac{[R_iY_j]_{\text{s.s.}}}{[Y_j]} = \frac{k_{ij}[R_iX]}{k_{-ij}} = \frac{a}{k_{-ij}} \qquad (5)$$

where $[R_iX]$ is the average level of the ultimate carcinogen and a is the increment of the adduct level per unit time (Ehrenberg *et al.*, 1986).

One drawback of DNA adducts as an end point in human dosimetry is that their lifetimes vary from hours to years due to chemical instability, repair, and cell turnover, and depend on the compound and cell type being studied. In order to utilize data on DNA-adduct levels in humans, steps should be taken to characterize the stability of the measured adducts by monitoring adduct levels after terminated exposures. In the case of chemically stable hemoglobin adducts, adduct levels increase in a predictable way during long-term exposure to reach the steady-state adduct level

$$\frac{[R_i Y_j]_{s.s.}}{[Y_j]} = \frac{at_{er}}{2} \tag{6}$$

where t_{er} is the lifetime of the erythrocytes in humans (i.e., the s.s. level is 9a if a is given as the weekly increment of the adduct level; Osterman-Golkar *et al.*, 1976; Segerbäck *et al.*, 1978) and measured adduct levels can be recalculated to obtain, for instance, the average weekly dose.

4.3. Blood Dose to Exposure Dose Ratio

High-to-low dose extrapolations and interspecies extrapolations of carcinogenicity data should be based on the target dose of ultimately reactive compounds. Carcinogen-DNA adduct formation is considered a critical event, and linearity of the dose-response curve at low doses is assumed.

The *in vivo* dose of a reactive compound is determined by the rate of uptake of the compound itself or of a precursor and by rates of chemical and/or enzyme-catalyzed reactions in tissues leading to formation and elimination of the compound (cf. Ehrenberg *et al.*, 1983). The relative rates of these processes may differ considerably between animal species and man (and between individuals), which results in a corresponding variation in the blood dose to exposure dose ratio.

It is also important to recognize that whereas exposures to humans generally are low and the processes follow first-order kinetics, this is not usually the case in laboratory experiments. Long-term assays of carcinogenicity are normally performed at extremely high concentrations in order to get a significant yield of tumors in a small number of animals.

The measurement of hemoglobin adducts as an end point in human epidemiological studies allows the classification of individuals with respect to blood dose of individual compounds and therefore considerably increases the power of detecting and quantifying an effect. Dosimetry, by means of hemoglobin adducts, in animal carcinogenicity studies would be helpful for the understanding of dose-response curves and would give some confidence in interspecies extrapolation.

Studies in experimental animals exposed to diverse carcinogens at low doses show that the degree of binding to macromolecules (DNA, RNA, and proteins) is directly proportional to exposure dose (cf. Farmer *et al.*, 1987). Human adduct data, compatible with a linear relationship between adduct levels and exposure dose, have been obtained for a few compounds, i.e., ethylene oxide (Duus *et al.*, 1989), propylene oxide (Högstedt *et al.*, 1989), and ethene in tobacco smoke (Törnqvist, 1989; Bailey *et al.*, 1988). For other compounds, such as polycyclic aromatic hydrocarbons (PAH), a greater variability may be anticipated.

4.4 Risk Estimation based on Target Dose and the Rad-Equivalence Approach

The dose-response or dose-risk relationship for the carcinogenic action of an initiator is determined by the (multiplicative) interaction of this initiator with promotive and cocarcinogenic conditions in the human population.

The rad-equivalence concept seeks to compensate for differences between humans and laboratory organisms with respect to factors that determine the expression of target dose in terms of mutation and cancer (e.g., DNA repair and general promotional pressure operating within individuals in a population). The prospective risk model we have proposed (Ehrenberg, 1980) is based on the quantitative determination of the respective effectiveness at low doses (defined in terms of target dose) of the test chemical and of penetrating ionizing radiation (usually acute γ radiation) to induce defined genetic damage in the same experimental species. In this way a numerical value (the quality factor Q) may be determined whereby the capacity of a defined target dose of the genotoxic chemical to induce genetic damage is expressed in terms of rad-equivalents, i.e., the number of rads giving the same response (or risk) as a unit of chemical dose (e.g., expressed in millimolar hour, mM · h). This chemical dose is in some cases suitably expressed in terms of the (cumulative) frequency of DNA adducts.

The significance of experimentally determined Q values hinges upon their possible extrapolative value. Thus the rad-equivalent value determined for a particular genotoxic chemical in a given experimental species may have a similar numerical value in other species including man (as indicated in the case of ethylene oxide; see below). Provided that this key proposition can be substantiated, experimentally determined rad-equivalence values may be used in conjunction with human radiation risk coefficients to assess the capacities of genotoxic agents to induce heritable damage in humans. Research on the theoretical background of the empirical rad-equivalence approach is required. It may be found necessary to replace γ radiation as a reference by chemical agents, representative of chemical classes, reaction mechanisms, and repair pathways.

5. AN EXAMPLE: RISK ESTIMATION OF ETHENE

5.1. The Model for Risk Estimation

The relationships between target dose, blood dose or tissue dose, and exposure dose can be illustrated as follows (cf. Törnqvist and Ehrenberg, 1989):

$$
\begin{array}{ccccc}
\text{Exposure} & \xrightarrow{\;f_1\;} & \text{Blood} & \xrightarrow{\;f_2\;} & \text{Target} \\
\text{dose} & & \text{dose} & & \text{dose} \\
D_{\text{exp}} & & D_{\text{blood}} & & D_{\text{targ}}
\end{array}
$$

The level of adducts to hemoglobin (or DNA in leukocytes) gives a measure of dose in blood. To the extent that an estimated risk should be based on exposure dose (D_{exp}) the relationship between blood dose and exposure dose (f_1) must be established through careful exposure assessment and modeling work. The relationship of target dose to blood dose (f_2) must be established primarily in animal experiments. The calculation of risk on the basis of exposure dose may accordingly be carried out by introducing the factors f_1 and f_2 and using the cancer risk radiation risk coefficient, k_γ, and the quality factor, Q:

$$
\text{Risk} = k_\gamma\, Q\, D_{\text{targ}}, \tag{7}
$$

$$
D_{\text{targ}} = f_1 f_2 D_{\text{exp}}.
$$

Ethene is converted to ethylene oxide in experimental animals as demonstrated directly by measurements of exhaled epoxide (Filser and Bolt, 1983) and indirectly by measurements of hydroxyethyl adducts to proteins (Ehrenberg *et al.*, 1977) and to DNA (Segerbäck, 1983, 1985). The metabolic conversion of ethene to ethylene oxide in humans is indicated through the observation of increased adduct levels in hemoglobin of cigarette smokers (Törnqvist *et al.*, 1986b) and confirmed in studies on ethene-exposed workers (Törnqvist *et al.*, 1989a). The risk estimation of ethene may be based on the risk estimation carried out for ethylene oxide using the values of Q and f_2 established for this compound.

5.2. Estimation of the Blood Dose to Exposure Dose Ratio (f_1)

Data on adduct levels in hemoglobin after exposure of rodents to ethene are compatible with an enzyme-catalyzed activation following Michaelis-Menten kinetics; that is, the dose of the reactive intermediate compound, ethylene oxide, is directly proportional to the air concentration of ethene at low concentrations ($\ll K_m$). At high concentrations the activating system is saturated and a constant amount of ethene is transformed per unit time (Segerbäck, 1983; cf. also Andersen *et al.*, 1980).

For an estimation of the cancer risk due to ethene a knowledge of the fraction (α) of the inhaled ethene metabolized at low air levels to ethylene oxide is essential. In rodents $\alpha \approx 0.08$ (Segerbäck, 1983; Törnqvist *et al.*, 1988) during continuous exposure. Adducts from ethylene oxide to *N*-terminal valine (HOEtVal) in hemoglobin have been determined in smokers in several studies (Törnqvist *et al.*, 1986b; Bailey *et al.*, 1988), and an average increment of 85 pmol HOEtVal/g Hb per 10 cigarettes/day has been calculated (Törnqvist, 1989). This figure corresponds to metabolism of about 6% of the ethene in the inhaled mainstream smoke (Törnqvist, 1989). In nonsmoking fruit-store workers exposed to ethene an incremental adduct level of HOEtVal corresponding to metabolism of ea. 3% of the inhaled ethene has been determined (Törnqvist *et al.*, 1989a). The α value of 6% is used in the tentative risk estimation below.

If alveolar ventilation (V_{alv}) is 0.11 liter $\cdot$ kg^{-1} $\cdot$ min^{-1} and the first-order rate constant for detoxification of ethylene oxide is estimated to be 3 h^{-1} (Osterman-Golkar and Bergmark, 1988; Duus *et al.*, 1989), we obtain

$$f_1 = \frac{0.06 \times 10^{-6} \text{ ppm}^{-1} \times 0.11 \text{ liter} \cdot \text{kg}^{-1} \cdot \text{min}^{-1} \times 60 \text{ min} \cdot \text{h}^{-1}}{25 \text{ liter} \cdot \text{mol}^{-1} \times 3 \text{ h}^{-1}}$$

$$= 5 \times 10^{-9} \text{ mol} \cdot \text{kg}^{-1} \cdot \text{h} \cdot (\text{ppm} \cdot \text{h})^{-1}$$
$$= 5 \times 10^{-6} \text{ (mM} \cdot \text{h)(ppm} \cdot \text{h})^{-1}.$$

(For details see Törnqvist, 1989.)

5.3. Target Dose to Blood Dose Ratio (f_2)

Segerbäck (1983, 1985) studied the dose distribution in mice and rats by means of DNA adducts after exposure to radiolabeled ethylene oxide or ethene. A variation by a factor less than 2 around the blood dose was shown. At the present state of knowledge we find it prudent to base the risk estimate on the assumption that the dose of ethylene oxide following exposure to ethylene oxide (or ethene) is evenly distributed in the body.

5.4. The Quality Factor Q

For the radiation-dose equivalent of the ethylene oxide dose, a mean value from measurements in various biological systems, $Q = 80$ (mM $\cdot$ h)$^{-1}$, is adopted (Ehrenberg *et al.*, 1983; Kolman *et al.*, 1988).

5.5　The Risk Coefficient for γ Radiation (k_γ)

We have found it appropriate to apply a risk coefficient $k_\gamma = 2 \cdot 10^{-4}$ and $4 \cdot 10^{-4}$ rad^{-1}, respectively, for total cancer risk mortality and morbidity risks. The cancer mortality value, considered to be valid at low dose rates, is 3 times lower than the coefficient given by the United Nations Scientific Committee on the Effects of Atomic Radiation (UNSCEAR, 1988) for high doses and high dose rates (cf. Törnqvist *et al.*, 1989b).

5.6.　Risk Estimation

The cancer (morbidity) risk coefficient for ethene is calculated from Eq. (7) according to

$$
\begin{aligned}
\text{Risk}/D_{exp} &= f_1 f_2 \, Q \, k_\gamma \\
&= 5 \times 10^{-6} \, (\text{mM} \cdot \text{h})(\text{ppm} \cdot \text{h})^{-1} \times 1 \\
&\quad \times 80 \, (\text{rad})(\text{mM} \cdot \text{h})^{-1} \times 4 \times 10^{-4} \, \text{rad}^{-1} \\
&= 0.16 \times 10^{-6} \, (\text{ppm} \cdot \text{h})^{-1}.
\end{aligned}
$$

An average exposure to 15 ppb in the environment (an estimated value for Sweden; Törnqvist, 1989) leads to a risk

$$
\begin{aligned}
&15 \times 10^{-3} \, \text{ppm} \times 365 \times 24 \, \text{h} \cdot \text{y}^{-1} \times 0.16 \times 10^{-6} \, (\text{ppm} \cdot \text{h})^{-1} \\
&= 2 \times 10^{-5} \, \text{y}^{-1}.
\end{aligned}
$$

During working hours V_{alv} may be assumed to be some 50% higher, i.e. 1 ppm $\cdot$ h ethene corresponds to $7.5 \cdot 10^{-6}$ mM $\cdot$ h or $0.6 \cdot 10^{-3}$ rad-equivalents. The dose limit for occupational exposure to radiation (50 mSv, or 5 rad γ radiation, per year, i.e., ca. 2000 h) would thus correspond to an average level of ethene of 4 ppm during working hours.

6.　SENSITIVITY OF HEMOGLOBIN-ADDUCT DETERMINATION

The sensitivity required of methods for dose monitoring may be defined by the rad-equivalence approach. The practical detection level of the N-alkyl Edman procedure developed at this laboratory for monitoring low-molecular-weight hemoglobin adducts (Törnqvist *et al.*, 1986a) is 1–10 pmol alkylvaline/g globin in human samples. This corresponds to the adduct level obtained from exposure to ethylene oxide at an average level of 0.5–5 ppb 40 h/week during more than 4 months, the life span of hemoglobin in humans. It appears that the method is sensitive enough to permit the determination of tissue doses in the general environment. In the case of ethylene oxide or alkylating agents with similar reaction patterns (Ehrenberg and Osterman-Golkar, 1980), the level 1–10 pmol N-alkylvaline/g globin caused by chronic or intermittent exposure is estimated to be associated with a cancer risk around 10^{-6}–10^{-5} per year. This means that the method permits monitoring at levels down to the range where risks to members of the public have been considered as acceptably small [International Commission on Radiological Protection (ICRP), 1977].

7. ADVANTAGES OF THE RAD-EQUIVALENCE APPROACH

Dose monitoring and application of the rad-equivalence approach have several advantages and eliminate a number of uncertainties encountered in efforts to estimate cancer risks from genotoxic agents, such as:

- Uncertainties are eliminated in interspecies extrapolations with respect to differences in metabolism and to the question of expression of dose (whether dose should be expressed per kg body weight or per m^2 body area and whether the daily dose or the lifetime dose is essential).
- The approach allows for the interindividual variation in humans.
- A major advantage is the possibility to identify and also estimate low cancer risks.
- The use of a common unit facilitates comparisons of risks in the evaluation step, and the adoption of general rules (such as those of ICRP) for judgement of acceptability.
- Long-term animal tests can be avoided.

ACKNOWLEDGMENTS
These studies were supported by the National Swedish Environment Protection Board, the Swedish Work Environment Fund, the Swedish Cancer Society, and Shell Internationale Research Maatschappij B.V.

REFERENCES

M. E. Andersen, M. L. Gargas, A. R. Jones, and L. J. Jenkins, Jr., Determination of the kinetic constants for metabolism of inhaled toxicants in vivo using gas uptake measurements, *Toxicol. Appl. Pharmacol.* **54**, 100–116 (1980).

E. Bailey, A. G. F. Brooks, C. T. Dollery, P. B. Farmer, B. J. Passingham, M. A. Sleightholm, and D. W. Yates, Hydroxyethylvaline adduct formation in haemoglobin as a biological monitor of cigarette smoke intake, *Arch. Toxicol.* **6**, 247–253 (1988).

J. T. Barr, The calculation and use of carcinogenic potency: A review, *Regul. Toxicol. Pharmacol.* **5**, 432–459 (1985).

P. Brookes, Quantitative aspects on the reaction of some carcinogens with nucleic acids and the possible significance of such reactions in the process of carcinogenesis, *Cancer Res.* **26**, 1994–2003 (1966).

U. Duus, S. Osterman-Golkar, M. Törnqvist, J. Mowrer, S. Holm, and L. Ehrenberg, Studies of determinants of tissue dose and cancer risk from ethylene oxide exposure. In: *Proceedings of the Symposium on Management of Risk from Genotoxic Substances in the Environment,* ed. L. Freij, Swedish National Chemicals Inspectorate, Solna, Sweden, pp. 141–153 (1989).

L. Ehrenberg, Purposes and methods of comparing effects of radiation and chemicals. In: *Radiobiological Equivalents of Chemical Pollutants,* IAEA, Vienna, Austria, pp. 23–36 (1980).

L. Ehrenberg, Aspects of statistical inference in testing for genetic toxicity. In: *Handbook of Mutagenicity Test Procedures,* eds. B. Kilbey, M. Legator, W. Nichols, and C. Ramel, Elsevier, Amsterdam, pp. 775–822 (1984).

L. Ehrenberg and S. Osterman-Golkar, Alkylation of macromolecules for detecting mutagenic agents, *Teratog. Carcinog. Mutagen.* **1**, 105–127 (1980).

L. Ehrenberg, E. Moustacchi, and S. Osterman-Golkar, Dosimetry of genotoxic agents and dose-response relationships of their effects, *Mutat. Res.* **123**, 121–182 (1983).

L. Ehrenberg, S. Osterman-Golkar, D. Segerbäck, K. Svensson, and C. J. Calleman, Evaluation of genetic risks of alkylating agents. III. Alkylation of haemoglobin after metabolic conversion of ethene to ethene oxide *in vivo, Mutat. Res.* **45**, 175–184 (1977).

L. Ehrenberg, S. Osterman-Golkar, D. Segerbäck, and M. Törnqvist, Power of methods for monitoring exposure to genotoxic chemicals by covalently bound adducts to macromolecules. In: *Environmental Mutagenesis and Carcinogenesis,* eds. N. K. Notani and P. S. Chauhan, Bhabha Atomic Research Centre, Bombay, pp. 155–166 (1986).

P. B. Farmer, H.-G. Neumann, and D. Henschler, Estimation of exposure of man to substances reacting covalently with macromolecules, *Arch. Toxicol.* **60**, 251–260 (1987).

J. G. Filser and H. M. Bolt, Exhalation of ethylene oxide by rats on exposure to ethylene, *Mutat. Res.* **120**, 57–60 (1983).

C. C. Harris, Interindividual variation among humans in carcinogen metabolism, DNA adduct formation and DNA repair, *Carcinog.* **10**, 1563–1566 (1989).

B. Högstedt, E. Bergmark, M. Törnqvist, and S. Osterman-Golkar, Chromosomal aberrations and micronuclei in lymphocytes in relation to alkylation of hemoglobin in workers exposed to ethylene oxide and propylene oxide. Hereditas, in press.

International Agency for Research on Cancer, IARC Monographs on the Evaluation of Carcinogenic Risks to Humans, Suppl. 7, *Overall Evaluations of Carcinogenicity: An Updating of IARC Monographs Volumes 1–42*, International Agency for Research on Cancer, Lyon, France (1987).

International Commission on Radiological Protection Publication No. 26, *Recommendations of the International Commission on Radiological Protection*, Ann. ICRP Vol. 1., No. 3, Pergamon Press, Oxford, (1977).

A. Kolman, D. Segerbäck, and S. Osterman-Golkar, Estimation of the cancer risk of genotoxic chemicals by the rad-equivalence approach. In: *Methods for Detecting DNA Damaging Agents in Humans: Applications in Cancer Epidemiology and Prevention*, IARC Scientific Publication No. 89, eds. H. Bartsch, K. Hemminki, and I. K. O'Neill, IARC, Lyon, France, pp. 258–264 (1988).

E. E. Miller and J. A. Miller, Mechanisms of chemical carcinogenesis: Nature of proximate carcinogens and interactions with macromolecules, *Pharmacol. Res.* **18**, 805–838 (1966).

T. O'Riordan, Environmental impact analyses and risk assessment in management perspective. In: *Energy Risk Management*, eds. G. T. Goodman and W. D. Rowe, Academic Press, New York, pp. 21–36 (1979).

S. Osterman-Golkar and E. Bergmark, Occupational exposure to ethylene oxide: Relation between in vivo dose and exposure dose, *Scand. J. Work Environ. Health* **14**, 372–377 (1988).

S. Osterman-Golkar, L. Ehrenberg, D. Segerbäck, and I. Hällström, Evaluation of genetic risks of alkylating agents. II. Haemoglobin as a dose monitor. *Mutat. Res.* **34**, 1–10 (1976).

S. Osterman-Golkar, D. Hultmark, D. Segerbäck, C. J. Calleman, R. Göthe, L. Ehrenberg, and C. A. Wachtmeister, Alkylation of DNA and proteins in mice exposed to vinyl chloride, *Biochem. Biophys. Res. Commun.* **76**, 259–266 (1977).

D. Segerbäck, Alkylation of DNA and hemoglobin in the mouse following exposure to ethene and ethene oxide, *Chem. Biol. Interact.* **45**, 139–151 (1983).

D. Segerbäck, In vivo dosimetry of some alkylating agents as a basis for risk estimation, Doctoral thesis, University of Stockholm, Stockholm (1985).

D. Segerbäck, C. J. Calleman, L. Ehrenberg, G. Löfroth, and S. Osterman-Golkar, Evaluation of genetic risks of alkylating agents. IV. Quantitative determination of alkylated amino acids in haemoglobin as a measure of the dose after treatment of mice with methyl methanesulfonate, *Mutat. Res.* **49**, 71–82 (1978).

K. Svensson, Alkylation of protein and DNA in mice treated with urethane, *Carcinog.* **9**, 2197–2201 (1988).

M. Törnqvist, Monitoring and cancer risk assessment of carcinogens, particularly alkenes in urban air, Doctoral thesis, University of Stockholm, Stockholm (1989).

M. Törnqvist and L. Ehrenberg, Comparative studies of cancer risks from fossil and nuclear energy generation. In: *Proceedings of the Symposium on Management of Risk from Genotoxic Substances in the Environment*, ed. L. Freij, Swedish National Chemicals Inspectorate, Solna, Sweden, pp. 204–219 (1989).

M. Törnqvist, J. Almberg, E. Bergmark, S. Nilsson, and S. Osterman-Golkar, Ethylene oxide doses in ethene exposed fruit store workers, *Scand. J. Work Environ. Health* **15**, 436–438 (1989a).

M. Törnqvist, A. Kautiainen, R. N. Gatz, and L. Ehrenberg, Hemoglobin adducts in animals exposed to gasoline and diesel exhausts. 1. Alkenes, *J. Appl. Toxicol.* **8**, 159–170 (1988).

M. Törnqvist, J. Mowrer, S. Jensen, and L. Ehrenberg, Monitoring of environmental cancer initiators through hemoglobin adducts by a modified Edman degradation method, *Anal. Biochem.* **154**, 255–266 (1986a).

M. Törnqvist, S. Osterman-Golkar, A. Kautiainen, S. Jensen, P. B. Farmer, and L. Ehrenberg, Tissue doses of ethylene oxide in cigarette smokers determined from adduct levels in hemoglobin, *Carcinog.* **7**, 1519–1521 (1986b).

M. Törnqvist, D. Segerbäck, and L. Ehrenberg, The "rad-equivalence approach" for assessment and evaluation of cancer risks, exemplified by studies of ethylene oxide and ethene, Biomonitoring and Carcinogen Risk Assessment Meeting, Cambridge, 27–28 July, 1989b, (in press).

UNSCEAR, *Sources, Effects and Risks of Ionizing Radiation*, United Nations Scientific Committee on the Effects of Atomic Radiation. Report to the General Assembly, United Nations, New York (1988).

J. C. Willey, A. Weston, A. Haugen, T. Krontiris, J. Resan, E. McDowell, B. Trump, and C. C. Harris, DNA restriction fragment length polymorphism analysis of human bronchogenic carcinoma, *Methods for Detecting DNA Damaging Agents in Humans: Applications in Cancer Epidemiology and Prevention*, IARC Scientific Publication No. 89, eds. H. Bartsch, K. Hemminki, and I. K. O'Neill, IARC, Lyon, France, pp. 439–450 (1988).

Analytical Techniques and Specific Applications

Human Tissue Culture Systems for Studies of Carcinogen Metabolism and Macromolecular Interactions

Gary D. Stoner and Michael J. Paolino
Department of Pathology
Medical College of Ohio
Toledo, Ohio

1. INTRODUCTION

Fifteen years ago, there was very little information on the ability of human tissues derived from different organ sites to metabolize environmental procarcinogens. The principal reason for this was the lack of methods to culture normal human tissues in a viable state prior to the early 1970s. Since then however, significant developments have been made in the culture of normal human tissues and cells and specifically epithelial tissues, which are the principal sites for cancer development in humans (Harris, 1987). As a result of these advancements, tissues from several human organs including the tracheobronchus (Barrett *et al.*, 1976; Trump *et al.*, 1980a), peripheral lung (Stoner *et al.*, 1978; Stoner, 1980), colon (Autrup, 1980), esophagus (Hillman *et al.*, 1980a), skin (Selkirk *et al.*, 1983), breast (Hillman *et al.*, 1980b), pancreas (Andersson and Hellerstrom, 1980; Jones *et al.*, 1980), kidney (Trump *et al.*, 1980b), endometrium (Kaufman *et al.*, 1980), uterine cervix (Wilbanks *et al.*, 1980), oral cavity (Buchner and Mlinek, 1975), bladder (Knowles *et al.*, 1980, 1983; Reedy and Heatfield, 1987), placenta (Stromberg, 1980), prostate (Heatfield, *et al.*, 1980b), liver (Autrup *et al.*, 1984), gastric mucosa (Donaldson and Kapadia, 1980), small intestine (Trier, 1980), and blood vessels (Kocan *et al.*, 1980) have been maintained for various periods in a viable state *in vitro*. Some of these tissues have been employed for investigations of the metabolism of several classes of carcinogens including aflatoxins (Autrup *et al.*, 1979, 1980a; Harries *et al.*, 1986; Mandal *et al.*, 1987; Stoner *et al.*, 1982b), aromatic amines (Harries *et al.*, 1986; Moore *et al.*, 1982, 1984; Schut *et al.*, 1984; Shivapurkar *et al.*, 1987; Stoner *et al.*, 1987), 1,2-dimethylhydrazine (Autrup *et al.*, 1980a; Harris *et al.*, 1977b), polycyclic aromatic hydrocarbons (Autrup *et al.*, 1977, 1978a, 1978b, 1980a, 1980b, 1982; Autrup and Harris, 1983; Dorman *et al.*, 1981; Harris *et al.*, 1976a, 1976b, 1977, 1978, 1979; Moore *et al.*, 1982; Selkirk *et al.*, 1983; Stoner *et al.*, 1978, 1982a, 1986; Teel *et al.*, 1986a,b; Yang, 1977), 6-nitrochrysene (Delclos *et al.*, 1989), and nitrosamines (Autrup and Stoner, 1982; Castonguay *et al.*, 1983; Harris *et al.*, 1977a, 1977b, 1979; Hecht *et al.*, 1982).

The objective of this chapter is to describe a method used in our laboratory, and in other laboratories (Autrup, 1980; Harris *et al.*, 1978; Trump *et al.*, 1980a), for the culture of fragments (explants) of normal human epithelial tissues. This method is one of several techniques available for the culture of human tissues. The advantages and disadvantages of the method for maintaining the functional state of different epithelial tissues will be discussed, as well as its applicability to investigations of chemical carcinogenesis.

2. HUMAN TISSUE CULTURE

2.1. Tissue Collection

The primary activity that must underlie all other features of investigations with human tissues is the collection of the tissues in a medically and ethically acceptable manner; i.e., with prior informed consent from either the subject or relatives of the subject and according to procedures approved by an institutional review board. The anonymity of the subject must be protected and, for this purpose, it is preferable not to keep the name of the subject on file in the laboratory. In addition, in no instance should names of the subjects be included in publications.

Tissues for carcinogenesis studies are obtained from either surgery or autopsy (Trump *et al.*, 1980a, 1980b). When collecting tissue from surgery, it is important that the surgeons be informed in advance to place the specimen

TABLE 7–1. Maximum Time After Death for Collection of Viable Human Tissues from Autopsies*

Tissue	Time (h)[†]
Colon	3
Esophagus	4
Oral Cavity	4
Peripheral lung	6
Tracheobronchus	18
Skin	> 24
Urinary Bladder	4

*Data in this table are confined to tissues cultured in our laboratory.
[†]These times are approximations and can vary from case to case.

in a suitable culture medium rather than in a preservative such as formalin. From autopsy, it is preferable to collect the tissues as soon as possible after death; i.e., within 1 hour, and to immediately place the tissue in a suitable culture medium to prevent it from drying. In most institutions, however, it is not possible to collect tissues from autopsy within 1 hour after death. This limitation does not preclude the development of a human tissue studies program since our laboratory and others (Resau *et al.,* 1987) have found that certain human tissues can be collected as long as 12–24 h after death and be propagated in a viable state *in vitro* (Table 1). The tissues may appear nonviable (by histopathological examination) at the time of collection, but the basal epithelial cells, in particular, can recover from ischemic damage and repopulate the epithelium after days to weeks in culture (Trump *et al.,* 1980a). The time period for recovery can vary with the type of tissue and from one specimen to another.

In our research, only grossly *normal* tissues are collected for carcinogenesis studies. With respect to surgical specimens, normal tissues are dissected from the diseased tissue such that there is an adequate amount of diseased tissue available for histological analysis and diagnosis. In the case of autopsy, care is taken not to collect tissues from individuals who have fungal or viral (including HIV) infections, a history of tuberculosis or active tuberculosis, metastatic cancer, or sepsis. In addition, lung specimens are not collected from patients maintained with life-support equipment (respirators) for extended periods (more than a few days), since the tracheobronchial epithelium of these individuals is usually heavily contaminated with microbes and is often nonviable.

Immediately after collection, tissue specimens from both surgery and autopsy are placed in ice-cold tissue culture medium for transport to the laboratory. We use L-15 medium (Leibovitz, 1963) for transport since it is internally buffered to equilibrate with the atmosphere and does not undergo extreme shifts in pH.

2.2 Explant Culture

Using aseptic technique, and in a laminar flow hood, the tissue specimens are removed from the transport vessel and placed in either 100- or 150-mm tissue culture dishes. The tissues are then cut into fragments (explants) and the explants are placed onto the etched surface of 60-mm tissue culture dishes as follows.

2.2.1. Tracheobronchus

Specimens of tracheobronchus are trimmed of adherent connective tissue and peripheral lung tissue and cut into flat 0.5-cm² explants (Barrett *et al.*, 1976). The explants are placed with the epithelium uppermost on opposite sides (2 per side) of each tissue culture dish.

2.2.2. Peripheral Lung

Specimens are cut into 0.1–0.2-cm² explants, and the explants are placed on opposite sides (10 per side) of each tissue culture dish. In our laboratory explants of peripheral lung are placed on the surface of Gelfoam sponge (UpJohn Company) to improve nutrient uptake into the lower portions of the tissue (Stoner *et al.*, 1978; Stoner, 1980).

2.2.3. Colon, Oral Cavity, Esophagus, Bladder

Mucosa from these tissues is dissected from the underlying muscularis and cut into 0.3–0.5-cm² explants (Autrup, 1980; Buchner and Mlinek, 1975, Hillman *et al.*, 1980b; Knowles *et al.*, 1980). The explants are placed with the epithelium uppermost on opposite sides (2 per side) of each tissue culture dish.

2.2.4. Pancreatic Duct

The main duct is opened along the entire length of the pancreas and the glandular portion dissected away from the duct (Jones *et al.*, 1980). The duct is cut into explants approximately 1 cm long, and the explants are placed into tissue culture dishes (3 per dish).

2.2.5. Skin

Epidermal mucosa is dissected from the dermis and trimmed of hair. The mucosa is cut into explants of 0.5 cm², and the explants are placed on opposite sides (2 per side) of each tissue culture dish (Selkirk *et al.*, 1983).

2.2.6. Prostate

Specimens are sliced 0.1 cm in thickness with razor blades on a dental wax plate (Heatfield *et al.*, 1980b). Slices are then placed in L-15 medium in tissue culture dishes (100 mm) using aseptic surgical technique. They are then trimmed into small $0.1 \times 0.1 \times 1$-cm^3 explants with a scalpel, washed in L-15 medium, and transferred to 60-mm dishes (5–10 explants per dish).

2.2.7. Breast

Ductal elements are dissected free from the surrounding fat and stroma with the aid of a dissecting microscope and trimmed into 0.5–1-cm^2 explants (Hillman *et al.*, 1980a). The explants are then placed on opposite sides of each tissue culture dish (4–6 per dish).

2.2.8. Placenta

The placenta is dissected into 1-cm^2 explants, and the explants are placed into tissue culture dishes (3–4 per dish) (Stromberg, 1980).

To each tissue culture dish is added 2.5 to 3 ml of CMRL-1066 medium, or another suitable culture medium (Stoner *et al.*, 1982a), supplemented with hydrocortisone hemisuccinate (0.1 μg per ml), insulin (1 μg per ml), glutamine (2 mM), β-retinyl acetate (0.1 μg per ml), gentamicin (50 μg per ml), and amphotericin B (0.25 μg per ml). Serum is not added to the medium.

The cultures are then placed in a controlled atmosphere chamber (Bellco Glass Co., Vineland, N.J.) and gassed with a mixture of either 50% O_2–45% N_2–5% CO_2 or 5% CO_2 in air for approximately 5 min. The survival of most tissues is comparable in either gaseous environment; however, that of the peripheral lung is optimal in 50% O_2–45% N_2–5% CO_2. The chamber is then placed on a rocker platform and rocked at approximately 10 cycles per minute to permit the medium to flow intermittently over the surface of the tissues. The explants are incubated at 36.5°C and the medium and atmosphere replaced after the first 24 h of incubation and every 48 h thereafter. Changing the medium after the first 24 h of incubation is important for reducing the risk of contamination.

2.3. Determination of Tissue Viability

Several procedures can be used to assess the viability of the tissues both before and during incubation *in vitro*. At the time of initiation of the explant cultures at least one explant per tissue is placed in 10% buffered formalin for light microscopy, and/or a solution of 4 parts formaldehyde and 1 part glutaraldehyde (McDowell and Trump, 1976) for both light and electron microscopy. This procedure allows an assessment of the morphological appearance of the tissue prior to culture. A lack of normal epithelium covering

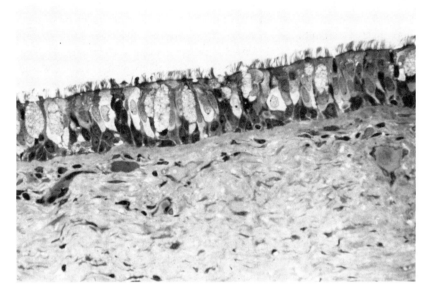

Figure 1. Explant of human tracheobronchus maintained for 7 days *in vitro.*

the surfaces of tracheobronchial, oral cavity, esophagus, bladder, colon, pancreatic duct, and skin specimens, or evidence of necrosis in peripheral lung, prostate, breast, and placental specimens would indicate poor viability. However, as indicated above, tissues may appear nonviable at the time of collection and recover from ischemic damage during the culture. A further evaluation is done by placing 1 or 2 explants per specimen in primary culture and observing the cultures for outgrowth of epithelial cells by phase contrast microscopy during a period of 3–4 weeks.

Other parameters of cell viability include both biochemical and autoradiographic determination of the incorporation of radiolabeled nucleotides into DNA (Autrup, 1980; Stoner, 1980) and radiolabeled amino acids into protein (Autrup, 1980). In our experience, there is a strong correlation between the morphological evaluation of the tissue and the uptake of radiolabeled precursors into cellular DNA and protein.

2.4. Maintenance of Differentiated Function

2.4.1. Tracheobronchus

When cultured in the above-described conditions, the normal morphology of the columnar epithelium is maintained for a period of 1–2 months (Fig. 1) (Barrett *et al.,* 1976). During this time, mucus is produced into the culture medium. After this, the epithelium becomes squamous and the cells continue to proliferate. Interestingly, when tracheobronchial explants are cultured in a simple medium; i.e., Eagle's minimal essential medium (MEM), the normal differentiation of the tissue is maintained for a period of at least one year (Trump *et al.*, 1980a). This maintainence of differentiated function appears to be related to a slower growth rate of the epithelial cells in MEM versus CMRL-1066 medium.

2.4.2. Peripheral Lung

The normal architecture of the peripheral lung is maintained for a period of 3–4 days (Fig. 2) (Stoner, 1980). After this, the structural integrity of the alveoli is lost in the central portions of the explant due to collapse of the alveolar walls. In addition, the central portions of the explant eventually become necrotic. In the outer portions of the explant, there is a marked thickening of the alveolar walls during the first few days in culture due to an increase in both the size and number of epithelial lining cells. Lamellar inclusion bodies are extruded into the alveolar spaces, indicative of the functional activity of the type 2 alveolar epithelial cells. After 7–25 days, the alveoli in the outer portion of the explant remain intact and the necrotic areas in the internal portion become filled with connective tissue cells. Bronchiolar epithelial cells and endothelial cells are well maintained during this period.

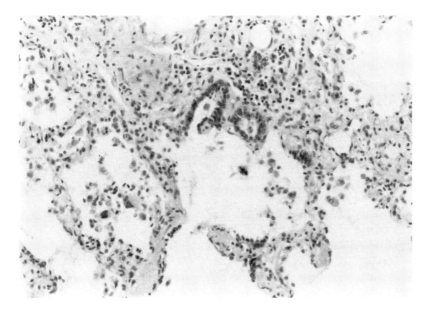

Figure 2. Explant of human peripheral lung maintained for 4 days *in vitro*.

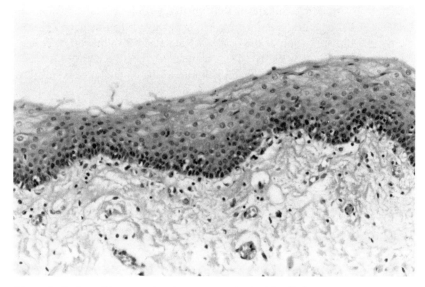

Figure 3. Explant of human oral cavity mucosa maintained for 10 days *in vitro*.

2.4.3. Oral Cavity

Explants of oral cavity mucosa can be maintained for a period of at least 14 days (Fig. 3) (Buchner and Mlinek, 1975). During this period the tissue remains morphologically normal.

2.4.4. Esophagus

Explants of human esophagus can be maintained for periods of 7 months (Fig. 4) (Hillman *et al.*, 1980b). During the first weeks in culture, the stratified, noncornified epithelium shows a rapid desquamation of the superficial layers. After 7 days, the esophageal epithelium gradually progresses from 1–2 cell layers in thickness to form a typical stratified squamous epithelium of several cell layers in thickness. The squamous epithelial characteristics of the cells are maintained during several months.

2.4.5. Bladder

The normal morphology of the bladder epithelium can be maintained for a period of 5–7 days *in vitro* (Fig. 5) (Knowles *et al.*, 1980). Some epithelial cells contain large quantities of glycogen. The epithelium then becomes thickened (hyperplastic) particularly along the edge of the explant adjacent to the culture dish. In our laboratory, bladder explants are maintained in a viable state for a period of at least 30 days (Stoner *et al.*, 1982a, 1987). Similar results were reported from other laboratories (Knowles *et al.*, 1983; Reedy and Heatfield, 1987). The morphology and cell kinetic parameters of

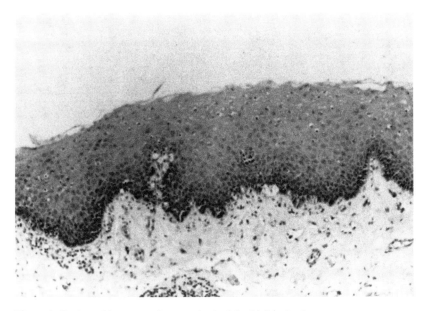

Figure 4. Explant of human esophagus maintained for 14 days *in vitro*.

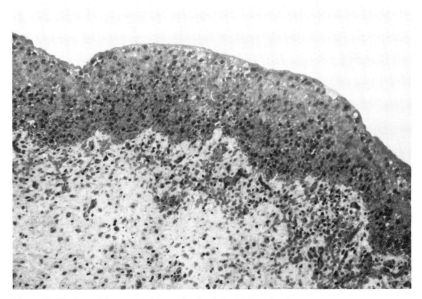

Figure 5. Explant of human bladder maintained for 30 days *in vitro*.

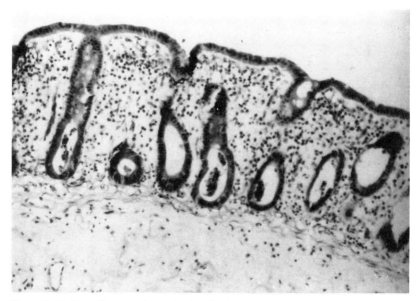

Figure 6. Explant of human colon maintained for 3 days *in vitro*.

cultured explants of human bladder maintained for several weeks in culture have been described in detail (Heatfield *et al.*, 1980a; Reedy and Heatfield, 1987).

2.4.6. Colon

Explants of human colon can be maintained for at least 28 days (Autrup, 1980). Survival of the tissue beyond 14 days is enhanced by the addition of 1.5% dimethylsulfoxide to the culture medium and the use of Gelfoam sponge as support for the explants. After 48–72 h of incubation, the epithelium of colonic explants is usually composed of a single layer of epithelium with few glands (Fig. 6). The normal structure of the crypts is lost, and there is a progressive reduction in the production of mucus into the medium. Autoradiograms revealed that the epithelial cells incorporated both [^{3}H]leucine and [^{3}H]thymidine to a greater extent than the stromal cells, and that the incorporation occurred predominately in the middle and lower parts of the crypt in early cultures (Autrup, 1980).

2.4.7. Pancreatic Duct

Viable human pancreatic ductal explants can be maintained *in vitro* for at least 2 months (Jones *et al.*, 1980). Alcian-blue dye-periodic-acid-Schiff (PAS) dye-positive material is observed in epithelial cells and in lumens of the secondary ducts during 5–6 weeks in culture. Any acinar or islet cells attached to the main duct prior to culture are lost during the incubation.

2.4.8. Skin

Explants of human skin can be maintained in a viable state for at least 30 days (Fig. 7) (Selkirk *et al.*, 1983). No attempts were made to culture skin for longer periods.

2.4.9. Prostate

Explants of human prostate can be maintained for periods of at least 24 weeks (Heatfield *et al.*, 1980b). Normal secretory cells of human prostate do not survive explant culture for more than a few days. However, epithelial cells with characteristics similar to basal cells do survive, proliferate, migrate around the edge of the explant, and differentiate into mucus-producing cells.

2.4.10. Breast

Normal-appearing breast epithelium can be maintained for at least 3–4 months *in vitro* (Hillman *et al.*, 1980a). During the first 2 weeks, the normal lobular structure of the breast is readily maintained. Within the central lumen, secretory material is often present. After 4 weeks, the acinar epithelium within the central portion of the explant remains normal, and the epithelial cells begin to migrate from the superficial acini onto the surface of the explant. With continued time in culture, the epithelial cells become flattened (squamous) and usually contain small secretory droplets. Treatment of the explants with insulin results in a marked improvement in maintaining the lobular architecture of the tissue (Hillman *et al.*, 1983). In addition,

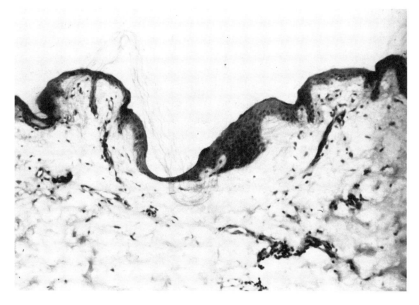

Figure 7. Explant of human skin maintained for 14 days *in vitro*.

treatment of explants with β-retinyl acetate inhibited squamous differentiation and increased the secretory activity of the glandular epithelial cells (Strum and Resau, 1986).

2.4.11. Placenta

The viability and functional activity of explant cultures of human placenta were determined by morphological studies and by measuring the secretion of human chorionic gonadotrophin (HCG) into the medium (Stromberg, 1980). The morphology of the placental villi remained normal during 3 weeks *in vitro*. The secretion of HCG rose to peak levels around the sixth day in culture and then declined in the manner of a bell-shaped curve.

2.5. Other Culture Systems

To our knowledge, the explant culture system described above has been used only for the maintainence and growth of lung, oral cavity, esophagus, bladder, colon, pancreatic duct, skin, prostate, breast, and placental tissues. However, other human tissues such as endometrium (Kaufman *et al.*, 1980), uterine cervix (Wilbanks *et al.*, 1980), pancreatic islets (Andersson and Hellerstrom, 1980), gastric mucosa (Donaldson Jr. and Kapadia, 1980), small intestine (Trier, 1980), blood vessels (Kocan *et al.*, 1980), and liver (Autrup *et al.*, 1984) have been maintained in explant culture, and the reader is referred to the referenced publications for descriptions of the methodology.

3. INVESTIGATIONS OF CARCINOGEN METABOLISM

Explant cultures of human tissue have been utilized extensively for investigations of the metabolism, DNA binding, and DNA-adduct formation of environmental procarcinogens (Table 2). In many of these studies, comparisons were made between the metabolism in human tissues and in the corresponding animal tissues in which the carcinogen was shown to induce tumors (Autrup *et al.*, 1980b; Autrup and Stoner, 1982; Daniel *et al.*, 1983; Delclos *et al.*, 1989; Moore *et al.*, 1982, 1984; Schut *et al.*, 1984; Shivapurkar *et al.*, 1987; Stoner *et al.*, 1982a, 1982b, 1986, 1987). The results of these investigations are described in detail in a series of review articles (Autrup and Harris, 1983; Daniel *et al.*, 1984; Harris *et al.*, 1978, 1982; Harris, 1987), and the reader is referred to these articles for additional information.

3.1. Methods

The methods employed in our laboratory for investigating the metabolism of environmental chemicals in cultured human tissues are as follows.

TABLE 7–2. Carcinogen Metabolism in Cultured Human Tissues

Carcinogen	Tissue	Reference
2-Acetylaminofluorene	Bladder	Moore *et al.*, 1982; Schut *et al.*, 1984
Aflatoxin B$_1$	Bladder	Daniel *et al.*, 1984; Stoner *et al.*, 1982b
	Colon	Autrup *et al.*, 1980a
	Stomach	Autrup *et al.*, 1984
	Tracheobronchus	Autrup *et al.*, 1979; Daniel *et al.*, 1984; Stoner *et al.*, 1982b
Benzidine	Bladder	Moore *et al.*, 1984
Benzo (a) pyrene	Bladder	Daniel *et al.*, 1984; Moore *et al.*, 1982; Stoner *et al.*, 1982a
	Bronchus	Autrup *et al.*, 1978a, 1978b, 1980a, 1980b, 1982; Daniel *et al.*, 1983, 1984; Harris *et al.*, 1976a, 1976b, 1977, 1979; Selkirk *et al.*, 1983; Stoner *et al.*, 1982a, 1986; Teel *et al.*, 1986; Yang *et al.*, 1977
	Colon	Autrup *et al.*, 1977, 1978b, 1980a, 1982
	Duodenum	Autrup *et al.*, 1982
	Esophagus	Autrup *et al.*, 1982, 1984; Harris *et al.*, 1979
	Liver	Autrup *et al.*, 1984

TABLE 7–2 (cont.).

Carcinogen	Tissue	Reference
	Pancreatic duct	Harris *et al.*, 1977
	Peripheral lung	Stoner *et al.*, 1978
	Pulmonary alveolar macrophage	Autrup *et al.*, 1978a
	Skin	Selkirk *et al.*, 1983
	Stomach	Autrup *et al.*, 1984
7,12-Dimethylbenz(a)anthracene	Pancreatic duct	Harris *et al.*, 1977
	Tracheobronchus	Stoner *et al.*, 1986
1,2-Dimethylhydrazine	Bronchus	Harris *et al.*, 1977b
	Colon	Autrup *et al.*, 1980a
4,4′-Methylenebis-(2-chloroaniline)	Bladder	Shivapurkar *et al.*, 1987; Stoner *et al.*, 1987
4-Methylnitrosamino-1-(3-pyridyl)-1-butanone	Bladder	Castonguay *et al.*, 1983
	Esophagus	Castonguay *et al.*, 1983
	Oral cavity	Castonguay *et al.*, 1983
	Peripheral lung	Castonguay *et al.*, 1983
	Tracheobronchus	Castonguay *et al.*, 1983
2-Naphthylamine	Bladder	Moore *et al.*, 1984
6-Nitrochrysene	Tracheobronchus	Delclos *et al.*, 1989
N-Nitrosobenzylmethylamine	Esophagus	Autrup and Stoner, 1982
N-Nitrosodimethylamine	Bronchus	Harris *et al.*, 1977a, 1977b
	Colon	Autrup *et al.*, 1977
	Esophagus	Harris *et al.*, 1979
	Liver	Autrup *et al.*, 1984

N-Nitrosodiethylamine	Bronchus	Harris *et al.*, 1977a
	Esophagus	Autrup and Stoner, 1982
	Liver	Autrup *et al.*, 1984
N-Nitrosonornicotine	Bladder	Castonguay *et al.*, 1983
	Esophagus	Castonguay *et al.*, 1983
	Oral cavity	Castonguay *et al.*, 1983
	Peripheral lung	Castonguay *et al.*, 1983
	Tracheobronchus	Castonguay *et al.*, 1983
N-Nitrosopiperidine	Bronchus	Harris *et al.*, 1977a
N,N'-Dinitrosopiperizine	Bronchus	Harris *et al.*, 1977a
N-Nitrosopyrrolidine	Esophagus	Harris *et al.*, 1979
	Bronchus	Harris *et al.*, 1977a

3.1.1. Toxicity Studies

The first step in metabolism studies is to identify nontoxic concentrations of the chemical for the appropriate human tissue. Explant cultures are initiated as described earlier and, after incubation for 24 h, the medium is changed and the tissues treated with four 2-fold dilutions of the chemical. Control explants are treated with the solvent for the chemical (usually dimethylsulfoxide at 0.2–0.5% in the medium). Twenty-four hours after treatment, 6–10 explants per concentration of test chemical are exposed for 4 h to [³H]thymidine and [³H]leucine (Stoner *et al.*, 1982a). After exposure to radiolabeled thymidine, 3–5 explants are taken for examination by high-resolution light microscopy and for autoradiography (ARG) and another 3–5 explants for biochemical analysis of the uptake of thymidine into cell DNA. All [³H]leucine-exposed explants are used for biochemical analysis of the uptake of leucine into protein. For ARG, the explants are fixed in a phosphate-buffered solution containing 4% formaldehyde: 1% glutaraldehyde (McDowell and Trump, 1976) and embedded in Epon. Sections 1 μm thick are coated with NTB-2 photographic emulsion, exposed in the dark for 2 weeks, and autoradiograms are prepared as described in Boren *et al.* (1974). The [³H]thymidine labeling index, i.e., percent of epithelial cells with labeled nuclei, is determined for each tissue and experimental variable and compared to solvent controls. Nuclei overlayed with 20 or more silver grains are considered to be labeled.

For biochemical studies, explants are placed in NaOH (for DNA determinations) or homogenized in 0.1M Tris-buffer (pH 7.4) for protein determinations. DNA and acid-insoluble protein are precipitated by addition of 1N perchloric acid. The pellets are washed with 0.2N perchloric acid until the radioactivity in the supernatant is negligible. The DNA concentration is then determined by the diphenylamine reaction (Burton, 1956) and protein by the Lowry method (Lowry *et al.*, 1951). Radioactivity is quantitated in a liquid scintillation spectrophotometer, and the data are expressed as disintegrations per minute (dpm)/μg of DNA or protein. The uptake of radiolabeled precursors into DNA and protein of chemically treated explants is then compared to solvent controls.

3.1.2. Incubation with Radiolabeled Chemicals and Sample Preparation

On the second or third day of explant culture, nontoxic concentrations (usually two) of radiolabeled test chemical are added to the culture medium for periods of 6, 12, and 24 h. After incubation, the media and explants are separated and the media stored immediately at −80°C. The explants are removed from the dishes, pooled, rinsed twice with cold phosphate buffered saline (PBS), and stored at −80°C. Depending upon the size of the specimen, explants from 4–8 cultures are pooled for each experimental variable, i.e., concentration of test chemical and time of incubation. Methanol (3–4 ml) is then added to each culture dish for a period of 1–2 h to remove any of the chemical and its metabolites that were adherent to the surface of the dish. The methanol washes are stored at −80°C. Incubations of the medium with radiolabeled chemical but without the explants serve as controls.

3.1.3. Isolation of Radioactive Chemical Bound to DNA

Explants are homogenized in 2-ml 0.05M sodium phosphate buffer, pH 6.5, containing 0.01M EDTA (ethylenediaminetetraacetic acid) and 0.01M EGTA [ethylene glycol bis(aminoethyl ether)tetraacetic acid], using an homogenizer at maximum speed for 30 s (Stoner *et al.*, 1982a). The homogenates are made 1.5% (vol/vol) in sodium dodecyl sulfate, treated with 0.2 mg autodigested pronase B, and incubated for 2.5 h at 37°C with mixing. The samples are made 7% (wt/vol) in 4-aminosalicyclate-NaCl (6:1) and extracted for 1 h with 1 volume of phenol reagent (Kirby, 1965), after which the phases are separated by centrifugation (40 min at 12,000 × g) and the aqueous phase removed. The phenol phase is back extracted with 1-ml buffer, and the resulting aqueous phases pooled. Hydroxylapatite (0.5 g), suspended in 2-ml 0.05M sodium phosphate buffer, pH 6.5, is added to the pooled aqueous layers in a 15-ml polystyrene centrifuge tube. The gel is washed four times with 10-ml volumes of a solution of 8M urea in 0.12M sodium phosphate buffer, pH 6.5, in a batch procedure, followed by two 5-ml washes of 0.05M sodium phosphate buffer. The DNA is then eluted from the hydroxylapatite with two 1.5-ml washes of 0.55 M sodium phosphate buffer, pH 6.5. The amount of DNA present in the aqueous samples is determined fluorimetrically after reaction of an aliquot (150 µl) with 3,5-diaminobenzoic acid (Kissane and Robbins, 1958). The radioactivity associated with the DNA is determined by liquid scintillation counting after digestion with perchloric acid. The binding levels are then calculated and expressed as µmol test chemical per mole deoxyribonucleotide.

An alternate method for the isolation of DNA for determination of the bound radioactivity has been described (Gupta, 1984). This method is less time consuming and also yields highly purified DNA.

3.1.4. High-Performance Liquid Chromatography

Methods for the identification of chemical-DNA adducts and of metabolites in the culture medium vary with the chemical under study. The reader is referred to articles referenced in Table 2 for descriptions of the methodology used for the chemicals listed.

3.2. Results

The principal observations from studies of the metabolism of chemical carcinogens in cultured human explant tissues are as follows.

1. Humans vary significantly in their ability to convert environmental procarcinogens into metabolites that interact with cellular DNA. Explants of human tracheobronchus from different individuals varied from 10-fold (Stoner *et al.*, 1982a) to 75-fold (Harris *et al.*, 1976a) in their ability to metabolize benzo(a)pyrene into forms that were bound to DNA. Similarly, we observed a 40-fold variation in the ability of human bladder explants from different individuals to convert benzo(a)pyrene into DNA binding forms (Stoner *et al.*, 1982a).

2. Different organs from the same individual vary significantly in their ability to metabolize environmental procarcinogens (Autrup *et al.*, 1982, 1984; Castonguay *et al.*, 1983; Daniel *et al.*, 1984; Harris *et al.*, 1977a,b, 1978, 1982; Hecht *et al.*, 1982; Selkirk *et al.*, 1983; Stoner *et al.*, 1982a, 1982b). For example, the binding of benzo(a)pyrene to the DNA of human bladder explants was significantly higher than that to the DNA of tracheobronchial tissues (Stoner *et al.*, 1982a).

3. The pathways for the metabolism of several procarcinogens including benzo(a)pyrene (Autrup *et al.*, 1980a, 1980b; Daniel *et al.*, 1983; Moore *et al.*, 1982; Stoner *et al.*, 1986), 2-acetylaminofluorene (Schut *et al.*, 1984), and aflatoxin B_1 (Stoner *et al.*, 1982b) in cultured human and animal model explant tissues are similar.

4. The profiles of carcinogen-DNA adducts observed in cultured human tissues are qualitatively similar to those in the corresponding animal tissues. However, quantitative differences between human and animals in the extents of formation of these adducts have been observed (Autrup *et al.*, 1980a, 1980b; Schut *et al.*, 1984; Stoner *et al.*, 1982b).

4. OTHER APPLICATIONS

In addition to investigations of carcinogen metabolism, explant cultures of human tissue have been used to determine if treatment with carcinogens leads to histopathological effects. For example, a single treatment of explants of human bronchus with amosite asbestos was found to induce focal hyperplasia and epidermoid metaplasia with cellular atypia (Haugen *et al.*, 1982). Intracytoplasmic and intranuclear fibers of amosite were identified in these focal lesions using x-ray microanalysis in combination with transmission electron microscopy. Similarly, treatment of explants of human bladder with

N-methyl-*N*-nitrosourea led to the development of hyperplastic and dysplastic lesions, the extent of which was dependent upon the number of treatments (El-Gerzawi *et al.*, 1982). To our knowledge, in no instance has treatment of cultured explant tissues with carcinogens led to the development of malignant lesions, even after the treated tissues were transplanted to athymic, nude mice.

Another potential application of cultured human tissues is the identification of putative inhibitors and promoters of chemical carcinogenesis. For example, ellagic acid, a naturally occurring inhibitor of chemical carcinogenesis in animals, has been shown in our laboratory to inhibit the DNA-damaging effects of benzo(a)pyrene (Teel *et al.*, 1986a, 1986b) and aflatoxin B_1 (Mandal *et al.*, 1987) in explants of human bronchus. In contrast, treatment of human bladder explants with saccharin, either alone or in combination with *N*-methyl-*N*-nitrosourea, resulted in a prolongation of hyperplastic and dysplastic lesions (El-Gerzawi *et al.*, 1982). These results are consistent with the concept of saccharin as a weak carcinogen or promoter of carcinogenesis.

In our opinion, the continued (and expanded) use of cultured human tissue for studies in toxicology and carcinogenesis is both warranted and desirable. Human tissues, both normal and malignant, are currently available through a variety of collection networks. Investigations with these tissues should assist in "bridging the gap" between the human and animals and make important contributions to our understanding of carcinogenesis.

REFERENCES

A. Andersson and C. Hellerstrom, Explant culture: pancreatic islets. In: *Methods in Cell Biology,* Vol. 21, Normal Human Tissue and Cell Culture. B. Endocrine, Urogenital and Gastrointestinal Systems, eds. C. C. Harris, B. F. Trump, and G. D. Stoner, Academic Press, New York, Chap. 6, pp. 135–151 (1980).

H. Autrup, Explant culture of human colon. In: *Methods in Cell Biology,* Vol. 21, Normal Human Tissue and Cell Culture. B. Endocrine, Urogenital, and Gastrointestinal Systems, eds. C. C. Harris, B. F. Trump, and G. D. Stoner, Academic Press, New York, Chap. 19, pp. 385–401 (1980).

H. Autrup and C. C. Harris, Metabolism of chemical carcinogens by cultured human tissues. In: *Human Carcinogenesis,* eds. C. C. Harris and H. Autrup, Academic Press, New York, pp. 169–194 (1983).

H. Autrup and G. D. Stoner, Metabolism of *N*-nitrosamines by cultured human and rat esophagus, *Cancer Res.* **42,** 1307–1311 (1982).

H. Autrup, J. M. Essigmann, R. G. Croy, B. F. Trump, G. W. Wogan, and C. C. Harris, Metabolism of aflatoxin B_1 and identification of the major aflatoxin B_1-DNA adducts formed in cultured human bronchus and colon, *Cancer Res.* **39,** 694–698 (1979).

H. Autrup, R. C. Grafstrom, M. Brugh, J. F. Lechner, A. Haugen, B. F. Trump, and C. C. Harris, Comparison of benzo(a)pyrene metabolism in bronchus, esophagus, colon, and duodenum from the same individual, *Cancer Res.* **42**, 934–938 (1982).

H. Autrup, C. C. Harris, G. D. Stoner, M. L. Jesudason, and B. F. Trump, Binding of chemical carcinogens to macromolecules in cultured human colon, *J. Natl. Cancer Inst.* **59**, 351–354 (1977).

H. Autrup, C. C. Harris, G. D. Stoner, J. K. Selkirk, P. W. Schafer, and B. F. Trump, Metabolism of [³H]benzo(a)pyrene by cultured human bronchus and cultured human pulmonary alveolar macrophages, *Lab. Invest.* **38**, 217–225 (1978a).

H. Autrup, C. C. Harris, B. F. Trump, and A. M. Jeffrey, Metabolism of benzo(a)pyrene and identification of the major benzo(a)pyrene-DNA adducts in cultured human colon, *Cancer Res.* **38**, 3689–3696 (1978b).

H. Autrup, C. C. Harris, S. M. Wu, L-Y. Bao, S. T. Pei, T. T. Sun, and C. C. Hsia, Activation of chemical carcinogens by cultured human fetal liver, esophagus and stomach, *Chem.-Biol. Interact.* **50**, 15–25 (1984).

H. Autrup, R. D. Schwartz, J. M. Essigmann, L. Smith, B. F. Trump, and C. C. Harris, Metabolism of aflatoxin B₁, benzo(a)pyrene and 1,2-dimethylhydrazine by cultured rat and human colon, *Teratog. Carcinog. Mutagen.* **1**, 3–13 (1980a).

H. Autrup, F. C. Wefald, A. M. Jeffrey, H. Tate, R. D. Schwartz, B. F. Trump, and C. C. Harris, Metabolism of benzo(a)pyrene by cultured tracheobronchial tissues from mice, rats, hamsters, bovines and humans, *Int. J. Cancer* **25**, 293–300 (1980b).

L. A. Barrett, E. M. McDowell, A. L. Frank, C. C. Harris, and B. F. Trump, Long-term organ culture of human bronchial epithelium, *Cancer Res.* **36**, 1003–1010 (1976).

H. Boren, E. Wright, and C. C. Harris, Quantitative light microscopic autoradiography. In: *Methods in Cell Physiology,* Vol. 8, ed. D. Prescott, Academic Press, New York, pp. 277–288 (1974).

A. Buchner and A. Mlinek, In Vitro cultivation of adult human gingiva. II. Light and electron microscopic observations of the explants, *J. Peridontal Res.* **10**, 346–356 (1975).

K. Burton, A study of the conditions and mechanism of the diphenylamine reaction for the colorimetric estimation of deoxyribonucleic acid, *Biochem. J.* **62**, 315–323 (1956).

A. Castonguay, G. D. Stoner, S. S. Hecht, H. A. J. Schut, and D. Hoffman, Metabolism of tobacco-specific nitrosamines by cultured human tissues, *Proc. Natl. Acad. Sci. USA* **80**, 6694–6697 (1983).

F. B. Daniel, H. A. J. Schut, D. W. Sandwisch, K. M. Schenck, C. O. Hoffman, J. R. Patrick, and G. D. Stoner, Interspecies comparisons of benzo(a)pyrene metabolism and DNA-adduct formation in cultured human and animal bladder and tracheobronchial tissues, *Cancer Res.* **43**, 4723–4729 (1983).

F. B. Daniel, G. D. Stoner, and H. A. J. Schut, Interindividual variation in metabolism-induced DNA binding of chemical genotoxins to human bladder and bronchus explants. In: *Individual Susceptibility to Genotoxic Agents in the Human Population,* eds. F. J. deSerres and R. W. Pero, Plenum, New York, pp. 177–199 (1984).

K. B. Delclos, K. El-Bayoumy, D. A. Casciano, R. P. Walker, F. F. Kadlubar, S. S. Hecht, N. Shivapurkar, S. Mandal, and G. D. Stoner, Metabolic activation of 6-nitrochrysene in explants of human bronchus and in isolated hepatocytes *Cancer Res.* **49**, 2909–2913 (1989).

R. M. Donaldson, Jr. and C. R. Kapadia, Organ culture of gastric mucosa: Advantages and limitations. In: *Methods in Cell Biology*, Vol. 21, Normal Human Tissue and Cell Culture. B. Endocrine, Urogenital and Gastrointestinal Systems, eds. C. C. Harris, B. F. Trump, and G. D. Stoner, Academic Press, New York, Chap. 17, pp. 349–363 (1980).

B. H. Dorman, V. M. Benta, M. J. Mass, and D. G. Kaufman, Benzo(a)pyrene binding to DNA in organ cultures of human endometrium, *Cancer Res.* **41**, 2718–2722 (1981).

S. El-Gerzawi, B. M. Heatfield, and B. F. Trump, *N*-methyl-*N*-nitrosourea and saccharin: Effects on epithelium of normal human urinary bladder *in vitro, J. Natl. Cancer Inst.* **69**, 577–583 (1982).

R. C. Gupta, Nonrandom binding of the carcinogen *N*-hydroxy-2-acetylaminofluorene to repetitive sequences of rat liver DNA *in vivo, Proc. Natl. Acad. Sci. USA* **81**, 6943–6947 (1984).

G. C. Harries, A. R. Boobis, N. Collier, and D. S. Davies, Interindividual differences in the activation of two hepatic carcinogens to mutagens by human liver, *Hum. Toxicol.* **5**, 21–26 (1986).

C. C. Harris, Human tissues and cells in carcinogenesis research, *Cancer Res.* **47**, 1–10 (1987).

C. C. Harris, H. Autrup, R. Connor, L. A. Barrett, E. M. McDowell, and B. F. Trump, Interindividual variation in binding of benzo(a)pyrene to DNA in cultured human bronchi, *Science* **194**, 1067–1069 (1976a).

C. C. Harris, H. Autrup, and G. D. Stoner, Metabolism of benzo(a)pyrene in cultured human tissues and cells. In: *Polycyclic Hydrocarbons and Cancer: Chemistry, Molecular Biology and Environment*, Vol. 2, eds. P. O. P. T'so and H. V. Gelboin, Academic Press, New York, Chap. 20, pp. 331–342 (1978).

C. C. Harris, H. Autrup, G. D. Stoner, E. M. McDowell, B. F. Trump, and P. Schafer, Metabolism of acyclic and cyclic *N*-nitrosamines in cultured human bronchi, *J. Natl. Cancer Inst.* **59**, 1401–1407 (1977a).

C. C. Harris, H. Autrup, G. D. Stoner, E. M. McDowell, B. F. Trump, and P. Schafer, Metabolism of dimethylnitrosamine and 1,2-dimethylhydrazine in cultured human bronchi, *Cancer Res.* **37**, 2309–2311 (1977b).

C. C. Harris, H. Autrup, G. D. Stoner, B. F. Trump, E. Hillman, P. W. Schafer, and A. M. Jeffrey, Metabolism of benzo(a)pyrene, *N*-nitrosodimethylamine, and *N*-nitrosopyrrolidine and identification of the major carcinogen-DNA adducts in cultured human esophagus, *Cancer Res.* **39**, 4401–4406 (1979).

C. C. Harris, H. Autrup, G. D. Stoner, S. K. Yang, J. C. Leutz, H. V. Gelboin, J. K. Selkirk, R. J. Connor, L. A. Barrett, R. T. Jones, E. M. McDowell, and B. F. Trump, Metabolism of benzo(a)pyrene and 7,12-dimethylbenz(a)anthracene in cultured human bronchus and pancreatic duct, *Cancer Res.* **37**, 3349–3355 (1977).

C. C. Harris, A. L. Frank, C. van Haaften, D. G. Kaufman, R. Conner, F. Jackson, L. A. Barrett, E. M. McDowell, and B. F. Trump, Binding of [³H]benzo(a)pyrene to DNA in cultured human bronchus, *Cancer Res.* **36**, 1011–1018 (1976b).

C. C. Harris, B. F. Trump, R. Grafstrom, and H. Autrup, Difference in metabolism of chemical carcinogens in cultured human epithelial tissues and cells, *J. Cell. Biochem.* **18**, 285–294 (1982).

A. Haugen, P. W. Schafer, J. F. Lechner, G. D. Stoner, B. F. Trump, and C. C. Harris, Cellular ingestion, toxic effects and lesions observed in human bronchial epithelial tissue and cells cultured with asbestos and glass fibers, *Int. J. Cancer* **30**, 265–272 (1982).

B. M. Heatfield, H. Sanefuji, S. El-Gerzawi, B. Urso, and B. F. Trump, Surface alterations in urothelium of normal human bladder during long-term explant culture, Scanning Electron. Microsc. **3**, 61–70 (1980a).

B. M. Heatfield, H. Sanefugi, and B. F. Trump, Long-term explant culture of normal human prostate, In: *Methods in Cell Biology,* Vol. 21, Normal Human Tissue and Cell Culture. B. Endocrine, Urogenital and Gastrointestinal Systems, eds. C. C. Harris, B. F. Trump, and G. D. Stoner, Academic Press, New York, Chap. 8, pp. 171–194 (1980b).

S. S. Hecht, A. Castonguay, F-L. Chung, D. Hoffman, and G. D. Stoner, Recent studies on the metabolic activation of cyclic nitrosamines. In: *Nitrosamines and Human Cancer,* Banbury Report 12, Cold Spring Harbor Laboratory, Cold Spring Harbor, NY, pp. 103–120 (1982).

E. A. Hillman, M. G. Valerio, S. A. Halter, L. A. Barrett-Boone, and B. F. Trump, Long-term explant culture of normal mammary epithelium, *Cancer Res.* **43**, 245–257 (1983).

E. A. Hillman, M. J. Vocci, J. W. Combs, H. Sanefuji, T. Robbins, D. H. Janns, C. C. Harris, and B. F. Trump, Human breast organ culture studies. In: *Methods in Cell Biology,* Vol. 21, Normal Human Tissue and Cell Culture. B. Endocrine, Urogenital and Gastrointestinal Systems, eds. C. C. Harris, B. F. Trump, and G. D. Stoner, Academic Press, New York, Chap. 4, pp. 79–106 (1980a).

E. A. Hillman, M. J. Vocci, W. Schurch, C. C. Harris, and B. F. Trump, Human esophageal organ culture studies. In: *Methods in Cell Biology,* Vol. 21, Normal Human Tissue and Cell Culture. B. Endocrine, Urogenital and Gastrointestinal Systems, eds. C. C. Harris, B. F. Trump, and G. D. Stoner, Academic Press, New York, Chap. 16, pp. 331–348 (1980b).

R. T. Jones, B. F. Trump, and G. D. Stoner, Culture of human pancreatic ducts. In: *Methods in Cell Biology,* Vol. 21, Normal Human Tissue and Cell Culture. B. Endocrine, Urogenital and Gastrointestinal Systems, eds. C. C. Harris, B. F. Trump, G. D. Stoner, Academic Press, New York, Chap. 21, pp. 429–439 (1980).

D. G. Kaufman, T. A. Adamec, L. A. Walton, C. N. Carney, S. A. Melin, V. M. Genta, M. J. Mass, B. H. Dorman, N. T. Rodgers, G. J. Photopulos, J. Powell, and J. W. Grisham, Studies of human endometrium in organ culture. In: *Methods in Cell Biology,* Vol. 21, Normal Human Tissue and Cell Culture. B. Endocrine, Urogenital and Gastrointestinal Systems, eds. C. C. Harris, B. F. Trump, and G. D. Stoner, Academic Press, New York, Chap. 1, pp. 1–27 (1980).

K. S. Kirby, Isolation and charcterization of ribosomal ribonucleic acid, *Biochem. J.* **96**, 266–269 (1965).

J. M. Kissane and E. Robbins, The fluorometric measurement of DNA in animal tissues with special reference to the central nervous system, *J. Biol. Chem.* **233**, 184–188 (1958).

M. A. Knowles, A. Finesilver, A. E. Harvey, R. J. Berry, and R. M. Hicks, Long-term organ culture of normal human bladder, *Cancer Res.* **43**, 374–385 (1983).

M. A. Knowles, R. M. Hicks, R. J. Berry, and E. Milroy, Organ culture of normal human bladder: Choice of starting material and culture characteristics. In: *Methods in Cell Biology,* Vol. 21, Normal Human Tissue and Cell Culture. B. Endocrine, Urogenital and Gastrointestinal Systems, eds. C. C. Harris, B. F. Trump, and G. D. Stoner, Academic Press, New York, Chap. 12, pp. 257–285 (1980).

R. M. Kocan, N. S. Moss, and E. P. Benditt, Human arterial wall cells and tissues in culture. In: *Methods in Cell Biology,* Vol. 21, Normal Human Tissue and Cell Culture. A. Respiratory, Cardiovascular and Intergumentary Systems, eds. C. C. Harris, B. F. Trump, and G. D. Stoner, Academic Press, New York, Chap. 10, pp. 153–166 (1980).

A. Leibovitz, The growth and maintenance of tissue cell cultures in free gas exchange with the atmosphere, *Am. J. Hyg.* **78**, 173–187 (1963).

O. H. Lowry, N. J. Rosebrough, A. L. Farr, and R. J. Randall, Protein measurement with the folin phenol reagent, *J. Biol. Chem.* **193**, 265–275 (1951).

E. McDowell and B. F. Trump, Histological fixatives suitable for diagnostic light and electron microscopy, *Arch. Pathol.* **100**, 405–414 (1976).

S. Mandal, A. Ahuja, N. M. Shivapurkar, S. J. Cheng, J. D. Groopman, and G. D. Stoner, Inhibition of aflatoxin B_1 mutagenesis in *Salmonella typhimurium* and DNA damage in cultured rat and human tracheobronchial tissues with ellagic acid, *Carcinog.* **8**, 1651–1656 (1987).

B. P. Moore, R. M. Hicks, M. A. Knowles, and S. Redgrave, Metabolism and binding of benzo(a)pyrene and 2-acetylaminofluorene by short-term organ cultures of human and rat bladder, *Cancer Res.* **42**, 642–648 (1982).

B. P. Moore, P. M. Potter, and R. M. Hicks, Metabolism of 2-naphthylamine and benzidine by rat and human bladder organ cultures, *Carcinog.* **5**, 949–954 (1984).

E. A. Reedy and B. M. Heatfield, Histomorphometry and cell kinetics of normal human bladder mucosa *in vitro, Urol. Res.* **15**, 321–327 (1987).

J. H. Resau, J. R. Cottrell, K. A. Elligett, and E. A. Hudson, Cell injury and regeneration of human epithelium in organ culture, *Cell Biol. Toxicol.* **3**, 441–458 (1987).

H. A. J. Schut, F. B. Daniel, K. M. Schenck, T. R. Loeb, and G. D. Stoner, Metabolism and DNA adduct formation of 2-acetylaminofluorene by bladder explants from human, dog, monkey, hamster and rat, *Carcinog.* **5**, 1287–1292 (1984).

J. K. Selkirk, A. Nikbakht, and G. D. Stoner, Comparative metabolism and macro-molecular binding of benzo(a)pyrene in explant cultures of human bladder, skin, bronchus and esophagus from eight individuals, *Cancer Lett.* **18**, 11–19 (1983).

N. Shivapurkar, T. A. Lehman, H. A. J. Schut, and G. D. Stoner, DNA binding of 4,4'-methylenebis(2-chloroaniline) (MCOA) in explant cultures of human and dog bladder, *Cancer Lett.* **8**, 41–48 (1987).

G. D. Stoner, Explant culture of human peripheral lung. In: *Methods in Cell Biology,* Vol. 21, Normal Human Tissue and Cell Culture. A. Respiratory, Cardiovascular and Integumentary Systems, eds. C. C. Harris, B. F. Trump, and G. D. Stoner, Academic Press, New York, Chap. 4, pp. 65–77 (1980).

G. D. Stoner, F. B. Daniel, K. M. Schenck, H. A. J. Schut, P. J. Goldblatt, and D. W. Sandwisch, Metabolism and DNA binding of benzo(a)pyrene in cultured human bladder and bronchus, *Carcinog.* **3,** 195–201 (1982a).

G. D. Stoner, F. B. Daniel, K. M. Schenck, H. A. J. Schut, D. W. Sandwisch, and A. F. Gohara, DNA binding and adduct formation of aflatoxin B₁ in cultured human and animal tracheobronchial and bladder tissues, *Carcinog.* **3,** 1345–1348 (1982b).

G. D. Stoner, C. C. Harris, H. Autrup, B. F. Trump, E. W. Kingsbury, G. A. Meyers, and R. Newkirk, Organ culture of human peripheral lung tissue. I. Metabolism of benzo(a)pyrene, *Lab. Invest.* **38,** 685–692 (1978).

G. D. Stoner, H. A. J. Schut, F. B. Daniel, and R. Dixit, A comparison of covalent DNA binding of benzo(a)pyrene and 7,12-dimethylbenz(a)anthracene in respiratory tissues from human, rat and mouse, *Cancer Lett.* **30,** 231–241 (1986).

G. D. Stoner, N. M. Shivapurkar, H. A. J. Schut, and T. Lehman, DNA binding and adduct formation of 4,4′-methylenebis(2-chloroaniline) in explant cultures of human and dog bladder. In: *Carcinogenic and Mutagenic Responses to Aromatic Amines and Nitroarenes,* eds. C. M. King, L. J. Romano, and D. Schuetzle, Elsevier Science Publishing Co., Inc., New York, pp. 237–240 (1987).

K. Stromberg, The human placenta in cell and organ culture. In: *Methods in Cell Biology,* Vol. 21, Normal Human Tissue and Cell Culture. B. Endocrine, Urogenital and Gastrointestinal Systems, eds. C. C. Harris, B. F. Trump, G. D. Stoner, Academic Press, New York, Chap. 10, pp. 227–252 (1980).

J. M. Strum and J. H. Resau, Effects of β-retinyl acetate on human breast epithelium in explant culture, *Am. J. Anat.* **175,** 35–48 (1986).

R. W. Teel, M. S. Babcock, R. Dixit, and G. D. Stoner, Ellagic acid toxicity and interaction with benzo(a)pyrene and benzo(a)pyrene 7,8-dihydrodiol in human bronchial epithelial cells, *Cell Biol. Toxicol.* **2,** 53–62 (1986a).

R. W. Teel, G. D. Stoner, M. S. Babcock, R. Dixit, and K. Kim, Benzo(a)pyrene metabolism and DNA-binding in cultured explants of human bronchus and in monolayer cultures of human bronchial epithelial cells treated with ellagic acid, *Cancer Detect. Prev.* **9,** 59–66 (1986b).

J. S. Trier, Organ culture of the mucosa of the human small intestine. In: *Methods in Cell Biology,* Vol. 21, Normal Human Tissue and Cell Culture. B. Endocrine, Urogenital and Gastrointestinal Systems, eds. C. C. Harris, B. F. Trump, and G. D. Stoner, Academic Press, New York, Chap. 18, pp. 365–384 (1980).

B. F. Trump, J. Resau, and L. A. Barrett, Methods of organ culture for human bronchus. In: *Methods in Cell Biology,* Vol. 21, Normal Human Tissue and Cell Culture. A. Respiratory, Cardiovascular and Integumentary Systems, eds. C. C. Harris, B. F. Trump, and G. D. Stoner, Academic Press, New York, Chap. 1, pp. 1–14 (1980a).

B. F. Trump, T. Sato, A. Trifillis, M. Hill-Craggs, M. W. Kahng, and M. W. Smith, Cell and explant culture of kidney tubular epithelium. In: *Methods in Cell Biology,* Vol. 21, Normal Human Tissue and Cell Culture. B. Endocrine, Urogenital and Gastrointestinal Systems, eds. C. C. Harris, B. F. Trump, and G. D. Stoner, Academic Press, New York, Chap. 14, pp. 309–326 (1980b).

G. D. Wilbanks, E. Leipus, and D. Tsurumoto, Tissue culture of the human uterine cervix. In: *Methods in Cell Biology,* Vol. 21, Normal Human Tissue and Cell

Culture. B. Endocrine, Urogenital and Gastrointestinal Systems, eds. C. C. Harris, B. F. Trump, and G. D. Stoner, Academic Press, New York, Chap. 2, pp. 29–50 (1980).

S. K. Yang, H. V. Gelboin, B. F. Trump, H. Autrup, and C. C. Harris, Metabolic activation of benzo(a)pyrene and binding to DNA in cultured human bronchus, *Cancer Res.* **37**, 1210–1215 (1977).

32P-Postlabeling Analysis of Mutagen- and Carcinogen-DNA Adducts and Age-Related DNA Modifications (I-Compounds)

Kurt Randerath and Erika Randerath
Department of Pharmacology
Baylor College of Medicine
Houston, Texas

1. INTRODUCTION

Numerous lines of evidence indicate that naturally occurring and human-made chemicals play a significant role in the etiology of human cancer (Doll and Peto, 1981). In experimental animals and cultured human cells, the majority, but not all, of chemical carcinogens are converted via chemical or metabolic activation to electrophiles; these reactive intermediates react with DNA, RNA, and proteins, resulting in the formation of covalent adducts (Miller and Miller, 1981; Hemminki, 1983). A number of observations suggest that DNA is the critical target molecule in the initiation of multistage chemical carcinogenesis. Methods for the detection and measurement of covalent DNA lesions are of crucial importance in the identification of potential carcinogens and mutagens and the study of mechanisms of carcinogenesis. The prevention or minimization of human exposures to genotoxic compounds depends on the judicious application of such methods.

Several excellent short-term *in vitro* assays employing bacteria, lower eukaryotes, or mammalian cells in culture are available to detect genotoxic activity of chemicals (Ames, 1979; DeSerres and Ashby, 1981; Stich and San, 1981). Since these systems are frequently unable to convert carcinogens to electrophilic species, exogenous sources of metabolic enzymes are added. However, such assays do not accurately mimic the *in vivo* situation, which entails complex biological interactions such as absorption, distribution, bioactivation, detoxication, DNA repair, and loss of genetically damaged cells by attrition. An added layer of complexity exists for intact animals in that these parameters are strongly dependent on species, age, sex, nutritional status, cell type, and tissue as well as on chronobiological effects, none of which can be adequately integrated into short-term *in vitro* tests. Therefore a need was apparent to us several years ago for the development of an improved *in vivo* assay system for genotoxic compounds, preferably one that would be capable of directly measuring covalent DNA damage. Radioactive carcinogens or antibodies specific to known adducts have been utilized for this purpose (Baird, 1979; Poirier, 1981); however, since these techniques cannot be readily applied to the large number of chemicals in the human environment, alternative techniques were sought.

^{32}P-postlabeling represents a general and sensitive approach for the measurement of DNA lesions formed with nonradioactive mutagens and carcinogens (K. Randerath *et al.*, 1981, 1985; Gupta *et al.*, 1982; Reddy *et al.*, 1984). To this end, normal and modified (adducted) nucleotides are labeled with ^{32}P and detected and quantified after thin-layer chromatography (TLC). ^{32}P-Adduct measurements have been applied to DNA preparations modified by over 100 test chemicals, comprising arylamines, arylamides, nitroaromatics, nitrosamines, azo compounds, dyestuffs, polycyclic aromatic hydrocarbons, heterocyclic polycyclic aromatics, epoxides, and methylating agents, as well as mycotoxins, alkenylbenzenes, antibiotics, estrogens, and polychlorinated dibenzo-*p*-dioxins. In addition, DNA alterations associated with exposures to complex mixtures and unidentified sources have been detected by ^{32}P-postlabeling. In this review, we focus on the basic principles of the assay and present an overview of some of its past and current applications. Examples given are mainly from the work of the authors' laboratory.

2. GENERAL METHOD

As shown in Fig. 1, the basic method (K. Randerath *et al.*, 1981) entails the nucleolytic digestion of DNA to 3'-mononucleotides, the subsequent conversion of these digestion products to 5'-^{32}P-labeled 3',5'-bisphosphate derivatives with [γ-^{32}P]ATP as the donor of label and T4 polynucleotide kinase as the catalyst, the separation of the labeled nucleotides by polyethyleneimine (PEI)-cellulose TLC, autoradiography to detect adducts and normal

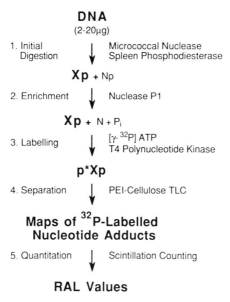

DNA
(2-20µg)

1. Initial   Micrococcal Nuclease
   Digestion   Spleen Phosphodiesterase

X p + Np

2. Enrichment   Nuclease P1

X p + N + P$_i$

3. Labelling   [γ-^{32}P] ATP
                    T4 Polynucleotide Kinase

p*Xp

4. Separation   PEI-Cellulose TLC

**Maps of ^{32}P-Labelled
Nucleotide Adducts**

5. Quantitation   Scintillation Counting

RAL Values

Figure 1. The various steps of the ^{32}P-postlabeling assay for DNA adducts. For further details, consult the text.

nucleotides, and scintillation counting for quantitation. A key feature of the assay is the class separation of aromatic, bulky, or hydrophobic adducts from the usually large excess of normal DNA nucleotides; this is readily accomplished by PEI-cellulose TLC employing nondenaturing solvents as the mobile phase (Gupta *et al.*, 1982; Reddy *et al.*, 1984). The various versions of the assay are illustrated in Fig. 2. In procedures 1 and 2, the adducts are labeled in the presence of the normal DNA nucleotides, while in procedures 3 and 4, the normal nucleotides are removed before ^{32}P-labeling. For this purpose, physical techniques such as butanol extraction (procedure 3) (Gupta, 1985) and enzymatic techniques such as nuclease P1 postdigestion (procedure 4) (Reddy and Randerath, 1986) have been developed. The latter two procedures afford the greatest sensitivity of adduct detection. A comparative study (Gupta and Earley, 1988) showed that recoveries of certain adducts in these enrichment procedures depend on adduct structure. Experimental details of these procedures and their applications have been reviewed recently (Gupta and Randerath, 1988).

The choice of the labeling procedure depends on the amounts and types of adducts present in the DNA under study. For procedure 1, the limit of detection of aromatic or hydrophobic adducts is 1 adduct in 3×10^7 nucleotides. This procedure, which affords quantitative labeling of both normal and adducted nucleotides with excess ATP, may serve as a means of calibration for the other procedures. Procedure 2 is more sensitive than procedure 1 and enables the detection of many adducts at levels as low as 1 in 10^9, provided that the adducts are labeled preferentially over the normal nucleotides (E. Randerath *et al.*, 1985). This procedure has been applied to adducts formed with a number of structurally diverse carcinogens, such as 7,12-dimethylbenz[a]anthracene, benzo[a]pyrene, dibenzo[c,g]carbazole, safrole, 4-aminobiphenyl, diethylstilbestrol, and 2-amino-3-methylimidazo[4,5-*f*]quinoline (IQ) (E. Randerath *et al.*, 1985; Schurdak and Randerath, 1985; Lu *et al.*, 1986; Liehr *et al.*, 1986; Schut *et al.*, 1988). The enrichment procedures (3 and 4) depend on two entirely different principles, and thus some differences in their performance are expected. For example, it appears that nucleotides containing carcinogen moieties attached at exocyclic nitrogen or oxygen atoms are recovered to similar extents by procedures 3 and 4, while the former procedure affords a better recovery of guanine-C8-substituted adducts. On the other hand, many I-compounds (see below) show higher recoveries after nuclease P1 enrichment.

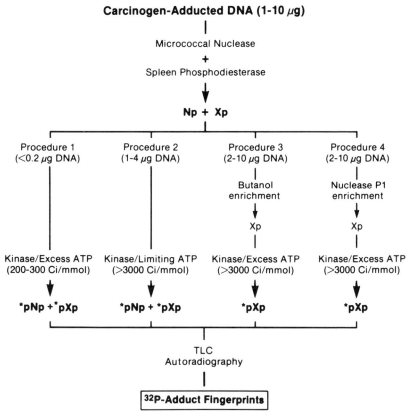

Figure 2. Scheme of various versions of the ^{32}P-postlabeling assay outlined in this article. N denotes normal DNA nucleotide moiety; X, adducted moiety. Asterisks denote ^{32}P-labeled groups.

Adduct Quantitation

1. Standard Procedure:

$$RAL = \frac{\text{cpm in adducts}}{\text{cpm in total N}}$$

2. Nuclease P1 Procedure:

$$RAL = \frac{\text{cpm in adducts}}{\text{spec. act. ATP [cpm/pmol] X pmol DNA-P}}$$

$$RAL \times 10^7 = \text{No. of adducts} / 10^7 \text{ N}$$

Figure 3. Adduct quantitation involves scintillation counting of excised nucleotide fractions. The RAL value measures adduct levels in the analyzed DNA sample.

The choice of the chromatographic procedure is of critical importance for optimal adduct resolution. TLC on PEI-cellulose has been used most extensively for this purpose. Concentrated electrolyte solutions containing 7–8.5 M urea (Gupta *et al.*, 1982) appear best suited as solvents in the separation of highly nonpolar adducts containing polycyclic aromatic moieties, while more polar adducts are resolved with solvents containing lower concentrations of electrolytes and urea (K. Randerath *et al.*, 1984; Reddy and Randerath, 1986; Gupta and Randerath, 1988). In two-dimensional chromatography, optimal resolution is usually achieved with acidic solvents (pH 3.3–4.0) in the first dimension, followed by an alkaline (pH 8) solvent in the second dimension (K. Randerath *et al.*, 1981; Gupta *et al.*, 1982). Derivatives of widely divergent polarities in a DNA sample may require several mapping systems (K. Randerath *et al.*, 1988a). Special procedures such as high-performance liquid chromatography have been reported to separate small alkylated (Wilson *et al.*, 1988) as well as polycyclic aromatic (Dunn and San, 1988) adducts in combination with ^{32}P-postlabeling. When the nature of the adducts in a sample of chemically modified DNA is unknown, it may be necessary to explore suitable conditions for their separation.

Adduct levels are calculated as relative adduct labeling (RAL) values, which represent the ratio of count rates of adducted nucleotides to count rates of total (adducted plus normal) nucleotides, as illustrated in Fig. 3. The radioactivity of normal nucleotides is evaluated by analyzing an aliquot of the labeled DNA digest (K. Randerath *et al.*, 1981; Gupta *et al.*, 1982). After nuclease P1 treatment (procedure 4), which removes normal nucleotides prior to labeling, the specific activity of [γ-^{32}P]ATP is determined and utilized for RAL calculations (Fig. 3). The RAL $\times$ 10^n is equal to the number of adducts in 10^n DNA nucleotides (K. Randerath *et al.*, 1985).

3. AUTHENTIC MUTAGENIC (GENOTOXIC) CARCINOGENS

A list of carcinogens whose DNA binding has been examined by ^{32}P-postlabeling is given in a recent review (Gupta and Randerath, 1988). For example, the assay was applied to DNA of mouse skin tissues after topical treatment with the polycyclic aromatic hydrocarbons, benzo[a]pyrene, 7,12-dimethylbenz[a]anthracene, and 3-methylcholanthrene (MC), respectively (E. Randerath *et al.*, 1983, 1985). As shown by the autoradiograms in Fig. 4, a large number of ^{32}P-labeled MC-DNA adducts was detected in digests of

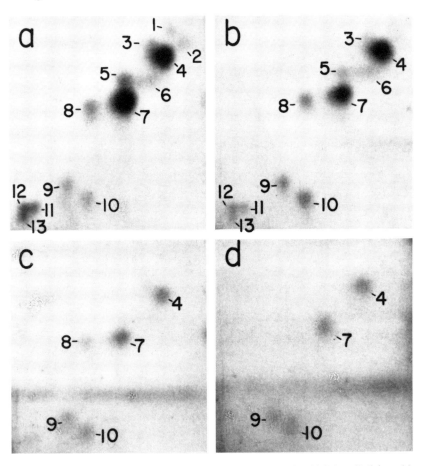

Figure 4. Maps of ^{32}P-labeled digests of mouse skin DNA isolated at (a) 1 day, (b) 6 days, (c) 14 days, and (d) 28 days after topical application of 3-methylcholanthrene (MC). Digests of DNA from control animals that had received vehicle (acetone) only did not exhibit any of the numbered spots. ^{32}P-Labeling was done by procedure 1, and the adducts were mapped as described by Gupta *et al.* (1982).

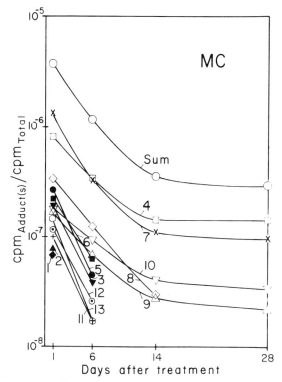

Figure 5. Time course of adduct labeling observed in the experiment illustrated in Fig. 4. Coefficients of variation were <15%.

DNA isolated from mouse skin at several time points after topical carcinogen treatment. Some of these adducts (spots 4, 7, 9, and 10) were highly persistent. As illustrated in Fig. 5, substantial removal of MC-DNA adducts occurred during the first two weeks after carcinogen application, while adducts remaining thereafter underwent little change. Some polycyclic aromatic hydrocarbon-DNA adducts have been shown to persist in mouse epidermis and dermis for at least 1 year after a single topical carcinogen exposure (E. Randerath *et al.*, 1985). Such results suggest that the persistent adducts occupy specific genomic sites in quiescent cells where they may be inaccessible to DNA repair enzymes. The persistence of adducts in certain organs, however, is not consistently associated with tumorigenicity (Wogan and Gorelick, 1985). Excellent associations have been observed by 32P-postlabeling, on the other hand, between the extent of initial binding to target organ DNA and hepatocarcinogenicity of dibenzo[c,g]carbazole (Schurdak and Randerath, 1985) and safrole (K. Randerath *et al.*, 1984; Reddy and Randerath, 1990). In the case of dibenzo[c,g]carbazole, adducts disappeared from target organ (liver) DNA at a much faster rate than from nontarget organ DNA (Schurdak and Randerath, 1988).

A unique and most important property of the ^{32}P-postlabeling assay is its ability to detect and measure chemically unidentified DNA adducts and especially adducts from unidentified or insufficiently characterized sources. In the following sections, we have chosen several examples to illustrate this point, i.e., (i) estrogen-induced adducts, (ii) I-compounds, (iii) carcinogen effects on I-compounds, and (iv) adducts formed as a consequence of exposures to complex mixtures.

4. ESTROGEN-INDUCED AND 2,3,7,8-TETRACHLORODIBENZO-*p*-DIOXIN-INDUCED DNA ALTERATIONS

An example for the application of ^{32}P-postlabeling to chemically unidentified adducts is provided by the analysis of kidney DNA from estrogen-treated male Syrian hamsters (Liehr *et al.*, 1986). These animals develop a high incidence of renal carcinoma after 6–8 months of exposure to natural or synthetic estrogens. ^{32}P-postlabeling showed that stilbene and steroid estrogens of widely divergent structures all led to the gradual formation of a set of chromatographically identical DNA adducts, as shown in Fig. 6. This was confirmed by isolating the adducts from the thin-layer plates and re-chromatographing them in various systems on PEI-cellulose, silica gel, and reversed-phase thin layers (Liehr *et al.*, 1986). It was concluded that the adducts did not contain estrogen moieties but rather were derived via the binding of an estrogen-induced unknown electrophilic metabolite (or metabolites) to kidney DNA; such adducts were not observed in nontarget organs. Liehr *et al.* (1986) postulated that such as yet unidentified metabolites and their DNA products are involved in the initiation of renal tumors in the hamster, as illustrated in Fig. 7. The estrogen-associated adducts were not detected in estrogen-induced kidney tumors (Lu *et al.*, 1988). Metabolic studies suggested that the formation of the adducts involves the activation of an endogenous metabolite by cytochrome P-450 but not prostaglandin endoperoxide synthase (Liehr *et al.*, 1987). Whether a similar mechanism of DNA damage also pertains to estrogen-induced tumorigenesis in other systems, including human tissues, has not yet been elucidated. The finding of I-compounds as discussed below also suggests that mechanisms exist by which "endogenous" electrophiles may lead to DNA adducts in animal and presumably human tissues.

In contrast to the estrogens, the potent nonmutagenic hepatocarcinogen, 2,3,7,8-tetrachlorodibenzo-*p*-dioxin (TCDD), was found by ^{32}P-postlabeling not to give rise to DNA adducts in the target tissue of female Sprague-Dawley rats; however, a drastic reduction in the level of I-compounds (see below) was noted in liver DNA after prolonged feeding of a carcinogenic dose of this compound (K. Randerath *et al.*, 1988b). This effect may be related to the hepatocarcinogenicity of the compound because kidney (nontarget) DNA of the animals was not affected similarly (K. Randerath *et al.*, 1988b).

Thus the ^{32}P-postlabeling assay has revealed that two mechanistically different types of DNA alterations may be induced by nonmutagenic carcinogens in target organs of carcinogenesis. Further work is needed to determine the biological consequences of these alterations and their possible relationship to tumorigenesis.

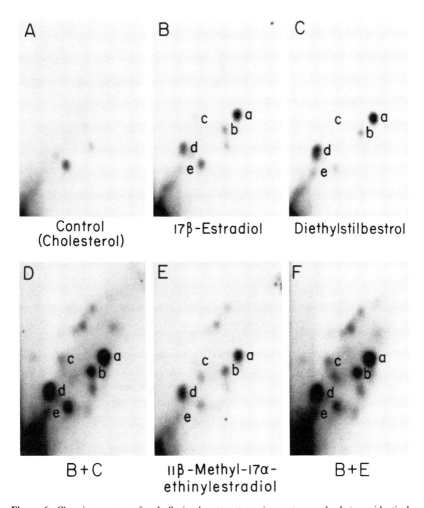

Figure 6. Chronic exposure of male Syrian hamsters to various estrogens leads to an identical set of DNA adducts (indicated by small letters) in kidney, the target organ of carcinogenesis; these adducts do not contain bound estrogen moieties. The maps shown are typical examples obtained by ^{32}P-postlabeling (Liehr *et al.*, 1986).

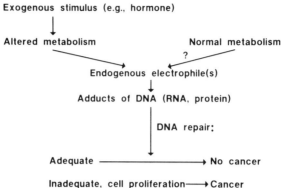

Exogenous stimulus (e.g., hormone)

Altered metabolism   Normal metabolism
                                 ?
        Endogenous electrophile(s)

        Adducts of DNA (RNA, protein)

                DNA repair:

    Adequate ─────────────→ No cancer

Inadequate, cell proliferation───→ Cancer

Figure 7. Formation of DNA adducts via "endogenous" electrophiles: In the hamster kidney model, estrogens are not part of the adduct structure but rather induce the binding of an unknown electrophile(s) specifically to kidney DNA. Likewise, DNA-reactive endogenous metabolites are postulated to arise in the course of normal metabolism, resulting in covalent DNA modifications. If these show an age-dependent increase, they are termed I-compounds (K. Randerath *et al.*, 1986a).

5. I-COMPOUNDS

Tissue DNA of untreated aging animals contains covalent modifications, termed I-compounds (K. Randerath *et al.*, 1986a), that were detected by ^{32}P-postlabeling (Fig. 8). The amounts and patterns of I-compounds were found to depend not only on age but also on animal species, strain or stock, tissue type, and sex (K. Randerath *et al.*, 1989). Furthermore, the composition of the animals' diet appears to play a major role in their formation (Li and Randerath, 1990). Analogous DNA modifications are not present or only marginally detectable in newborn animals (K. Randerath *et al.*, 1986a). While some I-compounds resemble aromatic DNA adducts in their chromatographic properties, more polar I-compounds have been detected recently by ^{32}P-postlabeling (K. Randerath *et al.*, 1988a). The origin and chemistry of the I-compounds are unknown, and no exogenous source of these DNA derivatives has been detected. In regenerating rat liver, I-compounds were found to resemble carcinogen adducts in their formation and not 5-methylcytosine, a normal enzymatic DNA modification, suggesting that DNA-modifying enzymes may not be directly involved in I-compound formation (K. Randerath *et al.*, 1988a). The observation of tissue-specific I-spots argues against a direct role of exogenous environmental carcinogens in their formation, because chemical carcinogens would be expected to elicit qualitatively identical adduct patterns in different organs. We have hypothesized that these DNA modifications are due to the binding to DNA of small amounts of reactive electrophilic by-products of normal metabolic activities, leading to the slow accumulation with age of I-compounds in tissue DNA.

As adduct analogues, I-compounds may play a role in natural aging and "spontaneous" tumor formation and may exert other as yet unidentified effects. Loss of I-compounds during chemical carcinogenesis (K. Randerath *et al.*, 1988b), for example, could conceivably facilitate DNA synthesis in preneoplastic and/or neoplastic cells. Consistent with this possibility, exposure to a number of nongenotoxic or nonmutagenic carcinogens, in addition to TCDD, has recently been found to lead to drastic losses of I-compounds in target tissue DNA (Li *et al.*, 1988; K. Randerath *et al.*, 1990).

6. CIGARETTE-SMOKE-INDUCED DNA LESIONS

Cigarette-smoke-induced DNA adducts were initially detected in both experimental animals (K. Randerath *et al.*, 1986b; E. Randerath *et al.*, 1986) and humans (E. Randerath *et al.*, 1986; Everson *et al.*, 1986, 1988) by procedure 2 (Fig. 2). The recently developed nuclease P1 version of the ³²P-postlabeling assay (procedure 4) (Reddy and Randerath, 1986) allowed the detection of much lower levels of smoking-related DNA adducts than heretofore possible. Results obtained by this procedure in human and mouse tissues (Fig. 9) (Everson *et al.*, 1986; E. Randerath *et al.*, 1988) provided evidence for the presence of DNA adducts. The cigarette-smoke-related ³²P-labeled material occupied two extensive diagonal radioactive zones (DRZ 1 and DRZ 2) on the autoradiographic maps. We have postulated (E. Randerath

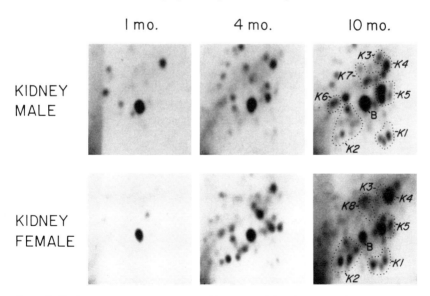

Figure 8. Typical I-spot patterns detected in kidney DNA of female and male Sprague-Dawley rats of different ages. DNA specimens were analyzed as described by K. Randerath *et al.* (1989).

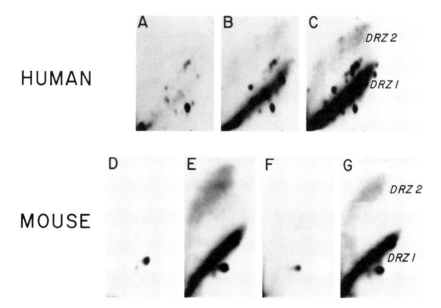

A   B   C

HUMAN

DRZ 2

DRZ I

D   E   F   G

MOUSE

DRZ 2

DRZ I

Figure 9. Autoradiograms of ³²P-labeled DNA adducts from lung of smokers and lung and heart of mice treated dermally with cigarette smoke condensate ("tar"). Adducts were purified by nuclease P1 treatment (Reddy and Randerath, 1986) and mapped as described by E. Randerath *et al.* (1988). A: nonsmoker; B: smoker (50 pack years); C: smoker (75 pack years); D: mouse lung control; E: lung after exposure to 6 daily doses of cigarette "tar" corresponding to 0.75 cigarette each; F: heart control; G: heart after treatment as in E. DRZ: diagonal radioactive zone (see text).

et al., 1988) that these results are indicative of the presence in the DNA samples of numerous, only partially resolved aromatic DNA adducts of varying polarities. This is in accord with the presence in cigarette smoke of a large number of polycyclic aromatic compounds containing known as well as unidentified tumor initiators and/or carcinogens [Hoffman *et al.,* 1976, 1987; U.S. Department of Health and Human Services, 1982; International Agency for Research on Cancer (IARC), 1985]. The adducts detected by ³²P-postlabeling thus appear to provide a valid dosimeter for initiator and/or carcinogen exposure. The similar patterns of DNA lesions in human and mouse tissues (Fig. 9) suggest common characteristics of tobacco-associated carcinogen activation and adduct formation in different mammalian species.

As shown in Fig. 10, smoking-associated DNA adducts were found to occur in a dose- and time-dependent manner in surgical specimens of lung, bronchus, and larynx from smokers with cancer of these organs (K. Randerath *et al.,* 1987). The available data best fit a power curve, $Y = aX^b$, especially if two outlying values from heavy alcohol users were omitted (Fig. 10). From our preliminary data, alcohol abuse appeared to lower the levels of smoking-associated DNA adducts in heavy smokers (our unpub-

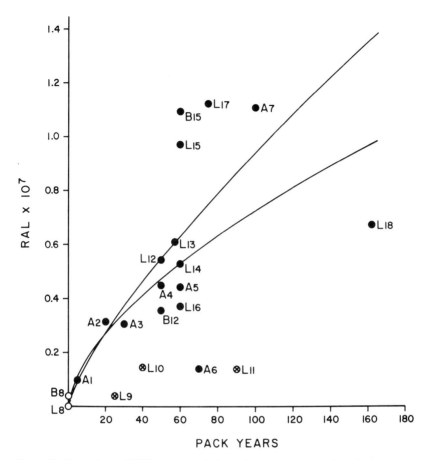

Figure 10. Dependence of DNA-adduct levels in smokers' target organs of carcinogenesis on smoking exposure. Tissues (A, aryepiglottic fold; B, bronchus; L, lung) were surgical specimens from patients with tumors of the larynx or lung, respectively. (Patient numbers are listed together with the tissue symbols.) ●, smoker; ○, nonsmoker; ⊗, ex-smoker. Ex-smokers ceased smoking 10 (L9, L10) or 14 (L11) years before the operation. Subjects smoked until at least 14 days before surgery, except for L14, which was from a patient with a smoke-free interval of 42 days. Smoking exposure is given in pack years, i.e. (packs of cigarettes/day) × years smoked. Smoking-associated DNA adducts were assayed as in Fig. 9. DNA-adducts levels, expressed as RAL × 10⁷ values, were calculated from the sum of the count rates of DRZ1 and DRZ 2 (see Fig. 9). Coefficients of variation of the mean RAL × 10⁷ values ranged from 5% to 25% in replicate (N = 3 or 4) DNA analyses. Data (excluding the ex-smokers) best fit a power curve, $Y = aX^b$. The lower curve ($Y = 0.042X^{0.616}$; $r = 0.69$) was calculated from the data for all smokers and the nonsmoker, while for the calculation of the upper curve ($Y = 0.027X^{0.772}$; $r = 0.87$) the outlying samples A6 and L18 (from heavy drinkers) were omitted.

lished experiments), but this needs further investigation. After cessation of smoking, DNA adduct levels declined only slowly; smoking-associated DNA adducts were still measurable after 10 or 14 years in ex-smokers with previous heavy exposure (Fig. 10, samples L10 and L11). Thus smoking-induced DNA lesions persisted in human tissues similarly to adducts elicited by authentic aromatic carcinogens in animals, where DNA adduct concentrations decline rapidly during the early phase after carcinogen exposure and subsequently stabilize at lower levels for extended periods of time (K. Randerath *et al.*, 1984; E. Randerath *et al.*, 1985). The extent of DNA damage in lung and larynx paralleled cancer risk and mortality, which are known to increase with the number of cigarettes smoked and the duration of smoking (U.S. Department of Health and Human Services, 1982; IARC, 1985). Similarly, the persistence of smoking-associated DNA adducts was in accord with the slow decline of lung and larynx cancer risk and mortality after cessation of smoking (U.S. Department of Health and Human Services, 1982; IARC, 1985). Our results, therefore, fulfilled the prediction that smoking-associated DNA adducts in major target organs of human carcinogenesis, as assayed by ^{32}P-postlabeling, provide a valid dosimeter for initiator and/or carcinogen exposure.

The tissue distribution of smoking-associated DNA adducts in autopsy samples from heavy smokers was found to parallel that seen in mice treated dermally with cigarette-smoke condensate (CSC) (E. Randerath *et al.*, 1988), i.e., adduct levels in lung and heart DNA of both species were high (cf. Fig. 9, panels E and G), in contrast to considerably lower levels in other human and mouse tissues, such as kidney, liver, and spleen. These similarities suggest that the mouse provides a suitable model system for studying DNA damage and mechanisms of smoking-associated carcinogenesis.

Our results imply that tobacco-related DNA lesions in target organs of carcinogenesis, such as lung and larynx, play a key role in cancer induction in these tissues. In addition, the DNA lesions detected by ^{32}P-postlabeling may be involved, also, in nonneoplastic lung and heart pathology, such as chronic obstructive lung disease and cardiomyopathy, in line with recent reports that DNA adducts induced by the carcinogenic food pyrolysis product, aminomethyl(α)carboline, are associated with degenerative tissue alterations (Takayama *et al.*, 1985; Yamashita *et al.*, 1986).

7. DNA ALTERATIONS IN WHITE BLOOD CELLS OF IRON-FOUNDRY AND COKE-OVEN WORKERS

White blood cells (WBC) provide an easily accessible source of DNA for biomonitoring of human exposure to environmental or occupational genotoxicants via a minimally invasive technique. In preliminary, still ongoing

**TABLE 8–1. DNA-Adduct Levels in White Blood Cells of
Iron-Foundry and Coke-Oven Workers**

Summary of Preliminary Results

1. Aromatic adducts detected in WBC DNA of most (53/68) foundry workers and most (15/22) coke-oven workers so far examined
2. Dose–response relationships established between estimated exposure and DNA adduct levels
3. Similar adduct patterns within each exposure group
4. Chromatographic identity of several adducts between the 2 groups
5. Chromatographic properties of adducts suggest PAH derivatives
6. Identification of some major adducts in progress
7. No clear effects due to age, sex, or smoking habits
8. ^{32}P-postlabeling assay of WBC DNA is useful in biomonitoring human exposure to occupational genotoxicants

experiments, we have collaborated with Dr. K. Hemminki of the Institute for Occupational Health in Helsinki, Finland, to detect and measure DNA lesions in WBC of iron-foundry and coke-oven workers. It is planned to relate the results of these measurements eventually to the exposure status of the subjects, but in the initial experiments, we have primarily attempted to answer the question whether adducts associated with occupational exposures can be detected by ^{32}P-postlabeling in WBC DNA.

The results are summarized in Table 1. (1) Aromatic adducts were detected in WBC DNA of most (53/68) foundry workers and most (15/22) coke-oven workers so far examined. (2) Dose-response relationships were established between estimated exposure and DNA-adduct levels. (3) The adduct patterns within each exposure group were similar, as shown in Fig. 11. Adduct levels in highly exposed individuals ranged from 1 adduct in 10^7 to 1 adduct in 3×10^5 DNA nucleotides (approximately $0.3–10$ fmol adduct/ μg DNA). (4) Several adducts appeared to be identical between the two groups of workers. (5) The chromatographic properties of the adducts suggested polycyclic aromatic hydrocarbon (PAH) derivatives. (6) The identification of some major adducts is in progress. (7) No clear effects due to age, sex, or smoking habits were observed. (8) These results indicate that ^{32}P-postlabeling of WBC DNA is a useful and generally applicable tool in biomonitoring of human exposure to occupational genotoxicants. How accurately WBC-DNA adduction reflects DNA lesions in internal organs is currently being studied. Notably, however, the levels of DNA adduction found in WBC of the exposed workers are similar to the total levels of persistent adducts in smokers' target tissues and in tissues of animals exposed to initiating or carcinogenic doses of authentic aromatic genotoxicants.

COKE OVEN WORKERS

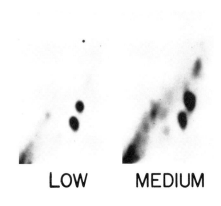

LOW   MEDIUM

HIGH

Figure 11. Typical maps of adducts in WBC DNA samples of coke-oven workers. Adducts were resolved two-dimensionally in $3.2M$ lithium formate, $7.7M$ urea, pH 4.0 (from bottom to top) and $0.7M$ sodium phosphate, $7M$ urea, pH 6.4 (from left to right).

8. CONCLUSIONS AND PROSPECTS

In this article, we have emphasized the use of ^{32}P-postlabeling in the analysis of covalent DNA alterations elicited by unknown electrophiles (e.g., estrogen-induced adducts and I-compounds) or incompletely characterized mixtures of genotoxicants (e.g., cigarette smoke components). When applying the assay to unidentified adducts one needs to keep in mind that recoveries of the DNA derivatives in ^{32}P-labeled form may be incomplete, and therefore, quantitative data represent minimum estimates. Nevertheless, it is possible to discern via postlabeling factors that affect adducts *in vivo*, such as the animal age versus I-compound levels, time of estrogen exposure versus adduct buildup in hamster kidney DNA, duration and intensity of cigarette smoking versus adduct levels in target tissues, etc. Our work with I-compounds, for example, has demonstrated in animal models that complex mixtures of unknown DNA adducts can be reproducibly quantified and related to environmental (e.g., diet), genetic (e.g., species, sex, tissue), and other (e.g., age) factors. The data so far obtained strongly suggest that DNA adducts assayed by ³²P-postlabeling may also become useful as dosimeters of human exposures to genotoxicants. Future efforts need to be directed at the structural identification and characterization of unknown DNA modifications detected by ³²P-adduct assay. It is hoped that a combination of the assay with improved sensitive techniques for structural analysis of small amounts of adducts will become a powerful tool in the characterization of the origin and nature of human adducts from both exogenous and endogenous sources.

ACKNOWLEDGMENTS

The authors' work cited in this review has been supported by U.S. Public Health Service Grants, No. CA 25590, No. CA 32157, and No. CA 43263 awarded by the National Cancer Institute and Grant No. AG 07750 awarded by the National Institute on Aging. The contributions to the development of the ³²P-postlabeling assay by our collaborators, especially Dr. Ramesh C. Gupta and Dr. M. Vijayaraj Reddy, are acknowledged with gratitude.

REFERENCES

B. N. Ames, Identifying environmental chemicals causing mutations and cancer, *Science* **204**, 587–593 (1979).

W. M. Baird, The use of radioactive carcinogens to detect DNA modifications by chemical carcinogens. In: *Chemical Carcinogens and DNA,* Vol. 1, ed. P. L. Grover, CRC Press, Boca Raton, FL, pp. 59–83 (1979).

F. J. De Serres and J. Ashby, *Evaluation of Short-Term Tests for Carcinogens: Reports of the International Collaborative Program,* Elsevier, New York (1981).

R. Doll and R. Peto, *The Causes of Cancer,* Oxford University Press, Oxford (1981).

B. P. Dunn and R. H. C. San, HPLC enrichment of hydrophobic DNA-carcinogen adducts for enhanced sensitivity of ³²P-postlabeling analysis, *Carcinog.* **9**, 1055–1060 (1988).

R. B. Everson, E. Randerath, R. M. Santella, R. C. Cefalo, T. A. Avitts, and K. Randerath, Detection of smoking related covalent DNA adducts in human placenta, *Science* **231**, 54–57 (1986).

R. B. Everson, E. Randerath, R. M. Santella, T. A. Avitts, I. B. Weinstein, and K. Randerath, Quantitative associations between DNA damage in human placenta and maternal smoking and birth weight, *J. Natl. Cancer Inst.* **80**, 567–576 (1988).

R. C. Gupta, Enhanced sensitivity of ³²P-postlabeling analysis of aromatic carcinogen-DNA adducts, *Cancer Res.* **45**, 5656–5662 (1985).

R. C. Gupta and K. Earley, ^{32}P-adduct assay: comparative recoveries of structurally diverse DNA adducts in the various enhancement procedures, *Carcinog.* **9,** 1687–1693 (1988).

R. C. Gupta and K. Randerath, Analysis of DNA adducts by ^{32}P-labeling and thin-layer chromatography. In: *DNA Repair, A Laboratory Manual of Research Procedures,* Vol. 3, eds. E. C. Friedberg and P. C. Hanawalt, Marcel Dekker, New York, pp. 399–418 (1988).

R. C. Gupta, M. V. Reddy, and K. Randerath, ^{32}P-Postlabeling analysis of nonradioactive aromatic carcinogen-DNA adducts, *Carcinog.* **3,** 1081–1092 (1982).

K. Hemminki, Nucleic acid adducts of chemical carcinogens and mutagens, *Arch. Toxicol.,* **52,** 249–285 (1983).

K. Hemminki, F. P. Perera, D. H. Phillips, K. Randerath, M. V. Reddy, and R. M. Santella, Aromatic DNA adducts in white blood cells of foundry workers. In: *Detection Methods for DNA-Damaging Agents in Man: Applications in Cancer Epidemiology and Prevention,* eds. H. Bartsch and K. Hemminki, Li, IARC Sci. Publ. No. 89, International Agency for Research on Cancer, Lyon, France, pp. 190–195 (1988).

D. Hoffmann, I. Schmeltz, S. S. Hecht, and E. L. Wynder, Chemical studies on tobacco smoke. XXXIX. On the identification of carcinogens, tumor promoters, and cocarcinogens in tobacco smoke. In: *DHEW Publication (NIH)-76-1221, Smoking and Health.* Vol. 1, *Modifying the Risk for the Smoker,* United States Department of Health, Education, and Welfare, Rockville, MD, pp. 125–146 (1976).

D. Hoffmann, E. L. Wynder, S. S. Hecht, K. D. Brunnemann, E. J. LaVoie, and N. J. Haley, Chemical carcinogens in tobacco. In: *Cancer Risks. Strategies for Elimination,* ed. P. Bannasch, Springer Verlag, New York, pp. 101–113 (1987).

IARC Monographs on the Evaluation of the Carcinogenic Risk of Chemicals to Humans, Vol. 38, *Tobacco Smoking.* International Agency for Research on Cancer, Lyon, France (1985).

D. Li, N. Chandar, B. Lombardi, and K. Randerath, Loss of I-compounds in liver DNA of rats fed a choline-devoid diet. *Proc. Am. Assoc. Cancer Res.* **29,** 110 (1988).

D. Li and K. Randerath, Association between diet and age-related DNA modifications (I-compounds) in rat liver and kidney, *Cancer Res.,* **50,** 3991–3996 (1990).

J. G. Liehr, T. A. Avitts, E. Randerath, and K. Randerath, Estrogen-induced endogenous DNA adduction: Possible mechanism of hormonal cancer, *Proc. Natl. Acad. Sci. USA* **83,** 5301–5305 (1986).

J. G. Liehr, E. R. Hall, T. A. Avitts, E. Randerath, and K. Randerath, Localization of estrogen-induced DNA adducts and cytochrome P-450 activity at the site of renal carcinogenesis in the hamster kidney, *Cancer Res.* **47,** 2156–2159 (1987).

L.-J. W. Lu, R. M. Disher, M. V. Reddy, and K. Randerath, ^{32}P-Postlabeling assay of transplacental DNA damage induced by the environmental carcinogens safrole, 4-aminobiphenyl, and benzo(a)pyrene, *Cancer Res.* **46,** 3046–3054 (1986).

L.-J. W. Lu, J. G. Liehr, D. A. Sirbasku, E. Randerath, and K. Randerath, Hypomethylation of DNA in estrogen-induced and -dependent hamster kidney tumors, *Carcinog.* **9,** 925–929 (1988).

E. C. Miller and J. A. Miller, Searches for ultimate chemical carcinogens and their reactions with cellular macromolecules, *Cancer* **47**, 2327–2345 (1981).

M. C. Poirier, Antibodies to carcinogen-DNA adducts, *J. Natl. Cancer Inst.* **67**, 515–519 (1981).

E. Randerath, H. P. Agrawal, M. V. Reddy, and K. Randerath, Highly persistent polycyclic aromatic hydrocarbon-DNA adducts in mouse skin: Detection by ^{32}P-postlabeling analysis, *Cancer Lett.* **20**, 109–114 (1983).

E. Randerath, H. P. Agrawal, J. A. Weaver, C. B. Bordelon, and K. Randerath, ^{32}P-Postlabeling analysis of DNA adducts persisting for up to 42 weeks in the skin, epidermis and dermis of mice treated with 7,12-dimethylbenz(a)anthracene, *Carcinog.* **6**, 1117–1126 (1985).

E. Randerath, T. A. Avitts, M. V. Reddy, R. H. Miller, R. B. Everson, and K. Randerath, Comparative ^{32}P-analysis of cigarette smoke induced DNA damage in human tissues and mouse skin, *Cancer Res.* **46**, 5869–5877 (1986).

E. Randerath, D. Mittal, and K. Randerath, Tissue distribution of covalent DNA damage in mice treated dermally with cigarette "tar": Preference for lung and heart DNA, *Carcinog.* **9**, 75–80 (1988).

K. Randerath, M. V. Reddy, and R. C. Gupta, ^{32}P-Labeling test for DNA damage, *Proc. Natl. Acad. Sci. USA* **78**, 6126–6129 (1981).

K. Randerath, R. E. Haglund, D. H. Phillips, and M. V. Reddy, ^{32}P-Postlabeling analysis of DNA adducts formed in the livers of animals treated with safrole, estragole and other naturally-occurring alkenylbenzenes. I. Adult female CD-1 mice, *Carcinog.* **5**, 1613–1622 (1984).

K. Randerath, E. Randerath, H. P. Agrawal, R. C. Gupta, M. E. Schurdak, and M. V. Reddy, Postlabeling methods for carcinogen-DNA adduct analysis, *Environ. Health Persp.* **62**, 57–65 (1985).

K. Randerath, M. V. Reddy, and R. M. Disher, Age- and tissue-related DNA modifications in untreated rats: Detection by ^{32}P-postlabeling assay and possible significance for spontaneous tumor induction and aging, *Carcinog.* **7**, 1615–1617 (1986a).

K. Randerath, M. V. Reddy, T. A. Avitts, R. H. Miller, R. B. Everson, and E. Randerath, ^{32}P-Postlabeling test for smoking-related DNA adducts in animal and human tissues. In: *Mechanisms in Tobacco Carcinogenesis*, Banbury Report 23, eds. D. Hoffmann and C. C. Harris, Cold Spring Harbor Laboratory, Cold Spring Harbor, NY, pp. 85–98 (1986b).

K. Randerath, T. A. Avitts, R. H. Miller, and E. Randerath, ^{32}P-Postlabeling analysis of cigarette smoking-related DNA adducts in target tissues of human carcinogenesis, *Proc. Am. Assoc. Cancer Res.* **28**, 98 (1987).

K. Randerath, L.-J. W. Lu, and D. Li, A comparison between different types of covalent DNA modifications (I-compounds, persistent carcinogen adducts and 5-methylcytosine) in regenerating rat liver, *Carcinog.* **9**, 1843–1848 (1988a).

K. Randerath, K. L. Putman, E. Randerath, G. Mason, M. Kelley, and S. Safe, Organ-specific effects of long term feeding of 2,3,7,8-tetrachlorodibenzo-p-dioxin and 1,2,3,7,8-pentachlorodibenzo-p-dioxin on I-compounds in hepatic and renal DNA of female Sprague-Dawley rats, *Carcinog.* **9**, 2285–2289 (1988b).

K. Randerath, J. G. Liehr, A. Gladek, and E. Randerath, Age-dependent covalent DNA alterations (I-compounds) in rodent tissues: species, tissue and sex specificities, *Mutat. Res.* **219**, 121–133 (1988c).

K. Randerath;, D. Li, and E. Randerath, Age-related DNA modifications (I-compounds): Modulation by physiological and pathological processes, *Mutat. Res.* **238,** 245–253 (1990).

M. V. Reddy, R. C. Gupta, E. Randerath, and K. Randerath, ^{32}P-Postlabeling test for covalent DNA binding of chemicals *in vivo*: Application to a variety of aromatic carcinogens and methylating agents, *Carcinog.* **5,** 231–243 (1984).

M. V. Reddy and K. Randerath, Nuclease P_1-mediated enhancement of sensitivity of ^{32}P-postlabeling test for structurally diverse DNA adducts, *Carcinog.* **7,** 1543–1551 (1986).

M. V. Reddy and K. Randerath, A comparison of DNA adduct formation in white blood cells and internal organs of mice exposed to benzo[a]pyrene, dibenzo[c,g]carbazole, satrole and cigarette smoke condensate, *Mutat. Res.* **241,** 37–48 (1990).

M. E. Schurdak and K. Randerath, Tissue-specific DNA adduct formation in mice treated with the environmental carcinogen, 7H-dibenzo(c,g)carbazole, *Carcinog.* **6,** 1271–1274 (1985).

M. E. Schurdak and K. Randerath, Comparison of the persistence of DNA adduction by the environmental carcinogen 7H-dibenzo(c,g)carbazole (DBC) in various mouse tissues, *Proc. Am. Assoc. Cancer Res.* **29,** 95 (1988).

H. A. J. Schut, K. L. Putman, and K. Randerath, DNA adduct formation of the carcinogen 2-amino-3-methylimidazo[4,5-*f*]quinoline (IQ) in target tissues of the F-344 rat, *Cancer Lett.* **41,** 345–352 (1988).

H. F. Stich and R. H. C. San, *Short-Term Tests for Chemical Carcinogens,* Springer-Verlag, New York (1981).

S. Takayama, Y. Nakatsura, H. Ohgaki, S. Sato, and T. Sugimura, Atrophy of salivary glands and pancreas of rats fed on diet with amino-methyl-α-carboline, *Proc. Jpn. Acad. Ser. B* **61,** 277–280 (1985).

U.S. Department of Health and Human Services, *The Health Consequences of Smoking: Cancer,* A Report of the Surgeon General, Office on Smoking and Health, Rockville, MD (1982).

V. L. Wilson, A. K. Basu, J. M. Essigmann, R. A. Smith, and C. C. Harris, O^6-Alkyldeoxyguanosine detection by ^{32}P-postlabeling and nucleotide chromatographic analysis, *Cancer Res.* **48,** 2156–2161 (1988).

G. N. Wogan and N. J. Gorelick, Chemical and biochemical dosimetry of exposure to genotoxic chemicals, *Environ. Health Perspect.* **62,** 5–18 (1985).

K. Yamashita, S. Takayama, M. Nagao, S. Sato, and T. Sugimura, Amino-methyl-α-carboline induced DNA modification in rat salivary glands and pancreas detected by ^{32}P-postlabeling method, *Proc. Jpn. Acad. Ser. B* **62,** 45–48 (1986).

DNA-Adduct Analysis by 32P-Postlabeling in the Study of Human Exposure to Carcinogens

David H. Phillips
Haddow Laboratories
Institute of Cancer Research
Surrey, United Kingdom

1. INTRODUCTION

The majority of chemical carcinogens share the property of covalent interaction of electrophilic reactive intermediates with cellular macromolecules and are thought to mediate their carcinogenic activity through their binding to DNA. With the advent of sensitive techniques for the detection of carcinogen-DNA adducts that do not require the use of radiolabeled test compounds [the means by which adducts were initially detected and analyzed in studies involving laboratory animals (Baird, 1979)] has come the possibility of monitoring DNA isolated from human tissue for evidence of prior exposure to chemical carcinogens and mutagens (Phillips, 1990). Occupational exposure to carcinogenic substances is believed to contribute to a small but significant percentage of human cancers in industrial societies (Doll and Peto, 1981). In most cases, however, human exposure is to complex mixtures of chemicals, with the result that it is difficult to differentiate the components that are deleterious to human health from those that are insignificant. In addition, the causative agents responsible for most of the commonest cancers remain unidentified although they are believed, in many cases, to be chemical. Effective strategies for cancer prevention are likely to be assisted by reliable monitoring of biologically significant human exposure to environmental carcinogens that are present in food, air, and water, or others that are formed endogenously.

^{32}P-Postlabeling analysis is a method of DNA-adduct detection that is readily applicable to monitoring human DNA for the presence of covalently bound carcinogen moieties for several reasons (Phillips, 1990). First, the method is extremely sensitive and can, in some circumstances, detect adducts present at frequencies as low as 1 adduct in 10^9–10^{10} nucleotides. Second, the assay can be performed on small quantities of DNA (0.1–10 μg), making it readily applicable to the analysis of human DNA, often obtainable only in very small amounts. Third, the assay is applicable to adducts with a wide range of different chemical structures (although it is most sensitive for aromatic and other hydrophobic moieties) and can be used to study adduct formation by complex mixtures of carcinogenic compounds. Fourth, it is not a requirement that an adduct be characterized by physicochemical methods, or prepared synthetically in large quantities for the ^{32}P-postlabeling method to be adapted for the detection of that adduct in biological samples. In this chapter I describe studies in our laboratory on the detection of aromatic adducts in DNA isolated from individuals exposed occupationally, medicinally, or socially to polycyclic aromatic hydrocarbons (PAHs) and corroborating studies on the formation of adducts in experimental animals treated with these PAH mixtures. Preliminary results from the surveillance of DNA isolated from different tissues of people not known to be exposed occupationally to chemical carcinogens are also described.

2. EXPERIMENTAL PROCEDURES

The principle of ^{32}P-postlabeling, as outlined by Gupta *et al.* (1982), is that (i) DNA containing adducts is digested enzymatically to nucleoside 3'-monophosphates, (ii) the DNA digest is incubated with [γ-^{32}P]ATP and T4 polynucleotide kinase, (iii) the labeled digest containing [5'-^{32}P]-labeled nucleoside 3',5'-bisphosphates is subjected to multidirectional chromatography, first to separate the adducted nucleotides from nonadducted nucleotides and then to resolve them, and (iv) the adducts are detected by autoradiography and quantitated by scintillation or Čerenkov counting of the radioactive spots on the chromatograms.

The experimental conditions adopted for these procedures in our laboratory are essentially those described by Randerath and co-workers (Gupta *et al.*, 1982; Reddy *et al.*, 1984; Everson *et al.*, 1986) with modifications that we have described (Phillips *et al.*, 1986) to allow for the use of reagents and materials obtained from different sources.

Several modifications of the standard ^{32}P-postlabeling procedure dramatically improve the sensitivity of the assay. If limiting quantities of high-specific-activity ATP are used instead of an excess of low-specific-activity material, then an enhancement of adduct labeling can be achieved without increasing the amount of radioactivity used (Phillips *et al.*, 1984; Randerath *et al.*, 1985). However, the degree of enhancement is variable and dependent in an unpredictable way on the nature of the adduct, and thus absolute quantitation requires a standard sample of DNA containing the adduct at a level sufficiently high for it to be detected by the standard procedure (in practice 1 adduct in 10^6–10^7 nucleotides). Other methods of improving sensitivity involve the separation of adducted nucleotides from normal ones prior to labeling, either by phase-transfer extraction (Gupta, 1985) or by reverse-phase high-pressure liquid chromatography (HPLC) (Dunn *et al.*, 1987; Dunn and San, 1988). Alternatively, incubation of DNA digests with nuclease P_1 prior to ^{32}P-labeling dephosphorylates normal nucleotides so that they are no longer substrates for the kinase, but not many types of adduct (Reddy and Randerath, 1986). These modifications, which are best suited to hydrophobic adducts, can make possible the detection of adducts present at frequencies as low as 1 in 10^9–10^{10} nucleotides in a sample of 5–10 µg DNA.

For the majority of our studies, we have found the nuclease P_1 digestion method most suitable for studying aromatic adducts present at low levels in human and animal DNA. However, care must be taken to ensure that the optimum concentration and incubation conditions are used. Underdigested samples will contain unmodified nucleotides that will compete with the nucleotide adducts for the available [γ-^{32}P]ATP; overdigestion will cause a partial dephosphorylation of the adducts and lead to reduced recovery. In practice, we have found that it is necessary to determine the concentration of nuclease P_1 with a standard incubation time of 1 h, that leads to the highest determination of adduct levels for a series of samples. The optimum concentration of enzyme may thus be different for coke-oven workers' white blood cell DNA than it is for, say, lung DNA from smokers. For some studies we have also used butanol extraction to concentrate adducts, but have found in some cases a 2–3-fold lower recovery of adducts. This may be due, in part, to methodological factors and, in part, to the nature of the adducts detected (Gupta and Earley, 1988; Gallagher *et al.*, 1989).

It should be noted that the use of the high-sensitivity modifications of ^{32}P-postlabeling can lead to the appearance of artifactual radioactive spots on the chromatograms, for example, ''spot X'', which we have detected in human, animal, bacterial, and plasmid DNA that is not known to be carcinogen-modified (Phillips *et al.*, 1986). Thus it is essential for these studies that DNA from a suitable control source be analyzed alongside the "exposed" samples, and only those spots or areas of radioactivity that are present in the latter and consistently absent from the former be considered as adducts.

3. DNA ADDUCTS IN LYMPHOCYTES OF IRON-FOUNDRY WORKERS

The iron and steel industries employ approximately 2 million workers worldwide. Foundry workers are exposed to a variety of hazardous materials that include carbon monoxide, silica and other mineral dusts, metal fumes and dusts, organic binder chemicals, and polycyclic aromatic compounds [International Agency for Research on Cancer (IARC), 1984b]. The use of coal-tar products for binding the sand from which molds are constructed is a major source of PAHs. Contact of molten metal with the molds results in the release of airborne PAHs and other pyrolysis products, and those procedures within the foundry that result in greatest concentrations of airborne mutagens are associated with the greatest carcinogenic risk to the workers.

The functions of workers in a Finnish iron foundry have been classified as involving high, medium, or low exposure to airborne PAHs according to the atmospheric concentrations of a representative PAH, benzo[a]pyrene (BP). High exposure (>0.2 μg BP/m³ air) was representative of workers involved in casting and sand preparation, medium exposure (0.05–0.2 μg BP/m³) by those involved in core blowing, mold changing and finishing, and low exposure (<0.05 μg BP/m³) by workers employed in melting, transport of molten iron, core making, sand drying, tempering, spraying, emering, and heat treating. Blood samples were obtained from healthy individuals in these three exposure groups, and DNA was isolated from their white blood cells and analyzed by ^{32}P-postlabeling (Phillips *et al.*, 1988a). Control DNA was obtained from blood samples from nonfoundry workers whose occupations did not indicate occupational exposure to PAHs.

Of nine samples from the unexposed control group, all but one contained no DNA adduct spots other than spot X. In the low-exposure group, 3/18 samples showed evidence of aromatic adducts. Eight out of 10 samples from the medium-exposure group were positive, and 3/4 samples from the high-exposure group were positive (Phillips *et al.*, 1988a) (Fig. 1).

One noticeable feature of these results was that the positive samples showed interindividual variation in their patterns of adduct spots. Also, in a number of cases where a second blood sample was obtained from the same individual, differences in the adduct patterns were observed. Calculation of the total levels of adducts in the samples indicated a correlation between exposure and the mean levels of adducts in the high- and medium-exposure

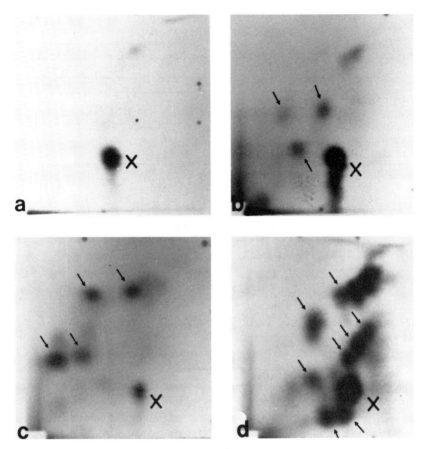

Figure 1. Autoradiographs of PEI-cellulose thin-layer chromatography maps of ³²P-labeled digests of DNA isolated from peripheral white blood cells from Finnish iron-foundry workers and control subjects. An example of an individual from each of the exposure catagories is shown: control (*a*), low exposure (*b*), medium exposure (*c*), and high exposure (*d*). Adducts are indicated by arrows. The small black dots visible at the top and right-hand edges of the maps are due to radioactive ink used to align the chromatograms for quantitation.

groups (combined because of the small number of samples obtained) compared to the control group (Table 1). A difference between the low-exposure and control groups was not, however, evident. No significant influence of age, sex, or smoking habits of the subjects was seen on the occurrence of adducts.

Although the workers were classified according to their exposure to airborne BP, for the most part their DNA-adduct profiles did not correspond to those of adducts formed from BP. Immunological methods of DNA-adduct detection, using antibodies raised against BP-DNA adducts (but known to cross-react with DNA modified by other PAHs), also showed a correlation

TABLE 9–1. ^{32}P-Postlabeling of White Blood Cell DNA
From Iron-Foundry Workers

Exposure Group (BP)	Mean No. Adducts/10^8 Nucleotides	Number of Individuals (No. of Samples)	Range
High-medium	1.8	8 (12)	0–10.0
Low	0.06	16 (17)	0–0.6
All exposed	0.8	24 (29)	0–10.0
After vacation	0.7	3 (3)	0.2–1.0
Controls	0.2	9 (9)	0–1.9

Source: Phillips *et al.* (1988a).

of exposure with adduct levels among iron-foundry workers (Perera *et al.*, 1988). Comparison of results obtained with the two methods indicated a relatively high correlation (Hemminki *et al.*, 1988), although there were variations in the actual positive values obtained.

A second study on a larger group of Finnish iron-foundry workers produced broadly similar results (Savela *et al.*, 1989). Elevated levels of adducts were found in white blood cell DNA from 53 workers compared to levels in control subjects. Large interindividual variations in adduct levels were again found, and in this case no discernible difference among exposure groups categorized according to airborne BP concentrations was evident. No effect of smoking was observed in this study, as in the first study, and independent analysis of the samples in three different laboratories gave correlations that were statistically highly significant.

These results indicate that assays such as ^{32}P-postlabeling can be used as dosimeters of human exposure to carcinogenic compounds, but also demonstrate the need for analyzing several samples, taken at different times, from each individual. It is not yet clear whether the differences among individuals is the consequence of qualitative and quantitative differences in exposure, or whether the variability in different samples from the same individual is a reflection of different recent exposure.

4. DNA ADDUCTS IN LYMPHOCYTES OF COKE-OVEN WORKERS

Coke production is an important industry producing fuel for iron-ore smelting and other uses. The destructive distillation of coal produces volatile fractions and the coke residue. Coke-oven emissions are complex mixtures of dusts, vapors, and gases. Many PAHs have been identified in the particulate matter of atmospheric samples taken from coke ovens. Levels of BP have been determined to be in the range 14–135 $\mu g/m^3$ at the battery top (where emission concentrations are highest) (Bjørseth *et al.*, 1978), and these levels represent exposures considerably higher than those observed in, for example, iron foundries (see above). Coke-oven emissions have been shown to produce tumors when applied to mouse skin (Nesnow *et al.*, 1982, 1983), and many case reports and epidemiological studies have identified that exposures in coke-oven plants are carcinogenic to humans, giving rise to lung cancer (IARC, 1984b).

The DNA isolated from the white blood cells of workers in a Polish coke-oven plant was examined for the presence of aromatic adducts by ^{32}P-postlabeling. Control samples were obtained by volunteers living in a rural region of Poland. In contrast to the chromatograms obtained with DNA from iron-foundry workers, a reproducible pattern of adduct spots was obtained with coke-oven worker DNA from different individuals. This consisted of a complex pattern of adduct spots and regions of diffuse radioactive bands, indicating the formation of adducts by a large number of different chemicals (Fig. 2). Of 22 samples studied, the mean level of adducts was equivalent to a total of 59 adducts/10^8 nucleotides, the highest level found being 202 in one individual (Table 2). Among 14 controls a very much weaker diagonal band of radioactivity was observed in the chromatograms of some of the samples, the mean levels of radioactivity being equivalent to a level of adducts approximately 10-fold lower than the mean level seen in the coke-oven workers. The highest value for a control sample was also 10-fold lower than the highest worker sample value.

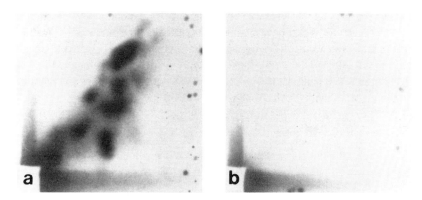

Figure 2. Autoradiographs of PEI-cellulose thin-layer chromatography maps of ^{32}P-labeled digests of DNA isolated from peripheral white blood cells from a Polish coke-oven worker (*a*) and from a control individual from a rural area of Poland (*b*).

5. DNA ADDUCTS IN SMOKERS AND NONSMOKERS

Lung cancer is among the most common cancers in man, and its rapid increase in incidence in the 20th century has been causally linked to the increase in cigarette smoking. Chemical analysis of cigarette smoke has revealed the presence of more than 50 carcinogenic chemicals, including several nitrosamines and PAH (IARC, 1986).

^{32}P-Postlabeling analysis, using the nuclease P_1 digestion method of sensitivity enhancement, was carried out on a series of DNA samples isolated from the nontumorous lung tissue of patients undergoing surgery for lung cancer (Phillips *et al.*, 1988b). The smoking histories of the patients were obtained, and chromatograms of ^{32}P-labeled digests of DNA from smokers' lungs revealed a strong band of radioactive material (Fig. 3*a*) and a weaker, faster migrating band (Fig. 3*b*). The adduct pattern observed for an ex-smoker of more than five years abstinence (Fig. 3*c*) was qualitatively similar but the radioactive bands were much weaker. A former smoker who had given up cigarettes three months prior to surgery had a pattern with adduct intensities similar to those of the smokers. The patterns of the nonsmokers contained no strong adduct bands (Fig. 3*d*), although low levels of radioactive material were seen in the same region of the chromatogram as band 1 in the smokers' samples.

The levels of adducts present in the DNA from 5 nonsmokers were in the range 1.7–4.9 adducts/10^8 nucleotides. Total adduct levels in 17 smokers' DNA samples ranged from 1.5 to 34.3 adducts/10^8 nucleotides, with 12 of the samples having levels in excess of 5 adducts/10^8 nucleotides. Of 7 former smokers, 2 had given up smoking only 1–3 months prior to surgery and had adduct levels typical of smokers (13.5 and 12.5 adducts/10^8 nucleotides); however 3 out of 4 who had given up more than 5 years previously had levels similar to the nonsmokers. Regression analysis of the data from the smokers revealed a linear relationship between the number of cigarettes smoked per day and the DNA-adduct levels (correlation coefficient = 0.724,

TABLE 9–2. ^{32}**P-Postlabeling of White Blood Cell DNA from Coke-Oven Workers**

Group	Mean No. Adducts/10^8 Nucleotides	Number of Individuals	Range
Coke-oven workers	59.1	22	4.7–202.7
Controls	6.2	14	1.5– 19.5

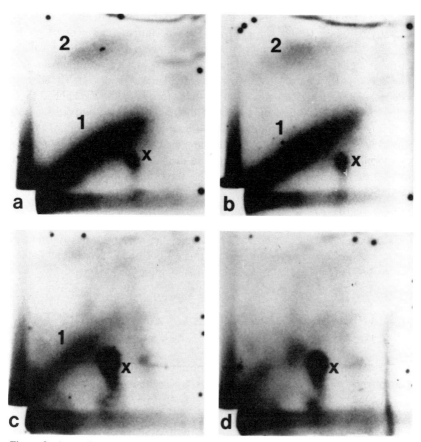

Figure 3. Autoradiographs of PEI-cellulose thin-layer chromatography maps of ^{32}P-labeled digests of DNA isolated from peripheral lung tissue of cigarette smokers (a,b), an ex-smoker (c), and a nonsmoker (d).

$P<$ 0.001). There was also a linear relationship between adduct levels and estimated lifetime cigarette consumption, although the correlation is not as strong (correlation coefficient = 0.663, $P < 0.001$) (Phillips *et al.*, 1988b).

Analysis of DNA isolated from human bronchial epithelium, that is, the primary tissue site of tobacco-induced lung tumor formation, revealed a diagonal band of radioactivity on the thin-layer chromatograms similar to that seen with peripheral lung DNA (Phillips *et al.*, 1990b,c). The intensity of the band was related to the smoking status of the individuals from whom the samples were obtained, although some overlap between the groups was evident; highest in current smokers (37 samples were in the range 2.2–9.6 adducts/10^8 nucleotides), intermediate in exsmokers (6 samples were in the range 2.9–6.0 adducts/10^8 nucleotides) and lowest in nonsmokers (8 samples had levels in the range 1.3–4.3 adducts/10^8 nucleotides).

The findings of these studies are broadly compatible with results reported recently by others (Randerath *et al.*, 1989) who found a multitude of smoking-induced adducts in the lung, bronchus, and larynx of smokers with cancer of these organs. Adduct profiles similar to those obtained with the smokers' lung DNA have been obtained when cigarette-smoke condensate (Randerath *et al.*, 1988) or other complex PAH mixtures (see below) were applied topically to mouse skin. The broadness of the adduct bands observed in each case suggest that a large number of different compounds were covalently bound to DNA. Although the chromatographic conditions employed and the use of nuclease P_1 digestion in analyzing the human lung DNA could lead to the loss of some types of adducts, e.g., those formed by aromatic amines or simple alkylating agents, nevertheless the demonstration of a linear relationship between adduct levels and numbers of cigarettes smoked daily supports the epidemiological finding that the more cigarettes smoked the greater the risk of lung cancer. Furthermore, the reduced levels of adducts in former smokers of several years' abstinence is compatible with the reducing risk of lung cancer after cessation of smoking (Doll, 1978).

6. EXPERIMENTAL STUDIES ON ENVIRONMENTAL PAH MIXTURES

Apart from those special groups occupationally exposed to chemical carcinogens, a wider potential health hazard is the exposure of the general population to fossil fuel products that contain PAH. Coal-tars, by-products of coal distillation, are used medicinally in the treatment of psoriasis and other chronic skin ailments, and also as roofing materials and in road surfacing. Creosote is a blend of coal-tar fractions that is widely used as a wood preservative and contains as much as 2g/kg BP. Bitumen is a product of oil refining that is used for roofing and road surfacing and in paints to protect metals from corrosion. Other sources of exposure to PAH include emissions from gasoline- and diesel-fueled vehicles, cigarette smoke, and carbon blacks used in inks and rubber tire manufacture.

There is evidence that many of these PAH mixtures are carcinogenic in experimental animals, and that some of them are carcinogenic in humans (IARC, 1985). For example, there is considered to be sufficient evidence that coal-tar and some mineral oils are human carcinogens, limited evidence for creosote, and inadequate evidence for bitumen and carbon blacks, although for all these mixtures there is evidence of carcinogenic activity in experimental animals and/or mutagenic activity in short-term tests (IARC, 1985).

Application to mouse skin of coal-tar, creosote, or bitumen at doses similar to those used in carcinogenicity tests results in the formation of DNA adducts by components of the mixtures. These adducts were detected by nuclease P_1-digestion enhanced ^{32}P-postlabeling (Schoket *et al.*, 1988a). The chromatograms contained a diagonal band of radioactivity similar to that seen with DNA from human lung or mouse skin exposed to cigarette smoke or its condensate, again indicating the formation of several (many?) different adducts. Single doses of the mixtures resulted in approximately 0.4 fmol total adducts/μg DNA (13 adducts/10^8 nucleotides) 24 h after treatment with coal-tar or creosote; treatment with bitumen resulted in approximately 4-fold lower levels of adducts, reflecting the lower PAH content of bitumen compared to the other two mixtures. Adduct persistence exhibited a rapid phase of removal, in which one-half to two-thirds of the adducts were removed in the first 7 days after treatment, followed by a slower removal of most, but not quite all, of the remaining adducts in the succeeding 25 days. This biphasic pattern adduct removal is similar to that observed with numerous pure carcinogens in a variety of tissues. Although the reasons for this pattern of adduct removal have not been fully elucidated, the fact that a proportion of the adducts are observed to persist for long periods after exposure suggests that human tissue, if similarly exposed *in vivo* to carcinogens, will bear evidence of that exposure for some time subsequently.

For the most part, human exposure to carcinogens is in the form of multiple, chronic doses rather than single acute doses, and multiple treatment of mouse skin with low doses of coal-tar, creosote, or bitumen was studied to mimic more closely conditions of human exposure (Schoket *et al.*, 1988a). Treatment of mouse skin twice weekly led to an accumulation of adducts over the five weeks of the experiment, with a steady-state level being approached towards the end of the period in some instances. A similar accumulation was evident in lung DNA, demonstrating the absorption of PAHs and their activation in organs distant from the site of application. The levels of adducts detected in lung DNA were approximately half those present in the skin. In contrast, application of cigarette-smoke condensate to mouse skin leads to the formation of higher levels of adducts in lung DNA than in skin DNA (Randerath *et al.*, 1988). We have also applied pharmaceutical formulations of PAH mixtures to mouse skin at doses used to treat psoriasis patients (Schoket *et al.*, 1990). Coal-tar ointment and juniper tar (cade oil) produced adducts in skin DNA at levels very similar to those produced by crude coal-tar. However, while treatment with coal-tar ointment resulted in only very low levels of adducts in the lungs of topically treated animals, higher levels were detected in the lungs of juniper tar-treated mice than in the skin. These results indicate that the formulation of the PAH mixtures has a significant influence on the absorption of the material and on the tissue distribution of adducts.

Application of samples of used gasoline and diesel engine lubricating oils to mouse skin gave rise to the formation of aromatic adducts in the skin and lungs of treated animals (Schoket *et al.*, 1989). This is consistent with the property of lubricating oils to accumulate PAHs in diesel and gasoline engines and attests to the potential carcinogenic activity of used oils (IARC, 1984a). Treatment of mouse skin with exhaust condensates from such engines also gave rise to PAH-DNA adducts (Schoket *et al.*, 1989). A significant proportion (31% and 48%, respectively) of the adducts formed by the gasoline and diesel exhaust condensates cochromatographed with the major BP-DNA adduct, but with the lubricating oils, only used gasoline engine oil, not diesel engine oil, produced significant amounts of an adduct (22% of total) that cochromatographed with the BP-DNA adduct.

Other ^{32}P-postlabeling studies on the exposure of experimental animals to PAH mixtures have been reported. Exposure of rats to diesel engine exhaust fumes has been demonstrated to result in the formation of a large number of different DNA adducts in the lungs of the animals (Wong *et al.*, 1986). Also, two studies of liver DNA from fish dwelling in waters heavily polluted with PAHs showed a complex pattern of adducts that was not observed with the DNA isolated from aquarium-dwelling fish (Dunn *et al.*, 1987) or from fish from a reference, unpolluted area (Varanasi *et al.*, 1989).

7. HUMAN SKIN EXPOSED TO PAH MIXTURES

In order to assess the potential hazard to humans of exposure to PAH mixtures, doses of coal-tar, creosote, and bitumen similar to those applied to mouse skin were applied to pieces of human skin freshly obtained from patients undergoing mastectomy or reduction mammoplasty and maintained in short-term organ culture (Schoket *et al.*, 1988b). Single doses of the materials resulted in the formation of adducts at levels similar to those formed in mouse skin, although interindividual variations that may reflect genetic differences in ability to metabolically activate PAH were evident. Application of the materials to human fetal skin in culture also produced DNA adducts, indicating that fetal skin has the capacity to metabolize and activate PAH, and that there is, therefore, a risk from exposure to these materials *in utero*. Treatment of human skin *in vitro* with the coal-tar ointment and juniper tar pharmaceutical preparations produced levels of adducts similar to those produced by these materials in mouse skin (Schoket *et al.*, 1990).

It has also been possible to study directly the formation of PAH-DNA adducts *in vivo* by ^{32}P-postlabeling DNA isolated from skin biopsies of psoriasis patients during a course of treatment with coal-tar ointment and/or juniper tar. The chromatograms of the labeled DNA digests revealed the presence of adducts at levels comparable to those seen in the human *in vitro* and mouse *in vivo* studies (Schoket *et al.*, 1990). The *in vivo* adduct patterns showed a pattern of adducts similar to those that were obtained in the mouse and short-term culture experiments, although somewhat more discrete spots were generally observed, suggesting that a lesser variety of covalently bound moieties were formed *in vivo* (Fig. 4).

8. DNA ADDUCTS IN THE GENERAL POPULATION

Aside from those individuals known to be exposed occupationally or medicinally to carcinogens, and apart from the well-documented exposure to carcinogens by the practice of smoking tobacco (IARC, 1986), there is reason to suppose that ^{32}P-postlabeling of human tissue DNA may provide some insight into the environmental causes of common human tumors for which prior exposure to known agents has not yet been implicated but where cancer incidences are nevertheless attributable to "lifestyle."

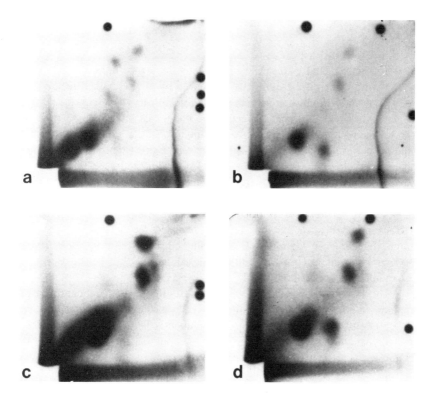

Figure 4. Autoradiographs of PEI-cellulose thin-layer chromatography maps of ^{32}P-labeled digests of DNA isolated from skin biopsy samples taken from psoriasis patients receiving treatment with coal-tar ointment (*a,b*) or juniper tar (*c,d*).

8.1. Bone Marrow

DNA from normal human bone marrow cells was analyzed by ^{32}P-postlabeling for the presence of aromatic or other hydrophobic adducts. Of 10 individuals examined, all were found to possess adducts in DNA isolated from both mononuclear and nonmononuclear cells (Phillips *et al.*, 1986). The adducts, whose patterns were subject to some interindividual variation, were present at levels in the range 1–9 adducts/10^8 nucleotides; no evidence for such adducts was found in DNA from human fetal bone marrow. The presence of the adducts in adult bone marrow would therefore suggest exposure to an as-yet-unidentified genotoxic agent, conceivably resulting from ingestion or inhalation of such a compound or from its environmentally-induced endogenous formation. Similar adducts were also observed in the white blood cells of human volunteers, albeit at lower levels and regardless of whether the individual was a smoker or nonsmoker. Thus the origins and identities of these adducts are at present unknown.

8.2. Colon

Epidemiological studies indicate that the incidence of colorectal cancer in the western hemisphere may be influenced by environmental factors, and diet is strongly implicated as being a significant factor. Tumors could conceivably be initiated either by genotoxic chemicals present in food, or by the endogenous formation of active compounds in the gut. Samples of normal-appearing human colonic mucosa were obtained from patients undergoing surgery for colorectal cancer, and from adult and fetal controls. ^{32}P-Postlabeling analysis of DNA from 44 cancer patients revealed the presence of a fast-migrating spot in 43% (19/44) of the samples at levels of up to 1 adduct in 10^8 nucleotides (Phillips *et al.*, 1988c). Analysis of colonic mucosa DNA from noncancer patients has not revealed the presence of the fast-migrating adduct in any of the 21 cases studied so far (Phillips *et al.*, 1988c, and unpublished results). In addition, the adduct was not detected in 10 of 11 samples of human fetal DNA.

The fast-migrating adduct was observed slightly less frequently in patients with poorly differentiated tumors than in those with moderately or well differentiated tumors, and slightly less frequently in patients with adenomas (5/15) than in patients without (12/29), although in neither case was the difference statistically significant in a study sample of this size. Nevertheless, it has been demonstrated that there is an excess of poorly differentiated large-bowel cancers in low-risk populations (Elmasry and Boulos, 1975) in whom inherited factors may assume greater importance than environmental ones. However, not all poorly differentiated colon cancers are genetically determined, nor is the appearance of adenomas in addition to carcinomas an indication of a genetically determined tumor in all cases. Thus the significance of the occurrence of the fast-migrating adduct in colonic DNA and the origin and possible role of this adduct in tumor initiation remain to be established.

9. SOURCES OF DNA FOR MONITORING DNA-ADDUCT FORMATION IN HUMANS

For routine analysis of populations for DNA-adduct formation, it is necessary to monitor DNA from a readily obtainable source. It has already been seen that white blood cell DNA can be used to monitor PAH exposure among foundry and coke-oven workers. Other sources of cells that might be exploited include buccal cavity scrapings and hair roots. Because use of such sources means that DNA from a tissue other than the target tissue for tumor formation by the carcinogenic exposure is being monitored, it raises the question as to the extent to which DNA-adduct formation in the tissue examined reflects what is occurring in the target tissue.

In studies to date on the formation of adducts as a result of cigarette smoking, there is a dose-dependent formation of adducts in the lung DNA of smokers (Phillips *et al.,* 1988b), and smoking-related adducts have also been detected in human placentas (Everson *et al.,* 1986). However, in several studies no effect of smoking was seen on the levels of adducts in white blood cells (Phillips *et al.,* 1986, 1988a; Everson *et al.,* 1987; Perera *et al.,* 1988). A large study of white blood cell DNA from heavy smokers and nonsmokers confirms this finding (Phillips *et al.,* 1990b). A significant difference between lung cancer patients who were smokers and noncancer current smoker control subjects has been seen in only one study (Perera *et al.,* 1989). Also, exfoliated mucosal cells of the oral cavity did not show any differences among nonsmokers, tobacco chewers, and inverted smokers when analyzed for aromatic adducts by ³²P-postlabeling (Dunn and Stich, 1986). Studies on cigarette smokers have established that they excrete metabolites of cigarette-smoke components (Putzrah *et al.,* 1981), have mutagens in their urine (Yamasaki and Ames 1977), have increased sister chromatid exchanges, have chromosomal aberrations in circulating peripheral lymphocytes (Obe *et al.,* 1982; Livingston and Fineman 1983; Perera *et al.,* 1987), and have elevated levels of hydroxyethylated (Tornquist *et al.,* 1986) and aminobiphenylated hemoglobin (Bryant *et al.,* 1987). It is thus unclear why smoking-related hydrophobic DNA adducts should fail to be detected in the blood. This finding suggests that the smoking habits of industrially-exposed workers will not interfere with the interpretation of data on the levels of hydrophobic adducts in their peripheral lymphocytes. However, the studies on adducts in smokers indicate that the most readily obtainable sources of human DNA may not be the most appropriate material to monitor for evidence of carcinogen exposure, and further work is needed to establish the relationships between adduct formation in nontarget and target tissues. The possibility that adducts can arise in white blood cell DNA as a result of ingestion and skin adsorption as well as of inhalation, and that these alternative routes may be significant in observed incidences of occupational exposure, will clearly need to be thoroughly investigated (Phillips *et al.*, 1990a).

10. CONCLUDING REMARKS

^{32}P-Postlabeling is the most sensitive assay available at present for the detection of aromatic and other hydrophobic adducts, although it is not as sensitive for simple alkyl adducts. Only small (microgram) quantities of DNA are required to perform the analysis, and it can readily be applied to the analysis of adducts formed by complex mixtures. It has the advantage of not requiring an adduct's identity to be known in order to detect it. Its potential for monitoring occupational exposure to carcinogens has already been demonstrated. Thus if the incidence of the majority of human cancers is influenced by environmental agents, then ^{32}P-postlabeling will be a useful tool in determining what role DNA-damaging chemicals may play in the etiology of common human cancers.

The requirement for relatively large quantities of ^{32}P means that care must be exercised when performing the assay. High-quality reagents must be used, particularly for the DNA isolation procedures, in order to obtain chromatograms with low backgrounds. Because hydrolysis of DNA samples is required, problems may be encountered if digestion is incomplete, leading to errors in adduct quantitation. Difficulties may also be encountered if an adduct is unstable as a mononucleotide. Although chemically synthesized standards are not obligatory, assessment of these effects cannot be reliably made without them. Nevertheless, in most cases where standards are available it has been found that near-quantitative labeling of adducts can be achieved. There will be uncertainty in the quantitation of very low levels of adducts where enhancement techniques such as nuclease P_1 digestion are used unless standardization is possible with more highly modified DNA samples. This emphasizes the need for control samples from individuals in low-risk or unexposed populations for comparison with the study group in order to assess the significance of adducts detected in the latter.

ACKNOWLEDGMENTS
Research in the author's laboratory is supported by grants to the Institute of Cancer Research from the Cancer Research Campaign, the U.K. Medical Research Council, and the U.S. National Cancer Institute. The studies described here have been carried out in collaboration with Alan

Hewer, Bernadette Schoket, Phil Grover, Kari Hemminki, Jeremy Jass, Kurt Randerath, and Colin Garner, whose contributions are gratefully acknowledged.

REFERENCES

W. M. Baird, The use of radioactive carcinogens to detect DNA modifications. In: *Chemical Carcinogens and DNA*, ed. P. L. Grover, CRC Press, Boca Raton, FL, pp. 59–83 (1979).

A. Bjørseth, O. Bjørseth, and P. E. Fjeldsted, Polycyclic aromatic hydrocarbons in the work atmosphere. II. Determination in a coke plant, *Scand. J. Work Environ. Health* **4**, 224–236 (1978).

M. S. Bryant, P. L. Skipper, S. R. Tannenbaum, and M. Maclure, Hemoglobin adducts of 4-aminobiphenyl in smokers and nonsmokers, *Cancer Res.* **47**, 602–608 (1987).

R. Doll, An epidemiological perspective of the biology of cancer, *Cancer Res.* **38**, 3573–3583 (1978).

R. Doll and R. Peto, *The Causes of Cancer*, Oxford University Press, London, 1981.

B. P. Dunn and R. H. C. San, HPLC enrichment of hydrophobic DNA adducts for enhanced sensitivity of ^{32}P-postlabeling analysis, *Carcinog.* **9**, 1055–1060 (1988).

B. P. Dunn and H. F. Stich, ^{32}P-Postlabeling analysis of aromatic DNA adducts in human oral mucosal cells. *Carcinog.* **7**, 1115–1210 (1986).

B. P. Dunn, J. J. Black, and A. Maccubbin, ^{32}P-Postlabeling analysis of aromatic DNA adducts in fish from polluted areas, *Cancer Res.* **47**, 6543–6548 (1987).

S. H. Elmasry and P. B. Boulos, Carcinoma of the large bowel on the Sudan, *Br. J. Surg.* **62**, 284–286 (1975).

R. B. Everson, E. Randerath, R. M. Santella, R. C. Cefalo, T. A. Avitts, and K. Randerath, Detection of smoking-related covalent DNA adducts in human placenta, *Science* **231**, 54–57 (1986).

R. B. Everson, E. Randerath, T. A. Abitts, H. A. J. Schut, and K. Randerath, Preliminary investigations of tissue specificity, species specificity, and strategies for identifying chemicals causing DNA adducts in human placenta, *Prog. Exp. Tumor Res.* **31**, 86–103 (1987).

J. E. Gallagher, M. A. Jackson, M. H. George, J. Lewtas, and I. G. C. Robertson, Differences in detection of DNA adducts in the ^{32}P-postlabeling assay after either 1-butanol extraction or nuclease P1 treatment, *Cancer Lett.* **45**, 7–12 (1989).

R. C. Gupta, Enhanced sensitivity of ^{32}P-postlabeling analysis of aromatic carcinogen: DNA adducts, *Cancer Res.* **45**, 5656–5662 (1985).

R. C. Gupta and K. Earley, ^{32}P-adduct assay: comparative recoveries of structurally diverse DNA adducts in the various enhancement procedures, *Carcinog.* **9**, 1687–1693 (1988).

R. C. Gupta, M. V. Reddy, and K. Randerath, ^{32}P-Postlabeling analysis of non-radioactive aromatic carcinogen-DNA adducts, *Carcinog.* **3**, 1081–1092 (1982).

K. Hemminki, F. P. Perera, D. H. Phillips, K. Randerath, M. V. Reddy, and R. M. Santella, Aromatic DNA adducts in white blood cells of foundry workers. In: *Methods for Detecting DNA Damaging Agents in Humans: Applications in Cancer*

Epidemiology and Prevention, (IARC Scientific Publication No. 89), ed. H. Bartsch, K. Hemminki, and I. K. O'Neill, International Agency for Research on Cancer, Lyon, France, pp. 190–195 (1988).

IARC, *IARC Monographs on the Evaluation of the Carcinogenic Risk of Chemicals to Humans,* Vol. 33, *Polynuclear Aromatic Compounds,* Part 2, *Carbon Blacks, Mineral Oils and Some Nitroarenes,* IARC, Lyon, France (1984a).

IARC, *IARC Monographs on the Evaluation of the Carcinogenic Risk of Chemicals to Humans,* Vol. 34, *Polynuclear Aromatic Compounds,* Part 3, *Industrial Exposures in Aluminum Production, Coal Gasification, Coke Production, and Iron and Steel Founding,* IARC, Lyon, France (1984b).

IARC, *IARC Monographs on the Evaluation of the Carcinogenic Risk of Chemicals to Humans,* Vol. 35, *Polynuclear Aromatic Compounds,* Part 4, *Bitumens, Coaltars and Derived Products, Shale-oils and Soots,* IARC, Lyon, France (1985).

IARC, *IARC Monographs on the Evaluation of the Carcinogenic Risk of Chemicals to Humans,* Vol. 38, *Tobacco Smoking,* IARC, Lyon, France (1986).

G. K. Livingston and R. M. Fineman, Correlation of human lymphocyte SCE frequency with smoking history, *Mutat. Res.* **119**, 59–64 (1983).

S. Nesnow, L. L. Triplett, and T. J. Slaga, Comparative tumor-initiating activity of complex mixtures from environmental particulate emissions on SENCAR mouse skin, *J. Natl. Cancer Inst.* **68**, 829–834 (1982).

S. Nesnow, L. L. Triplett, and T. J. Slaga, Mouse skin tumor initiation-promotion and complete carcinogenesis bioassays: mechanisms and biological activities of emission samples, *Environ. Health Perspect.* **47**, 255–268 (1983).

G. Obe, H.-J. Vogt, S. Madle, A. Fahning, and W. D. Heller, Double blind study on the effect of cigarette smokers' urine by high performance liquid chromatography, *Mutat. Res* **92**, 309–319 (1982).

F. P. Perera, K. Hemminki, T. L. Young, D. Brenner, G. Kelly, and R. M. Santella, Detection of polycyclic aromatic hydrocarbon-DNA adducts in white blood cells of foundry workers, *Cancer Res.* **48**, 2288–2291 (1988).

F. P. Perera, J. Mayer, A. Jaretzki, S. Hearne, D. Brenner, T. L. Young, H. K. Fischman, M. Grimes, S. Grantham, M. X. Tang, W.-Y. Tsai, and R. M. Santella, Comparison of DNA adducts and sister chromatid exchange in lung cancer cases and controls, *Cancer Res.* **49**, 4446–4451 (1989).

F. P. Perera, R. M. Santella, D. Brenner, M. C. Poirier, A. A. Munshi, H. K. Fischman, and J. Van Ryzin, DNA adducts, protein adducts, and sister chromatid exchange in cigarette smokers and nonsmokers, *J. Natl. Cancer Inst.* **79**, 449–456 (1987).

D. H. Phillips, Modern methods of DNA adduct determination. In: *Chemical Carcinogenesis and Mutagenesis 1,* eds. C. S. Cooper and P. L. Grover, Springer-Verlag, pp 503–546 (1990).

D. H. Phillips, K. Hemminki, A. Alhonen, A. Hewer, and P. L. Grover, Monitoring occupational exposure to carcinogens: detection by ^{32}P-postlabeling of aromatic DNA adducts in white blood cells from iron foundry workers, *Mutat. Res.* **204**, 531–541 (1988a).

D. H. Phillips, A. Hewer, and P. L. Grover, Aromatic DNA adducts in human bone marrow and peripheral blood leukocytes, *Carcinog.* **7**, 2071–2075 (1986).

D. H. Phillips, A. Hewer, P. L. Grover, and J. R. Jass, An aromatic DNA adduct in colonic mucosa from patients with colorectal cancer. In: *Methods for Detecting DNA Damaging Agents in Humans: Applications in Cancer Epidemiology and Prevention* (IARC Scientific Publications No. 89), eds. H. Bartsch, K. Hemminki, and I. K. O'Neill, IARC, Lyon, France, pp. 368–371 (1988c).

D. H. Phillips, A. Hewer, C. N. Martin, R. C. Garner, and M. M. King, Correlation of DNA adduct levels in human lung with cigarette smoking, *Nature* **336**, 790–792 (1988b).

D. H. Phillips, M. V. Reddy, and K. Randerath, ^{32}P-Postlabeling analysis of DNA adducts formed in the livers of animals treated with safrole, estragole and other naturally-occurring alkenylbenzenes. II. Newborn male B6C3F$_1$ mice, *Carcinog.* **5**, 1623–1628 (1984).

D. H. Phillips, B. Schoket, and A. Hewer, DNA adducts in white blood cells versus other human tissues. In: *Proceedings of Biomonitoring and Carcinogen Risk Assessment Meeting,* 27–29 July, 1989, Queen's College, Cambridge, in press (1990a).

D. H. Phillips, B. Schoket, A. Hewer, and P. L. Grover, Human DNA adducts due to smoking and other exposures to carcinogens. In: *Mutation and the Environment, Part C: Somatic and Heritable Mutation, Adduction and Epidemiology,* eds. M. L. Mendelsohn and R. T. Allertini, New York, Wiley-Liss, pp 283–292 (1990c)

D. H. Phillips, B. Schoket, A. Hewer, E. Bailey, S. Kostic and I. Vincze, Influence of cigarette smoking on the levels of DNA adducts in human bronchial epithelium and white blood cells, *Int. J. Cancer*, in press (1990b).

R. M. Putzrath, D. Langden, and E. Eisenstadt, Analysis of mutagenic activity in cigarette smokers' urine by high performance liquid chromatography, *Mutat. Res.* **85**, 97–108 (1981).

E. Randerath, K. Randerath, H. P. Agrawal, J. A. Weaver, and C. B. Bordelon, ^{32}P-Postlabeling analysis of DNA adducts persisting for up to 42 weeks in the skin, epidermis and dermis of mice treated topically with 7,12-dimethylbenz[a]anthracene, *Carcinog.* **6**, 1117–1126 (1985).

E. Randerath, K. Randerath, R. H. Miller, D. Mittal, T. A. Avitts, and H. A. Dunsford, Covalent DNA damage in tissues of cigarette smokers as determined by ^{32}P-postlabeling assay, *J. Natl. Cancer Inst.* **81**, 341–347 (1989).

E. Randerath, D. Mittal and K. Randerath, Tissue distribution of covalent DNA damage in mice treated dermally with cigarette "tar": preference for lung and heart DNA, *Carcinog.* **9**, 75–80 (1988).

M. V. Reddy and K. Randerath, Nuclease P1-mediated enhancement of sensitivity of ^{32}P-postlabeling test for structurally diverse DNA adducts, *Carcinog.* **7**, 1543–1551 (1986).

M. V. Reddy, R. C. Gupta, E. Randerath, and K. Randerath, ^{32}P-Postlabeling test for covalent DNA binding of chemicals *in vivo*: application to a variety of aromatic carcinogens and methylating agents, *Carcinog.* **5**, 231–243 (1984).

K. Savela, K. Hemminki, A. Hewer, D. H. Phillips, K. L. Putman, and K. Randerath, Interlaboratory comparison of the ^{32}P-postlabeling assay for aromatic DNA addicts in white blood cells of iron foundry workers, *Mutat. Res.,* **224**, 485–492 (1989).

B. Schoket, A. Hewer, P. L. Grover, and D. H. Phillips, Covalent binding of components of coal-tar, creosote and bitumen to the DNA of the skin and lungs of mice following topical application, *Carcinog.* **9**, 1253–1258 (1988a).

B. Schoket, A. Hewer, P. L. Grover, and D. H. Phillips, Formation of DNA adducts in human skin maintained in short-term organ culture and treated with coal-tar, creosote or bitumen, *Int. J. Cancer* **42**, 622–626 (1988b).

B. Schoket, A. Hewer, P. L. Grover, and D. H. Phillips, ^{32}P-Postlabeling analysis of DNA adducts in the skin of mice treated with petrol and diesel engine lubricating oils and exhaust condensates, *Carcinog.* **10**, 1485–1490 (1989).

B. Schoket, I. Horkay, A. Kosa, L. Paldeak, A. Hewer, P. L. Grover, and D. H. Phillips, Formation of DNA adducts in the skin of psoriasis patients, in human skin in organ culture and in mouse skin and lung following topical application of coal-tar and juniper tar, *J. Invest. Dermatol.*, **94**, 241–246 (1990).

M. Tornquist, S. Osterman-Golkar, A. Kautiainen, S. Jensen, P. B. Farmer, and L. Ehrenberg, Tissue doses of ethylene oxide in cigarette smokers determined from adduct levels in hemoglobin, *Carcinog.* **7**, 1519–1521 (1986).

U. Varanasi, W. L. Reichert, and J. E. Stein, ^{32}P-Postlabeling analysis of DNA adducts in liver of wild English sole (*parophys vetulus*) and winter flounder (*Pseudopleuronectes americanus*), *Cancer Res.* **49**, 1171–1177 (1989).

D. Wong, C. E. Mitchell, R. K. Wolff, J. L. Manderly, and A. M. Jeffrey, Identification of DNA damage as a result of exposure of rats to diesel engine exhaust, *Carcinog.* **7**, 1595–1597 (1986).

E. Yamasaki and B. N. Ames, Concentration of mutagens from urine by adsorption with the non polar resin XAD-2: cigarette smokers have mutagenic urine, *Proc. Natl. Acad. Sci. USA* **7**, 3555–3559 (1971).

Quantitative Analysis of DNA Adducts: the Potential for Mass Spectrometric Techniques

Norbert Fedtke* and James A. Swenberg[†]
Chemical Industry Institute of Toxicology
Department of Biochemical Toxicology and Pathobiology
Research Triangle Park
North Carolina

1. DNA ADDUCTS AND CARCINOGENESIS: MOLECULAR DOSIMETRY

Chemically induced carcinogenesis is regarded as a multistage process including initiation, promotion, and progression. Current evidence suggests that mutating events participate in all of these stages. One mechanism for the induction of mutations is the covalent interaction of a chemical carcinogen or its metabolites with DNA. The DNA lesions that are formed after chemical reaction of carcinogenic agents or their electrophilic metabolites with DNA constituents are commonly referred to as DNA adducts. Specific DNA adducts are formed and repaired at different rates in different cell types. If promutagenic DNA adducts are not repaired prior to replication,

*Present address: Hüls AG, PS Biologie/Toxikologie, P.O. Box 1320, D-4370 Marl, FRG.
[†]To whom correspondence should be addressed: Dr. James A. Swenberg, University of North Carolina, Campus Box 7095, Chapel Hill, NC 27599.

they can cause errors during replication, leading to insertions, deletions, base-pair substitutions, and recombination.

Factors that affect the frequency of initiating events in a given cell population exposed to a carcinogen are as follows: (1) the molecular dose of each type of DNA adduct, (2) the mutational efficiencies of these adducts, i.e., the efficiency for causing base-pair mismatch, (3) the site of DNA-adduct formation, and (4) the extent of cell proliferation (Swenberg *et al.*, 1987). DNA-adduct formation is only one of the determinants in initiation; however, quantitation of this factor has received considerable attention in recent years. The molecular dose of DNA adducts reflects the amount of the carcinogen that reacts with and persists in DNA, integrating the complex pharmacokinetic mechanisms involved in absorption of the compound, distribution, metabolic activation, detoxication, reaction kinetics with nucleic acids, spontaneous loss of the adduct, and enzymatic DNA repair (Swenberg *et al.*, 1987, 1990; Zeise *et al.*, 1987). Thus, compared to approaches that relate dose–response relationships for carcinogenicity to external exposure concentrations, the determination of DNA adducts should provide more accurate predictions of the biological response to carcinogen exposure.

The determination of DNA adducts can establish dose–response relationships for chemical carcinogens ranging from the high-exposure concentrations used in animal bioassays to doses relevant to human exposure. More accurate risk estimation for humans should be facilitated by using information on the actual target doses in appropriate animal models (Swenberg *et al.*, 1987, 1990; Belinsky *et al.*, 1987; Lutz, 1987). DNA-adduct measurement has been suggested as a means for monitoring the actual dose of a genotoxic agent in individuals exposed to mutagenic chemicals (Farmer *et al.*, 1987).

Two principal approaches for DNA-adduct measurement are the detection of excised bases in urine and the detection of the modified bases in DNA extracted from tissues or cells. The determination of adducts in urine may be useful for demonstrating actual exposures: however, since it measures DNA adducts that have been lost or repaired, its use in quantitative risk assessment remains to be established. Furthermore, for each adduct in urine, it must be shown that the adduct was originally present in the host's DNA and not of dietary origin.

For the determination of DNA adducts in man or in laboratory animals exposed to low doses of carcinogens, methods must be developed that are capable of detecting 1 adduct in 10^7–10^{12} unmodified bases with a high degree of specificity. Analytical methods currently used for DNA-adduct quantification include high-pressure liquid chromatography (HPLC), immunological assays, ^{32}P-postlabeling, and assays based on mass spectrometry (Farmer *et al.*, 1987; Bolt *et al.*, 1988). The choice of the method depends on the sensitivity and specificity required and the properties of the DNA adduct to be measured. Methods using HPLC coupled with UV or fluorescence detectors are limited by high detection limits and the lack of specificity. ^{32}P-postlabeling methods provide extreme sensitivity for bulky adducts, but the adducts are usually not characterized structurally and are only detected after the enzymatic attachment of ^{32}P. Immunological assays require thorough characterization of the antibodies and whenever possible the combination of multiple methods, e.g., HPLC and enzyme-linked immunoabsorbent assay (ELISA), in order to determine the specificity of the response. Mass spectrometric methods utilize expensive equipment and internal standards that may require custom synthesis, but are regarded as the most accurate and precise methods.

This chapter focuses on current approaches using mass spectrometric techniques for the quantification of DNA adducts. The recently developed electrophore labeling technique of DNA adducts prior to detection by negative-ion chemical ionization mass spectrometry is described in detail.

2. THE CURRENT ROLE OF MASS SPECTROMETRY IN DNA-ADDUCT RESEARCH

Mass spectrometry (MS) is one of the most useful tools for structure elucidation of nucleic acid adducts. Since Biemann (1962) and Biemann and McCloskey (1962) reported the first mass spectra of free bases and nucleosides, tremendous progress has been made in the area of characterization of nucleic acid constituents. Overviews of techniques used in qualitative MS analyses of bases, nucleosides, and nucleotides have been published by Farmer *et al.* (1988), McCloskey (1986), Hogg *et al.* (1986), McCloskey (1985), Pang *et al.* (1982), and Hignite (1980). Examples of DNA-adduct identification by MS include those formed by chemical or physical carcinogens such as aromatic amines (reviewed by Mitchum *et al.*, 1985), 4-nitroquinoline 1-oxide (Galiegue-Zoutina *et al.*, 1986), pyrrolizidine alkaloids (Tomer *et al.*, 1986), fluoranthene (Babson *et al.*, 1986), free oxygen radicals (Dizdaroglu, 1986; Dizdaroglu and Bergtold, 1986), UV radiation (Kumar *et al.*, 1987), and γ radiation (Dizdaroglu *et al.*, 1987).

MS has been used to complement the ^{32}P-postlabeling methods of Randerath *et al.* (1981) and Gupta *et al.* (1982). After separation of the ^{32}P-postlabeled adducts of styrene oxide by HPLC and on polyethyleneimine (PEI) cellulose plates, the adducts were recovered and identified using LC-MS (Kaur *et al.*, 1988). A similar approach was taken by Dirr *et al.* (1989), who isolated and characterized the main DNA adducts of IQ (2-amino-3-methylimidazo[4,5—f]quinoline) by ^{32}P-postlabeling and MS, suggesting that similar approaches might become powerful analytical tools for future DNA-adduct research.

In contrast to the widespread use of MS techniques for qualitative identification of DNA adducts, there have been few reports of quantitative MS techniques in DNA-adduct research. Methods developed for quantitative determination of natural and modified nucleobases, nucleosides, nucleotides, and related compounds in DNA, cell extracts, and body fluids are listed in Table 1. Because of great differences in abundance, the sensitivity and selectivity required are different for naturally occurring nucleosides, endogenous modified nucleobases such as 5-methylcytosine, or chemically induced DNA adducts such as 7-(2'-hydroxyethyl)guanine. The techniques and resulting limits of detection listed in Table 1 reflect these different requirements.

Most methods for DNA-adduct analysis cited in Table 1 use gas chromatography (GC) for introduction of the sample into the mass spectrometer. This combination has been advanced by the development of inert, high-resolution capillary columns coupled with on-column injection. Since the constituents of nucleic acids are not very volatile, they are not amenable to direct GC-MS analysis. Derivatization techniques have been developed that result in analytes combining good gas chromatographic properties and mass spectra containing valuable structural information and often enhanced sensitivity (Blau and King, 1978; Knapp, 1978; Anderegg, 1988; Saha *et al.,* 1989).

All but two of the methods listed in Table 1 use single-ion monitoring (SIM) as the operation mode. In SIM, the mass spectrometer is programmed to monitor only a narrow mass range specific for the analyte, instead of scanning a wide mass range. This increases the signal-to-noise ratio and can effectively filter out unwanted responses from other chemicals present in the sample (Halpern, 1981). Although most of the structural information in a complete mass spectrum is lost, SIM is regarded as a specific method, especially when molecular ions or ions with high masses are monitored.

An internal standard that can correct variation due to sample preparation, sample introduction, ionization, or instrument performance is crucial in quantitative MS techniques. Many of the methods in Table 1 use the analyte labeled at two or more positions with stable isotopes such as ^{13}C or ^{2}H. Other quantitative analytical techniques for DNA adducts cannot use these ideal standards because their detection methods lack the selectivity necessary to distinguish the analyte and its stable labeled analogue. The use of stable labeled analogues in quantitative MS techniques is referred to as isotope dilution mass spectrometry (IDMS). The advantages of this technique are discussed in detail in several reviews (Garland and Powell, 1981; Ballard *et al.,* 1986; Heumann, 1986; Colby, 1983; DeLeenheer *et al.,* 1985) and include such attributes as identical extraction, derivatization, and elution for the analyte and the internal standard. In addition to stable isotopes, isomers and other structural analogues can be used as internal standards.

Seven of the methods listed in Table 1 were specifically developed for detection of adducts formed by interaction of exogenous chemicals with nucleic acids, but only four have been applied under *in vivo* conditions. The excretion of 7-[^{2}H$_3$]-methylguanine was measured as a heptafluorobutyryl-pentafluorobenzyl derivative using electron-impact ionization with a sensitivity of 0.5 mg/ml in urine after exposure of rats to deuterated methylating agents (Farmer *et al.*, 1986). The formation of 7-(2'-hydroxyethyl)guanine and 7-(2'-oxoethyl)guanine was measured after trimethylsilylation and using electron-impact ionization in liver DNA from rats exposed to ethylene oxide and vinyl chloride, respectively (Föst *et al.*, 1988, 1989). The limits of detection for these methods are based on the injection of dilutions of authentic compounds and do not reflect the sensitivity characteristic for the entire assay on tissues, e.g., extraction and derivatization yields. The formation and persistence of N^2,3-ethenoguanine was measured in DNA from various rat tissues after vinyl chloride exposure (Fedtke *et al.*, 1990b). Sixty femtomole DNA adducts per mmol unmodified base were detected by this method, demonstrating the excellent sensitivity achieved by the combination of electrophore labeling and negative-ion chemical ionization MS. The concept of this technique is discussed in detail in the following paragraphs.

3. ELECTROPHORE LABELING: A NEW APPROACH TO DNA-ADDUCT QUANTIFICATION

Electron-capture detection takes advantage of the electronegative properties (affinity for free electrons) of compounds (electrophores). Such compounds readily attach low-energy electrons and stabilize the negative charge forming negative ions. Highly sensitive detection is possible by either an electron-capture detector (ECD) or negative-ion chemical ionization MS (NICI-MS). General reviews on these detection techniques were published by Pellizzari (1974) and Budzikiewicz (1986).

The sensitive detection of DNA constituents after derivatization with electron-capturing groups (electrophores) was suggested by Gelijkens *et al.* (1981), who developed a derivatization method for purine and pyrimidine bases using trifluoroacetylation for the introduction of the electrophore. Giese and co-workers (Nazareth *et al.*, 1984; Mohamed *et al.*, 1984; Adams and Giese, 1985; Fisher *et al.*, 1985; Adams *et al.*, 1986; Trainor *et al.*, 1988; Fisher and Giese, 1988; Saha *et al.*, 1989) studied the properties of several electrophoric derivatives of nucleobases and nucleotides with regard to ultratrace analysis. They found pentafluorobenzylation very suitable for DNA-adduct determination because structurally characteristic fragment ions with high relative abundance are formed in NICI-MS by the loss of one pentafluorobenzyl group at mass [M-181]$^-$ (M = molecular ion). Furthermore, these derivatives are very stable compared to silyl derivatives (Trainor *et al.*, 1988). Nazareth *et al.* (1984) termed the modification of nucleobases and nucleosides with electron-capturing groups "electrophore labeling."

TABLE 10–1. Mass Spectrometric Methods Used for the Quantification of Nucleic Acid Constituents and Related Compounds*

Compound	Method	Standard	Derivative	Limit of Detection	In Vivo Conditions	Reference
5-Fluorouracil, 6-Mercaptopurine	GC-MS PICI, SIM	Imipramine	Permethyl	100 ng/ml Serum	Yes	Pantaretto et al., 1974
Cytosine Arabinoside	GC-MS EI, SIM	SLA	Acetyl Methyl	0.5 mg/ml Plasma	Yes	Boutagy and Harvey, 1978
5-Methyl Cytosine	GC-MS EI, SIM	Cytosine	Trimethyl-silyl	1/5000 Bases	No	Singer et al., 1979
Uracil, Thymine	GC-MS EI, SIM	SLA	Trimethyl-silyl	20 pmol/ml Plasma	Yes	Finn et al., 1979
cAMP	GC-MS EI, SIM	SLA	Trimethyl-silyl	3 pmol/ml†	Yes	Johnson et al., 1980
6-Mercapto-purinriboside	GC-MS EI, SIM	SLA	Permethyl	< 1 nmol/ml blood	Yes	Jardine and Weidner, 1980
5-Methyl cytosine	GC-MS EI, SIM	SLA	Butyldi-methylsilyl	0.3 pmol/mg DNA	Yes	Russel et al., 1983; Crain and McCloskey, 1983
Cytokinine nucleotides	GC-MS EI, SIM	SLA	Permethyl trimethyl silyl	1.8 pmol/g tissue	Yes	Scott and Horgan, 1984
7-[^{2}H$_3$]-Methyl guanine, 3-[^{2}H$_3$]-Methyl-adenine	GC-MS EI, SIM	7-Ethyl-guanine	Heptafluoro-buturyl/penta-fluorobenzyl	0.5 mg/ml urine	Yes	Shuker et al., 1984; Bailey et al., 1987; Farmer et al., 1986

Compound	Technique		Derivative	LOD	SLA	References
Methylated nucleobases	DIP-MS-MS PICI	SLA	…	nmol	No	Ashworth et al., 1984
5-Methyl cytosine	GC-MS NICI, SIM	Heptachlor	Heptafluoro-butyryl, penta-fluorobenzoyl	1 fg/ml†	No	Mohamed et al., 1984; Nazareth et al., 1984
5-Bromo-uracil	GC-MS EI, SIM	5-Chloro-uracil	Trimethyl-silyl	7 pmol/10 mg/DNA	Yes	Stetson et al., 1986; Maybaum et al., 1987
Adenosine	GC-MS EI, SIM	SLA	Butyldi-methylsilyl	5 ng/ml	Yes	Ballard et al., 1986
7-Ethylde-oxyguanosine	DIP-MS-MS PICI	SLA	…	78 pmol/ml†	No	Chang et al., 1986
O^4-Ethyl-thymidine	GC-MS NICI, SIM	SLA	Pentafluoro-benzyl	0.05 fmol/ml†	No	Turner et al., 1987; Koch et al., 1987
7-(2'-Hydroxy-ethyl)guanine	GC-MS EI, SIM	7-Methyl-guanine	Trimethyl-silyl	35 fmol/ml†	Yes	Föst et al., 1989; Bolt et al., 1988
7-(2'-Oxoethyl) guanine	GC-MS EI, SIM	7-Methyl-guanine	Oxime, trimethyl-silyl	300 fmol/ml†	Yes	Föst et al., 1988; Kaseman et al., 1988
N^2-3-Etheno-guanine	GC-MS NICI, SIM	SLA	Pentafluoro-benzyl	60 fmol/mmol guanine	Yes	Fedtke et al., 1990a, 1990b

*Abbreviations: EI, electron-impact ionization; PICI, positive-ion chemical ionization; NICI, negative-ion chemical ionization; DIP, direct insertion probe; SIM, single-ion monitoring; SLA, stable-labeled analogue.
†Limit of detection for diluted standard.

Figure 1. Structure of N^2,3-ethenoguanine (EG).

After derivatizing millimole quantities of O^4-ethylthymidine with pentafluorobenzyl bromide and diluting the resulting pentafluorobenzyl derivative, a detection limit of 45 amol/μl injected solution was determined using GC-ECD (Adams *et al.*, 1986). By removing the ribosyl and the ethyl moieties of O^4-ethylthymidine by acid hydrolysis and then derivatizing the resulting thymine with pentafluorobenzyl bromide, Turner *et al.* (1987) found a similar detection limit using GC-NICI-MS with SIM. These results demonstrated the tremendous potential of electrophore labeling combined with NICI-MS for DNA-adduct analysis. However, the cited detection limits reflect merely the amounts that can be measured in the detection process itself using diluted standards. This situation is very different from the "real world" situation where trace amounts of analyte have to be isolated from biological matrices, derivatized, and purified prior to the detection. Any of these steps is associated with sample loss, and it is usually not possible to inject the entire final solution onto the GC column. Hence, the detection limit, expressed as amount that can be quantitated per unit of biological material, can be significantly higher. Advances in sample handling are needed to bring the sensitivity for real samples closer to that of standards.

4. DETERMINATION OF N^2,3-ETHENOGUANINE AFTER ELECTROPHORE LABELING

N^2,3-Ethenoguanine (EG, Fig. 1) is a cyclic DNA adduct that has been tentatively detected in rats after exposure to radiolabeled vinyl chloride (VC; Laib *et al.*, 1985; Laib, 1986). A GC-NICI-MS assay combining electrophore labeling and isotope dilution was recently developed to quantify the formation and persistence of EG in preweanling rats (Fedtke *et al.*, 1990a, 1990b). Derivatization of milligram quantities of EG with pentafluorobenzyl bromide was used to obtain sufficient reaction products for an unequivocal characterization by nuclear magnetic resonance (NMR) and MS. The resulting two pentafluorobenzyl (PFB) derivatives $3,5\text{-}PFB_2\text{-}EG$ and $1,5\text{-}PFB_2\text{-}EG$ are shown in Fig. 2. The derivative $3,5\text{-}PFB_2\text{-}EG$ was formed at a higher yield, exhibited excellent gas chromatographic properties, and the fragmentation of the analyte and its stable isotope in NICI-MS were restricted to two major fragments (Figs. 3A and 3B). These characteristics rendered $3,5\text{-}PFB_2\text{-}EG$ an ideal compound for GC-NISI-MS using SIM. Routinely, 190-amol $3,5\text{-}PFB_2\text{-}EG$ standard could be detected with a signal-to-noise ratio of 10 when the fragment $[M\text{-}181]^-$ was monitored at the mass/charge ratio of 354. The diagram of the assay in Fig. 4 summarizes the sample preparation.

EG was cleaved from the DNA backbone by depurination using mild acid hydrolysis, resulting in an acidic mixture of purine bases and DNA

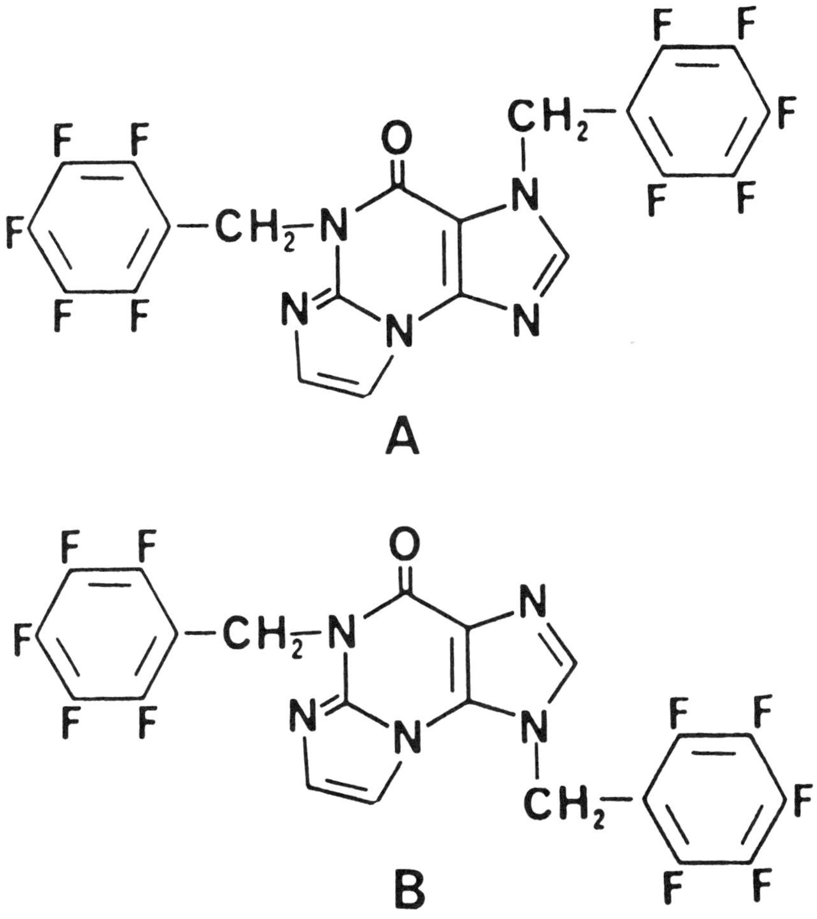

Figure 2. Structures of $3,5\text{-}PFB_2\text{-}EG$ (A) and $1,5\text{-}PFB_2\text{-}EG$ (B).

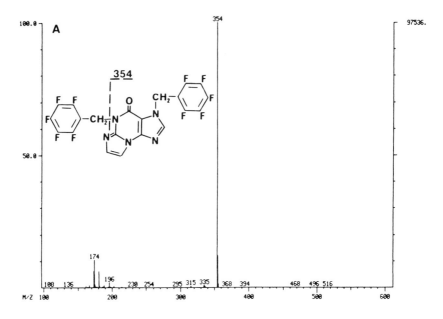

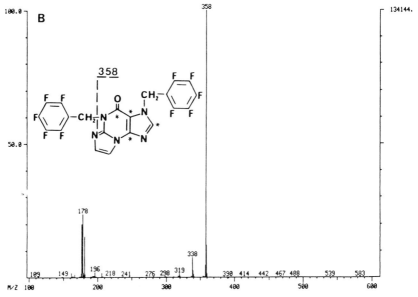

Figure 3. Negative ion chemical ionization mass spectra of 3,5-PFB$_2$-EG (A) and of 3,5-PFB$_2$-[2,3a,4,9a-^{13}C]-EG (B).

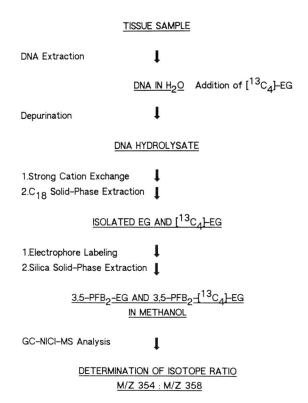

Figure 4. Isotope dilution mass spectrometry assay for EG.

backbone (Fedtke *et al.*, 1990a). Strong cation exchange chromatography was combined with C_{18} solid-phase extraction to isolate EG from the hydrolysates. EG was eluted with MeOH into a silanized reaction vessel, the MeOH was evaporated, and EG was electrophore-labeled by derivatization with pentafluorobenzyl bromide. Silica gel solid-phase extraction was used for the final purification of the reaction mixture. The internal standard, [2,3a,4,9a-^{13}C]-EG, was added to the hydrolysate prior to the EG isolation. The detection limit of the IDMS assay was 60 fmol EG per μmol guanine or 1 adduct in 1.7×10^7 guanine bases, i.e., in addition to the increased specificity, the method was 20-fold more sensitive than a previously developed HPLC method based on fluorescence detection of EG (Fedtke *et al.*, 1989). Figure 5 shows the ion intensity profiles obtained in an analysis of authentic 3,5-PFB$_2$-EG (Fig. 5A) and of liver DNA from a preweanling rat exposed to 600 ppm vinyl chloride for 5 days, 4 h per day (Figs. 5B1 and 5B2). The data on formation and persistence of EG obtained by application of the IDMS assay to various tissues of VC-exposed preweanling rats are discussed elsewhere (Fedtke *et al.*, 1990b).

Information on the molecular dose of EG can be combined with the information on its mutagenic efficiency. In addition, data on the other VC-derived DNA adducts 7-(2'-oxoethyl)guanine (Fedtke *et al.*, 1990b), 3,N^4-ethenodeoxycytidine, and 1,N^6-ethenodeoxyadenosine (Eberle *et al.*, 1989; Ciroussel *et al.*, 1989) were published recently. This unique data set is a prerequisite for a better understanding of VC-induced carcinogenesis, emphasizing the importance of the quantitative determination of DNA adducts.

ANALYSIS OF N^2,3-ETHENOGUANINE

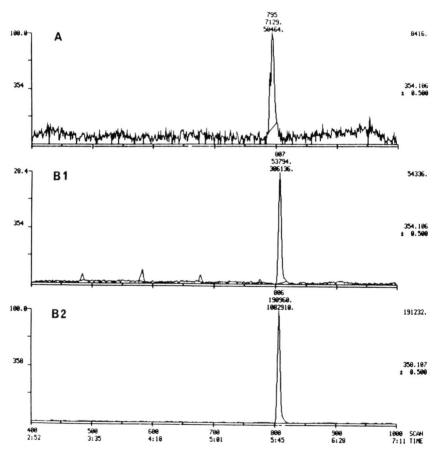

Figure 5. Analysis of EG. A: ion intensity profile for authentic 3,5-PFB$_2$-EG. B: ion intensity profiles for 3,5-PFB$_2$-EG (B1) and 3,5-PFB$_2$-[2,3a,4,9a-^{13}C]-EG (B2) obtained by analyzing liver DNA from VC exposed rats.

5. POTENTIAL AND LIMITATIONS

Quantitative MS can combine high specificity with sensitivities comparable to other analytical methods currently used for DNA-adduct quantification. GC-MS–based ultratrace analysis involving detection of amol (10^{-18}) quantities per injection became possible with the advance of analytical techniques and equipment in the last decade. To date, only a few efforts have been made to combine these techniques for the analysis of DNA adducts. The described EG assay is one approach for addressing problems associated with the analysis of a DNA adduct present in trace concentrations. A major limitation for many research laboratories is the availability of MS equipment. Furthermore, the time-consuming methods development process requires a considerable commitment of resources. Difficult steps include reproducible extraction of the analyte from biological matrices, derivatization reactions, sample clean-up, and final analysis. In most cases, sample preparation will probably be the limiting factor, not the detector.

It is difficult to fully assess future possibilities of quantitative MS methods with regard to DNA adducts. At this time the data on many aspects are scarce, e.g., their applicability to different classes of adducts or the overall difficulties of method development have not been elucidated. However, the success of MS methods in other biomedical research areas, especially in clinical chemistry, suggests that the same potential exists for its application to DNA-adduct analyses. In conclusion, MS techniques appear to be well suited for molecular dosimetry studies on DNA adducts at exposure concentrations relevant to humans.

REFERENCES

J. Adams and R. W. Giese, Cinnamoylation of the sugar hydroxyls of 3-methylthymidine. Formation of a relatively stable ester derivative, *J. Chromatogr.* **347**, 99–107 (1985).

J. Adams, M. David, and R. W. Giese, Pentafluorobenzylation of O^4-ethylthymidine and analogues by phase-transfer catalysis for determination by gas chromatography with electron capture detection, *Anal. Chem.* **58**, 345–348 (1986).

R. J. Anderegg, Derivatization in mass spectrometry: strategies for controlling fragmentation, *Mass Spectrom. Rev.* **7**, 395–424 (1988).

D. J. Ashworth, W. M. Baird, C.-J. Chang, J. D. Ciupek, K. L. Bush, and R. G. Cooks, Chemical modification of nucleic acids. Methylation of calf thymus DNA investigated by mass spectrometry and liquid chromatography, *Biomed. Mass Spectrom.* **12**, 309–318 (1984).

J. R. Babson, S. E. Russo-Rodriguez, W. H. Rastetter, and G. N. Wogan, *In vitro* DNA-binding of microsomally-activated fluoranthene: evidence that the major product is a fluoranthene N^2-deoxyguanosine adduct, *Carcinog.* **7**, 859–865 (1986).

E. Bailey, P. B. Farmer, and D. E. G. Shuker, Estimation of exposure to alkylating carcinogens by the GC-MS determination of adducts to hemoglobin and nucleic acid bases in urine, *Arch. Toxicol.* **60,** 187–191 (1987).

K. D. Ballard, T. D. Eller, J. G. Webb, W. H. Newman, and D. R. Knapp, Quantitative analysis of adenosine: statistical comparison of radioimmunoassay and gas chromatography–mass spectrometry-selected ion monitoring methods, *Biomed. Environ. Mass Spectrom.* **13,** 667–675 (1986).

S. A. Belinsky, C. M. White, T. R. Devereux, and M. W. Anderson, DNA adducts as a dosimeter for risk estimation, *Environ. Health Perspect.* **76,** 3–8 (1987).

K. Biemann, *Mass Spectrometry: Organical Chemical Applications,* McGraw-Hill, New York, p. 351 (1962).

K. Biemann and J. A. McCloskey, Jr., Application of mass spectrometry to structure problems. VI. Nucleosides, *J. Am. Chem. Soc.* **84,** 2005 (1962).

K. Blau and G. S. King, *Handbook of Derivatives for Chromatography,* Heyden, London (1978).

H. M. Bolt, H. Peter, and U. Föst, Analysis of macromolecular ethylene oxide adducts, *Int. Arch. Occup. Environ. Health* **60,** 141–144 (1988).

J. Boutagy and D. J. Harvey, Analysis of cytosine arabinoside and related pyrimidine nucleosides by gas chromatography and gas chromatography–mass spectrometry, *J. Chromatogr.* **156,** 153–166 (1978).

H. Budzikiewicz, Negative chemical ionization (NCI) of organic compounds, *Mass Spectrom. Rev.* **5,** 345–380 (1986).

C.-J. Chang, D. J. Ashworth, I. Isern-Flecha, X.-J. Jiang, and R. G. Cooks, Modification of calf thymus DNA by methyl methanesulfonate. Quantitative determination of 7-methyldeoxyguanosine by mass spectrometry, *Chem. Biol. Interact.* **57,** 295–300 (1986).

F. Ciroussel, A. Barbin, G. Eberle, and H. Bartsch, Investigations on the relationship between DNA ethenobase adduct levels in several organs of vinyl chloride–exposed rats and cancer susceptibility, *Biochem. Pharmacol.* **39,** 1109–1113 (1990).

B. N. Colby, Evaluation of stable labeled compounds as internal standards for quantitative GC/MS determinations, *U.S. Environmental Protection Agency, Office of Research and Development, Report No. EPA-600/2-83-127* (1983).

P. F.Crain and J. A. McCloskey, Analysis of modified bases in DNA by stable isotope dilution gas chromatography–mass spectrometry: 5-methylcytosine, *Anal. Biochem.* **132,** 124–131 (1983).

A. P. De Leenheer, M. F. Lefevere, W. E. Lambert, and E. S. Colinet, Isotope-dilution mass spectrometry in clinical chemistry, *Adv. Clin. Chem.* **24,** 111–161 (1985).

A. Dirr, I. Fasshauer, D. Wild, and D. Henschler, The DNA-adducts of the food mutagen and carcinogen IQ (2-amino-3-methylimidazo[4,5-f]quinoline, *Suppl. Arch. Toxicol.* **13,** 224–226 (1989).

M. Dizdaroglu, Characterization of free radical–induced damage to DNA by the combined use of enzymatic hydrolysis and gas chromatography–mass spectrometry, *J. Chromatogr.* **367,** 357–366 (1986).

M. Dizdaroglu and D. S. Bergtold, Characterization of free radical–induced base damage in DNA at biologically relevant levels, *Anal. Biochem.* **156,** 182–188 (1986).

M. Dizdaroglu, M.-L. Dirksen, H. Jiang, and J. H. Robbins, Ionizing-radiation-induced damage in the DNA of cultured human cells—identification of 8,5-cyclo-2-deoxyguanosine, *Biochem. J.* **241**, 929–932 (1987).

G. Eberle, A. Barbin, R. J. Laib, F. Ciroussel, J. Thomale, H. Bartsch, and M. F. Rajewsky, 1,N⁶-etheno-2'-deoxyadenosine and 3,N⁴-etheno-2'-deoxycytidine detected by mouse clonal antibodies in lung and liver DNA of rats exposed to vinyl chloride, *Carcinog.* **10**, 209–212 (1989).

P. B. Farmer, J. Lamb, and P. D. Lawley, Novel uses of mass spectrometry in studies of adducts of alkylating agents with nucleic acids and proteins. In: *Methods for Detecting DNA Damaging Agents in Humans: Application in Cancer Epidemiology and Prevention,* IARC Scientific Publication No. 89, eds. H. Bartsch, K. Hemminki, and I. K. O'Neill, Lyon, France (1988).

P. B. Farmer, H.-G. Neumann, and D. Henschler, Estimation of exposure of man to substances reacting covalently with macromolecules, *Arch. Toxicol.* **60**, 251–260 (1987).

P. B. Farmer, D. E. G. Shuker, and I. Bird, DNA and protein adducts as indicators of *in vivo* methylation by nitrosable drugs, *Carcinog.* **7**, 49–52 (1986).

N. Fedtke, J. A. Boucheron, M. J. Turner, Jr., and J. A. Swenberg, Vinyl chloride induced DNA adducts. I: Quantitative determination of N²,3-ethenoguanine based on electrophore labeling, *Carcinog.,* **11**, 1279–1285 (1990a).

N. Fedtke, J. A. Boucheron, V. E. Walker, and J. A. Swenberg, Vinyl chloride induced DNA adducts. II: Formation and persistence of 7-(2'-oxoethyl)guanine and N²,3-ethenoguanine in rat tissue DNA, *Carcinog.,* **11**, 1287–1292 (1990b).

N. Fedtke, V. E. Walker, and J. A. Swenberg, Determination of 7-(2'-oxo-ethyl)guanine and N²,3-ethenoguanine in DNA hydrosates by HPLC, *Suppl. Arch. Toxicol.* **13**, 214–218 (1989).

C. Finn, H.-J. Schwandt, and W. Sadee, Determination of uracil and thymine and their nucleosides and nucleotides in pmol amounts by gas chromatography mass spectrometry selected ion monitoring, *Biomed. Mass Spectrom.* **6**, 194–199 (1979).

D. H. Fisher and R. W. Giese, Determination of 5-methylcytosine in DNA by gas chromatography-electron-capture detection. *J. Chromatogr.,* **452**, 51–60 (1988).

D. H. Fisher, J. Adams, and R. W. Giese, Trace derivatization of cytosine with pentafluorobenzoyl chloride and dimethylsulfate, *Environ. Health Perspect.* **62**, 67–71 (1985).

U. Föst, B. Marczynski, R. Kasemann, and H. Peter, Bestimmung des Adduktes 7-(2'-oxoethyl)guanin mittels GC/MS aus Leber-DNA von Ratten nach Exposition gegen Vinylchlorid, *Naunyn Schmiedeberg's Arch. Pharmacol. Suppl. 2,* **338**, A249 (1988).

U. Föst, B. Marczynski, R. Kasemann, and H. Peter, Determination of 7-(2'-hydroxyethyl)guanine with GC/MS as a parameter for genotoxicity of ethylene oxide, *Suppl. Arch. Toxicol.* **13**, 250–253 (1989).

S. Galiegue-Zouitina, B. Bailleul, Y.-M. Ginot, B. Perly, P. Vigny, and M. H. Loucheux-Lefebre, N²-Guanyl and N⁶-adenyl arylation of chicken erythrocyte DNA by the ultimate carcinogen of 4-nitroquinoline 1-oxide', *Cancer Res.* **46**, 1858–1863 (1986).

W. A. Garland and M. L. Powell, Quantitative selected ion monitoring (QSIM) of drugs and/or drug metabolites in biological matrices, *J. Chromatogr. Sci.* **19**, 392–434 (1981).

C. F. Gelijkens, D. L. Smith, and J. A. McCloskey, Capillary gas chromatography of pyrimidines and purines: N,O-peralkyl and trifluoroacetyl-N,O-aklyl derivatives, *J. Chromatogr.* **225**, 291–299 (1981).

R. C. Gupta, M. V. Reddy, and K. Randerath, ^{32}P-Postlabeling analysis of nonradioactive aromatic carcinogen-DNA adducts, *Carcinog.* **3**, 1081–1092 (1982).

B. Halpern, Biomedical applications of gas chromatography–mass spectrometry, *CRC Crit. Rev. Anal. Chem.* **11**, 49–78 (1981).

H. G. Heumann, Isotope dilution mass spectrometry of inorganic and organic substances, *Fresenius Z. Anal. Chem.* **325**, 661–666 (1986).

C. Hignite, Nucleic acids and derivatives. In: *Biochemical Applications of Mass Spectrometry,* eds. G. R. Waller and O. C. Dermer, Wiley, New York, pp. 527–566 (1980).

A. M. Hogg, J. G. Kelland, and J. C. Vederas, Investigation of ribo- and deoxyribonucleosides and -nucleotides by fast-atom-bombardment mass spectrometry, *Helv. Chim. Acta* **69**, 908–917 (1986).

I. Jardine and M. M. Weidner, Approach to the quantitative analysis of nucleotides by gas chromatography–mass spectrometry, *J. Chromatogr.* **182**, 395–401 (1980).

L. P. Johnson, J. K. Macleod, R. E. Summons, and N. Hunt, Design of a stable isotope dilution gas chromatography/mass spectrometric assay for cAMP: comparison with standard protein-binding and radioimmunoassay methods, *Anal. Biochem.* **106**, 285–290 (1980).

R. Kasemann, U. Föst, and H. Peter, Selective GC/MS analysis of 7-(2'-oxoethyl)guanine in the presence of 7-(2'-hydroxyethyl)guanine by oximation, *Arch. Toxicol.* **61**, 245–246 (1988).

S. Kaur, K. Pongracz, S. F. Liu, A. Burlingame, and W. Bodell, Isolation and characterization of DNA adducts by ^{32}P-postlabeling and mass spectrometry, *Proc. Am. Assoc. Cancer Res.* **29**, 89 (1988).

S. A. M. Koch, M. J. Turner, Jr., and J. A. Swenberg, Quantitation of O^4-ethyldeoxythymidine using electrophore post-labelling, *Proc. Am. Assoc. Cancer Res.* **28**, 369 (1987).

D. R. Knapp, *Handbook of Analytical Derivatization Reactions,* Wiley, New York (1979).

S. Kumar, N. D. Sharma, R. J. H. Davies, D. W. Phillipson, and J. A. McCloskey, The isolation and characterization of a new type of dimeric adenine photoproduct in UV-irradiated deoxyadenylates, *Nucleic Acids Res.* **15**, 1199–1216 (1987).

R. J. Laib, The role of cyclic base adducts in vinyl chloride-induced carcinogenesis: Studies on nucleic acid alkylation in vivo. In: *The Role of Cyclic Nucleic Acid Adducts in Carcinogenesis and Mutagenesis,* IARC Scientific Publication No. 70, eds B. Singer and H. Bartsch, Oxford University Press, London, pp 101–108 (1986).

R. J. Laib, G. Doerjer, and H. M. Bolt, Detection of N^2,3-ethenoguanine in liver DNA hydrolysates after exposure of the animals to ^{14}C-vinyl chloride, *J. Cancer Clin. Oncol.* **109**, A7 (1985).

W. K. Lutz, Quantitative evaluation of DNA-binding data in vivo for low-dose extrapolations, *Suppl. Arch. Toxicol.* **11**, 66–74 (1987).

J. Maybaum, M. G. Kott, N. J. Johnson, W. D. Ensminger, and P. L. Stetson, Analysis of bromodeoxyuridine incorporation into DNA: Comparison of gas chromatographic/mass spectrometric, CsCl gradient sedimentation, and specific radioactivity methods, *Anal. Biochem.* **161**, 164–171 (1987).

J. A. McCloskey, Mass spectrometry of nucleic acid constituents and related compounds. In: *Mass Spectrometry in the Health and Life Sciences*, eds. A. L. Burlingame and N. Castagnoli, Elsevier, Amsterdam, pp. 521–546 (1985).

J. A. McCloskey, In: *Mass Spectrometry in Biomedical Research*, ed. S. J. Gaskell, Wiley, New York, pp. 75–95 (1986).

R. K. Mitchum, J. P. Freeman, F. A. Beland, and F. F. Kadlubar, Mass spectrometric identification of DNA adducts formed by carcinogenic aromatic amines. In: *Mass Spectrometry in the Health and Life Sciences*, eds. A. L. Burlingame and N. Castagnoli, Elsevier, Amsterdam, pp. 547–580 (1985).

G. B. Mohamed, A. Nazareth, M. J. Hayes, R. W. Giese, and P. Vouros, Gas chromatography–mass spectrometry characteristics of methylated perfluoroacyl derivatives of cytosine and 5-methylcytosine, *J. Chromatogr.* **314**, 211–217 (1984).

A. Nazareth, M. Joppich, S. Abdel-Baky, K. O'Connell, A. Sentissi, and R. W. Giese, Electrophore-labeling and alkylation of standards of nucleic acid pyrimidine bases for analysis by gas chromatography with electron capture detection, *J. Chromatogr.* **314**, 201–210 (1984).

C. Pantaretto, A. Martini, G. Belvedere, A. Bossi, M. G. Donelli, and A. Frigerio, Application of gas chromatography–chemical ionization mass fragmentography in the evaluation of bases and nucleoside analogues used in cancer chemotherapy, *J. Chromatogr.* **99**, 519–527 (1974).

H. Pang, D. L. Smith, P. F. Crain, K. Yamaizumi, S. Nishimura, and J. A. McCloskey, Identification of nucleosides in hydrolysates of transfer RNA by high resolution mass spectrometry, *Eur. J. Biochem.* **127**, 459–471 (1982).

E. D. Pellizzari, Electron capture detection in gas chromatography, *J. Chromatogr.* **90**, 323–361 (1974).

K. Randerath, M. V. Reddy, and R. C. Gupta, ^{32}P-labeling test for DNA damage, *Proc. Natl. Acad. Sci.* **78**, 6126–6129 (1981).

P. J. Russell, K. D. Rodland, E. M. Rachlin, and J. A. McCloskey, Differential DNA methylation during the vegetative life cycle of neurospora crassa, *J. Bacteriol.* **169**, 2902–2905 (1983).

M. Saha, G. M. Kresbach and R. W. Giese, Preparation and mass spectral characterization of pentafluorobenzyl derivatives of alkyl and hydroxyalkyl-nucleobase DNA adducts, *Biomed. Environ. Mass Spectrom.* **18**, 958–972 (1989).

I. Scott and R. Horgan, Mass-spectrometric quantification of cytokinin nucleotides and glycosides in tobacco crown-gall tissue, *Planta*, **161**, 345–354 (1984).

D. E. G. Shuker, E. Bailey, S. M. Gorf, J. Lamb, and P. B. Farmer, Determination of N-7-[^{2}H$_3$]methyl guanine in rat urine by gas chromatography–mass spectrometry following administration of trideuteromethylating agents or precursors, *Anal. Biochem.* **140**, 270–275 (1984).

J. Singer, W. C. Schnute, Jr., J. E. Shively, C. W. Todd, and A. D. Riggs, Sensitive detection of 5-methylcytosine and quantitation of the 5-methylcytosine/cytosine ratio by gas chromatography–mass spectrometry using mutiple specific ion monitoring, *Anal. Biochem.* **94**, 297–301 (1979).

P. L. Stetson, J. Maybaum, U. A. Shukla, and W. D. Ensminger, Simultaneous determination of thymine and 5-bromouracil in DNA hydrolysates using gas chromatography–mass spectrometry with selected ion monitoring, *J. Chromatogr.* **375**, 1–9 (1986).

J. A. Swenberg, N. Fedtke, T. R. Fennell, and V. E. Walker, Relationships between carcinogen exposure, DNA adducts and carcinogenesis. In: *Progress in Predictive Toxicology,* Elsevier, Amsterdam, pp. 161–184 (1990).

J. A. Swenberg, F. C. Richardson, J. A. Boucheron, F. H. Deal, S. A. Belinsky, M. Charbonneau, and B. G. Short, High- to low-dose extrapolation: critical determinants involved in the dose response of carcinogenic substances, *Environ. Health Perspect.* **76**, 57–63 (1987).

K. B. Tomer, M. L. Gross, and M. L. Deinzer, Fast atom bombardment and tandem mass spectrometry of covalently modified nucleosides and nucleotides: adducts of pyrrolizidine alkaloid metabolites, *Anal. Chem.* **58**, 2527–2534 (1986).

T. M. Trainor, R. W. Giese, and P. Vouros, Mass spectrometry of electrophore-labeled nucleosides: pentafluorobenzyl and cinnamoyl derivatives, *J. Chromatogr.* **452**, 369–376 (1988).

M. J. Turner, Jr., S. A. M. Koch, J. A. Boucheron, and J. A. Swenberg, Methods for quantitative determination of the DNA adduct O^4-ethylthymidine by electron capture negative ion chemical ionization mass spectrometry, presented at the 35th ASMS Conference on Mass Spectrometry and Allied Topics, Denver, Colorado, May 24–29, pp. 703–704 (1987).

L. Zeise, R. Wilson, and E. A. C. Crouch, Dose-response relationships for carcinogens: a review, *Environ. Health Perspect.* **73**, 259–308 (1987).

Analytical Approaches for the Determination of Protein–Carcinogen Adducts Using Mass Spectrometry

Peter B. Farmer
MRC Toxicology Unit
Surrey, United Kingdom

1. INTRODUCTION

The essence of human carcinogen–biomonitoring procedures is the sensitive and selective determination of covalently bound adducts formed between the genotoxic agent and reactive residues within the cell. The procedures that have been developed to date have varied extensively both in the analytical method used and in the biological site from which the adduct was extracted prior to its determination. Within this chapter the relative merits and disadvantages of the use of mass spectrometry as an analytical method for determining adducts of carcinogens with proteins (and in particular with hemoglobin) will be discussed.

2. USE OF PROTEINS FOR BIOMONITORING

An ideal biomonitoring procedure would measure the adducts formed by the carcinogen with the target site for its biological action. These sites are believed to be nucleophilic residues within DNA. No biomonitoring procedure that has been developed up to now has achieved this ideal aim. The reasons for this are twofold. First, the nature of the target site is in most cases unknown, and second, the acquisition of sufficient adduct for analysis from such target sites has proved difficult to achieve. An exception to this generality is the case with methylating carcinogens for which there is strong experimental evidence for the involvement of O^6-methylation of guanine (and possibly O^4-methylation of thymine) in DNA in the initiation stage of cancer. Immunoassay procedures for such adducts have been developed and used for human monitoring (Wild *et al.*, 1983, 1987; Foiles *et al.*, 1988). However, the nature of the target site within DNA for more structurally complex genotoxic agents is less certain.

There has consequently been a great deal of interest in the development of methods for determining *nontarget* site adducts of carcinogens as biomonitoring procedures. These sites may be within nucleic acids or proteins. Analytical techniques of this type have two possible functions. Their results may first be used as a monitor of exposure to the compound in question. Second, if the quantitative relationship between the binding of the carcinogen to the nontarget site and to the target site has been established, these techniques could also be used as a risk monitor for the carcinogen. The principles of this indirect assessment of carcinogen–target site interactions have largely been developed by Ehrenberg and colleagues (see review by Ehrenberg and Osterman-Golkar 1980).

The "nontarget site" that has received most attention over the past 10 years is hemoglobin. This protein is readily available from blood (see below for experimental details), resulting in up to gram quantities being available for analysis. Hemoglobin also has a long biological lifetime (120 days in humans). The stability of the carcinogen–hemoglobin adducts that have been studied experimentally generally approaches that of the protein, which thus allows evidence of exposure to be accumulated over this time period. A greater sensitivity therefore results for the monitoring of continuous low-level exposure to carcinogens.

All electrophilic genotoxic compounds that react with nucleic acids also appear to react with hemoglobin, although the relative extent of binding to the protein varies extensively from one compound to another. The extent of binding is measured by the "binding index," defined as binding (mmol/mol hemoglobin) divided by dose (mmol/kg). Any attempt at determining carcinogen risk factors from hemoglobin binding data must clearly take into account such differences in reactivity. A knowledge of the exact target site in

DNA is also a prerequisite if risk factor calculations are to be carried out. Thus in summary, the use of hemoglobin for carcinogen dose monitoring has the advantage of its ready accessibility and long lifetime, allowing highly sensitive analyses, especially for compounds with a high binding index. Additionally there is the possibility of using hemoglobin binding for carcinogenic risk monitoring in cases for which experimental DNA and hemoglobin binding studies have been carried out and for which the DNA target site is known.

Albumin also has many of hemoglobin's advantages. Although it has a more rapid turnover than hemoglobin ($t_{1/2}$ = about 19 days in humans), it is also abundant in blood (about 42 mg/ml plasma). Despite this, the use of albumin has up to now been limited to studies of carcinogen exposure in animals.

3. USE OF MASS SPECTROMETRY FOR CARCINOGEN-ADDUCT ANALYSIS

The justification for the use of mass spectrometry for carcinogen-adduct analysis rests in its sensitivity and "chemical specificity," (i.e., its ability to distinguish and characterize adducts from different carcinogenic chemicals), even if these have closely related structures. This structural specificity shown by mass spectrometry is an advantage of the technique that is not shown to the same extent by the other carcinogen-adduct monitoring methods that are in use (i.e., immunoassay, ^{32}P-postlabeling, and fluorescence measurements), in which one must be aware of possible cross-reactivity between different chemicals.

The sensitivity requirements for human carcinogen biomonitoring cannot precisely be defined at present, as adduct levels that would present an unacceptable risk for humans are unknown. However, in animals, DNA modification of 1 in 10^5–10^7 nucleotides is caused by tumorigenic doses of chemical carcinogens (Phillips, 1985). The human diploid genome contains on the order of 10^{10} nucleotides (Mahler and Cordes, 1971). One might assume therefore that analytical processes with sensitivity greater than 1 in 10^7 nucleotides, but probably not greater than 1 in 10^{10} nucleotides, may be required at present for human biomonitoring. The analytical procerures used to date (see above) have sensitivities for detecting modified nucleotides (1 in 10^6–10^{10}) generally within this range. Such techniques have been demonstrated to be capable of detecting occupational and environmental exposure to carcinogens (see the review by Farmer *et al.*, 1987), and in many cases have shown the presence of background levels of adducts in "unexposed" individuals. These observations support the conclusion that methods with detection limits greater than 1 modified nucleotide per 10^7 nucleotides are satisfactory for monitoring procedures.

The sensitivity required for protein-adduct monitoring will naturally depend on the protein to DNA binding ratio for the compound considered. For example, in the case of ethylene oxide it has been calculated (L. Ehrenberg, personal communication) that formation of 1 N-7-guanine adduct per 10^{10} bases in DNA corresponds to that of 10 pmol adduct per gram formed at the *N*-terminal position (valine) of hemoglobin. The analytical sensitivity for determining this adduct by gas chromatography–mass spectrometry (GC-MS) (see below) reaches this level, i.e., it can monitor the equivalent of about 1 DNA–N-7-guanine adduct per cell. The sensitivity requirements for measuring hemoglobin-adduct formation will be less severe for compounds that bind preferentially to hemoglobin compared to DNA, e.g., acrylamide (Hashimoto and Aldridge, 1970; Bailey *et al.*, 1986). The highest GC-MS analytical sensitivity for protein adducts so far achieved has been with hemoglobin–aromatic amine conjugates, where levels as low as 10 pg amine/g hemoglobin can be detected (Bryant *et al.*, 1987).

4. SPECIFICITY

The principle of mass spectrometric measurements is that an ion specific for the compound being analyzed is monitored, normally during elution of this compound from a chromatographic system [GC or high-pressure liquid chromatography (HPLC)]. The specificity of the measurement thus depends both on the nature of the ion chosen and on the chromatographic resolution of the compound.

Increased mass spectral specificity for the compound may be achieved by increasing the number of ions monitored or by repeating the analysis using a different form of ionization, e.g., by changing from electron impact (EI) to positive or negative chemical ionization (CI). An extra advantage of the use of a soft ionization technique (e.g., CI) is that higher mass ions are generally produced from the compound. These ions are less subject to contamination from background ions derived from impurities, which are normally of lower mass. Even more specific analyses may be carried out using high-resolution mass spectrometry, in which only compounds of the *exact* mass of (and hence the same empirical formula as) the adducts are detected. This technique, however, is not available on all mass spectrometers and also results in large losses in sensitivity for the analysis. It has consequently not been widely used for human biomonitoring, although its use in acrylamide exposure has been reported (Farmer *et al.*, 1988).

The chromatographic system most widely used in MS biomonitoring studies is high-resolution capillary GC. Techniques currently available for preparing capillary GC columns are so advanced that the theoretical maximum resolving power is nearly being achieved. Thus the scope for further increases in specificity of the analyses by improving GC column preparation is limited. However, the resolving power of the separation may be increased by using longer columns or by reducing internal diameter. Further confirmation of the characterization of a compound in a biomonitoring determination may be achieved by repeat analysis using a column coated with a different stationary phase or by employing a different derivative of the compound. The latter technique has the additional advantage of altering the ions to be monitored in the mass spectrometer, affording extra selectivity for the analysis. Although this procedure has been widely used in other forms of xenobiotic analysis, it has surprisingly not received much attention in the biomonitoring field. Further improvements in *sensitivity* may also be possible by selecting a derivative with a higher mass spectral response, e.g., in negative-ion CI.

5. GC-MS PROCEDURES FOR ANALYSIS OF HEMOGLOBIN–CARCINOGEN ADDUCTS

The procedures that have been used for GC-MS analysis of hemoglobin–carcinogen adducts fall into two general approaches: (a) analysis of a carcinogen–modified amino acid, derived from the hemoglobin polypeptide chain, and (b) analysis of the carcinogen, or derivative thereof, liberated and purified from a hemoglobin hydrolyzate. Approach (a) was the first to be used, and considerable improvements in its sensitivity have been achieved over the past 10 years. The more recent approach (b) shows even greater levels of sensitivity. A significant difference between approaches (a) and (b) is that the nature of the adduct is known with certainty for (a) but not for (b) although as will be shown below, good assumptions may be made for (b) based on chemical reaction models. This means that there will be a slightly greater uncertainty factor if one attempts to monitor risk (see Introduction) from approach (b) compared to (a). However, exposure monitoring may be carried out equally by either of the methods, subject of course to the requirement that the carcinogen under study binds to the appropriate nucleophilic sites in hemoglobin.

Summaries of the use and practical requirements of each approach will be given below.

5.1. Determination of Carcinogen-Modified Amino Acids from Hemoglobin

The procedure used initially for such determinations involved the hydrolysis of globin in strong acid in the presence of an appropriately labeled internal standard, followed by purification of the modified amino acid, derivatization, and quantitation by GC-MS using selected-ion recording. A summary of the amino acids determined in this way is given in Table 1. For example, the procedure may be illustrated in detail for the interaction of the carcinogen propylene oxide with the histidine residues in hemoglobin, which yields N^T-(2-hydroxypropyl)histidine (Fig. 1) (Farmer *et al.*, 1981). Blood samples (5–10 ml) from exposed individuals are centrifuged to separate the red cells, which are then washed and lysed. The lysate is subjected to high-speed centrifugation to remove cellular debris and the globin precipitated from it by decanting it into acetone containing 1% hydrochloric acid. The globin is washed with solvents and dried aliquots (10–50 mg) used for N^T-(2-hydroxypropyl)histidine determination. After addition of a synthetic deuterated internal standard [d_5-N^T-(2-hydroxypropyl)histidine], the protein is hydrolyzed *in vacuo* in 6N hydrochloric acid, and the resulting amino acid mixture chromatographed on a cation exchange column. The purified d_0- and d_5-N^T-(2-hydroxypropyl) histidine is derivatized by methylation of its carboxylic acid function and heptafluorobutyrylation of its amino and hydroxyl groups. The resulting derivative has a molecular weight of 619 (624 for the d_5 analogue) and yields in EI mass spectrometry an intense ion at *m/z* 560 (565 for the d_5 analogue), which is used for its monitoring. Comparison of the peak height of the signals of mass to charge ratio (*m/z*) 560 and 565, as the derivative is eluted from the capillary GC, allows calculation of the amount of unlabeled material present. A representative analysis is shown in Fig. 2. Further details of GC and MS conditions are given by Farmer *et al.* (1982). For human monitoring further purification of the alkylated amino acids and modification of the protein hydrolysis procedures may be necessary (Osterman-Golkar *et al.*, 1984). An analogous approach has been used to monitor the ethylene oxide adduct with histidine in hemoglobin [N^T-(2-hydroxyethyl)histidine], using a d_4-labeled internal standard (Calleman *et al.*, 1978; Osterman-Golkar *et al.*, 1983; Van Sittert *et al.*, 1985; Farmer *et al.*, 1986a).

Other adducts that have been monitored in this way include *S*-methylcysteine (Farmer *et al.*, 1980; Bailey *et al.*, 1981), N^T-methylhistidine (Törnqvist *et al.*, 1988a), and *S*-ethylcysteine (Farmer, 1982), which are formed following exposure of hemoglobin to methylating and ethylating carcinogens. Exposure to acrylamide results in the formation of *S*-(3-amino-3-oxopropyl)cysteine in hemoglobin; upon acidic hydrolysis this yields *S*-(2-carboxyethyl)cysteine, for which biomonitoring methods have also been developed (Bailey *et al.*, 1986). All of these cysteine adducts were determined by CI selective-ion monitoring as the EI spectra of their esterified acylated derivatives gave no intense high mass ions.

Although these methods for cysteine and histidine adducts were shown to be satisfactory for monitoring high-level exposures to the appropriate carcinogens, they show several practical disadvantages. These derive from the

TABLE 11-1. GC-MS Methods for Determining Alkylated Amino Acids*

Alkylating agents	Amino acid determined	Derivative	Ionization technique	Internal standard	Ions monitored	Reference
Analysis After Total Protein Hydrolysis						
Methylating agents	S-methylcysteine	n-Bu ester, HFB	+CI	d_3-methyl	388, 391	Farmer et al., 1980
	N^T-methylhistidine	Me ester, HFB	+EI	N^T-ethyl	320, 334	Törnqvist et al., 1988a
Ethylating agents	S-ethylcysteine	n-Bu ester, HFB	+CI	d_3-methyl	402, 391	Farmer, 1982
Acrylamide	S-(2-carboxyethyl)cysteine	Me ester, HFB	+CI	d_3-carboxyethyl	386, 389	Bailey et al., 1986
Ethylene/ethylene oxide	N^T-(2-hydroxyethyl)histidine	Me ester, HFB	+EI	d_4-hydroxyethyl	546, 550	Calleman et al., 1978
Propylene/propylene oxide	N^T-(2-hydroxypropyl)histidine	Me ester, HFB	+EI	d_5-hydroxypropyl	560, 565	Farmer et al., 1982
Analysis After Modified Edman Degradation						
Methylating agents	N-methylvaline	PFPTH	−CI	d_3-methyl	310, 313	Törnqvist et al., 1988a
Ethylene/ethylene oxide	N-(2-hydroxyethyl)valine	PFPTH	−CI	d_4-hydroxyethyl	348, 352	Törnqvist et al., 1986a
	valine	PFPTH, TMSi	+EI	d_4-hydroxyethyl	440, 444	Bailey et al., 1988
Styrene/styrene oxide	N-(2-hydroxy-2-phenyl-ethyl)valine	PFPTH	−CI	$d_7$valine analog	424, 431	Nördqvist et al., 1985

*Abbreviations: HFB, heptafluorobutyryl; PFPTH, pentafluorophenyl thiohydantoin; TMSi, trimethylsilyl; EI, electron impact; CI, chemical ionization; +, positive; −, negative.

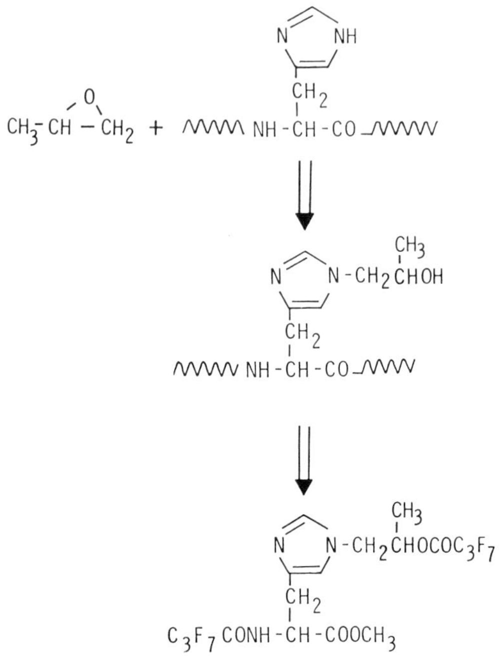

Figure 1. Interaction of propylene oxide with a histidine residue in hemoglobin, and derivatization of the adduct for GC-MS.

difficulty in chromatographically separating the modified amino acid from the normal amino acids (present in up to 10^6-fold greater quantities), and from the limitation in the amount of hemoglobin that may therefore consequently be used for analysis. A significant development in these techniques was made by Törnqvist *et al.* (1986a), who showed that carcinogen adducts with the *N*-terminal valine of hemoglobin could readily be extracted from the rest of the protein following a modified Edman degradation procedure. The consequent reduction in contamination with the adduct results in analytical sensitivities considerably higher than that achieved for the adducts derived from complete protein hydrolysis. An example of this approach is the analysis of *N*-(2-hydroxyethyl)valine, resulting from the interaction of hemoglobin with ethylene oxide (Fig. 3). Our approach (Bailey *et al.*, 1988), which is a modification of that of Törnqvist *et al.* (1986a), involves the following stages. Globin is prepared from blood samples as described above and is reacted with pentafluorophenyl isothiocyanate in formamide solution. An internal standard [globin reacted with stable isotope labeled (d_4-)ethylene oxide] is also added to this reaction mixture. The *N*-terminal valine adduct with ethylene oxide is converted by this procedure to a pentafluorophenylthiohydantoin, which is extracted into ether and washed. The product is then derivatized by trimethylsilylation and subjected to EI GC-MS selective-ion monitoring (see Fig. 4 for a representative trace and compound structure). The sensitivity of this analysis allows the detection of 10 pmol adduct/g hemoglobin. An analogous approach could be used to monitor N-terminal valine adducts with methylating agents (Törnqvist *et al.*, 1988a) propylene oxide, and styrene oxide (Törnqvist *et al.*, 1986a; Nördqvist *et al.*, 1985) (Table 1). [The procedure of Törnqvist *et al.* (1986a) differs from the procedure noted above in that the pentafluorophenylthiohydantoin is not derivatized and is quantitated by negative-ion CI mass spectrometry.]

5.2. Determination of Carcinogen-Derived Products after Mild Hydrolysis of Hemoglobin

A variety of carcinogen adducts with hemoglobin are labile to mild alkaline or acidic hydrolysis. As a result these adducts would not survive

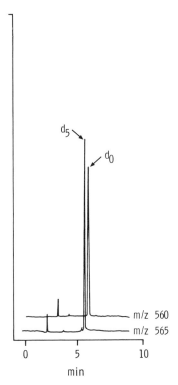

Figure 2. Selective-ion monitoring of derivatized N^τ-(2-hydroxypropyl)histidine (m/z 560) and its d_5 analogue (m/z 565). The sample was derived from human hemoglobin that had been exposed *in vivo* to propylene oxide. The amount of d_5 internal standard added was 100 ng. The concentration of d_0-N^τ-(2-hydroxypropyl)histidine in the globin was determined to be 13 nmol/g.

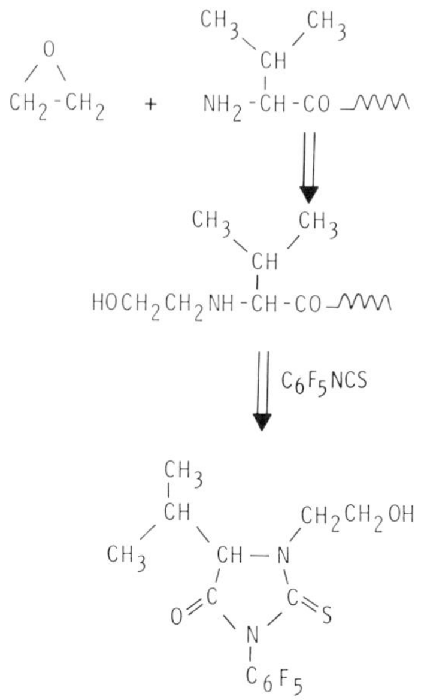

$$
\begin{array}{c}
\underset{\text{CH}_2\text{-CH}_2}{\overset{O}{\diagup\!\!\!\diagdown}} \quad + \quad \underset{\text{NH}_2\text{-CH-CO}\text{---}\backslash\!\backslash\!\backslash}{\overset{\text{CH}_3\diagdown\quad\diagup\text{CH}_3}{\underset{\text{CH}}{}}}
\end{array}
$$

$$
\Downarrow
$$

$$
\underset{\text{HOCH}_2\text{CH}_2\text{NH-CH-CO}\text{---}\backslash\!\backslash\!\backslash}{\overset{\text{CH}_3\diagdown\quad\diagup\text{CH}_3}{\underset{\text{CH}}{}}}
$$

$$
\Downarrow \quad C_6F_5NCS
$$

Figure 3. Interaction of ethylene oxide with an *N*-terminal valine residue in hemoglobin and conversion of the adduct to a pentafluorophenylthiohydantoin.

the strong acid hydrolysis used for degradation of the protein chain (see above), and they could also decompose if the acidic acetone procedure is used to remove the heme in preparation of globin. However, it is possible to analyze these labile adducts if one uses as the sample either an erythrocyte lysate or ethanol-precipitated hemoglobin. Two main classes of these adducts that have potential for this approach, carboxylic acid esters and cysteine-sulfinamides, are described below. Analogous approaches for determining tobacco-specific nitrosamines are discussed in chapter 18.

Little has so far been published on the use of carboxylic acid esters for biomonitoring, although the production of these adducts has been known for some time (Kim *et al.*, 1977). Mild acid hydrolysis of hemoglobin from animals treated with benzo[a]pyrene was shown by Shugart (1986a, 1986b), using HPLC and fluorescence detection, to yield benzo[a]pyrene tetrols. Acidic hydrolysis of benzpyrene-exposed hemoglobin, to release the tetrols, was also used by Santella *et al.* (1986) in their enzyme-linked immunosorbent assay for determining exposure to this hydrocarbon. An MS approach has also been developed for determining methanol derived from carboxylic acid esters (Gan *et al.*, 1989). Unpublished work from our laboratory suggests that this approach may also be feasible for other alcohols.

The situation with the cysteine–sulfinamide adducts is quite different, as MS has been used to study several human exposures to carcinogens. These adducts are formed by aromatic amines through a mechanism that involves the metabolism of the amine by hepatic cytochrome P-450 to a

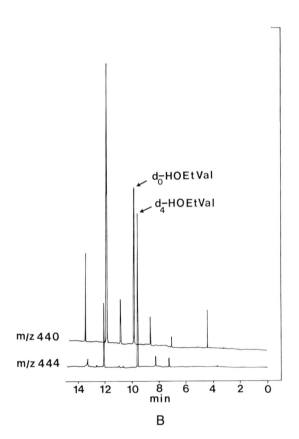

Figure 4. (a) Structures of trimethylsilyl derivatives of *N*-(2-hydroxyethyl)valine pentafluoro-phenylthiohydantoin and its d_4-labeled analogue. (b) Selective-ion monitoring of the trimethyl-silyl derivatives of *N*-(2-hydroxyethyl)valine pentafluorophenylthiohydantoin (*m/z* 440) and its d_4-labeled internal standard (*m/z* 444). The sample was derived from hemoglobin of a smoker of 20 cigarettes per day. The amount of d_4 internal standard added was 72 pmol. The concentration of d_0-*N*-(2-hydroxyethyl)valine in the globin was determined to be 225 pmol/g. Reproduced with permission from Springer-Verlag GmbH and Co.

hydroxylamine, which is subsequently further oxidized within the erythrocyte to the nitroso compound (with concomitant production of methemoglobin) (Kiese, 1974). The nitroso compound reacts directly with cysteine residues in hemoglobin to form an unstable intermediate that rearranges to the more stable sulfinamide. Such sulfinamides, whose potential as biomonitors of exposure was recognized by Neumann (see review, Neumann, 1988), may be hydrolyzed under mild acidic or alkaline conditions to yield the parent aromatic amine, which may be extracted, derivatized, and quantified by GC-MS. The procedure may be illustrated by the work of Bryant *et al.* (1987) on monitoring human exposure to 4-aminobiphenyl. Blood samples (10 ml) are centrifuged to isolate the red cells, which are lysed, centrifuged again to remove cell debris, and dialyzed. Hemoglobin is determined in the dialyzate, which is then hydrolyzed in the presence of an internal standard (4'-fluoro-4-aminobiphenyl) in $0.1M$ sodium hydroxide. Aromatic amines in the hydrolyzate are extracted with hexane, washed, and derivatized by treatment with pentafluoropropionic anhydride. GC-MS analyses were carried out using negative-ion CI selective-ion recording of the (M-HF)$^-$ ions. Levels of 4-aminobiphenyl less than 10 pg in the 10-ml blood sample are detectable with this approach. Similar approaches have been used to analyze hemoglobin cysteine adducts with *o*-, *m*-, and *p*-toluidines and 2-aminonaphthalene in human blood (Stillwell *et al.*, 1987).

Analytical variants of this approach are presented by Lewalter and Korallus (1985), who detected aniline adducts in acid-hydrolyzed erythrocytes by gas chromatography with the nitrogen-specific detection, by Birner and Neumann (1988), who analyzed a variety of monocyclic aromatic amines after $1N$ sodium hydroxide hydrolysis of ethanol-precipitated hemoglobin, and also by Farmer and Bailey (1988), who analyzed 4,4'-methylenedianiline as its heptafluorobutyryl derivative (using a d_8-labeled internal standard) following its hydrolysis from dialyzed exposed hemoglobin with $0.1N$ sodium hydroxide.

6. APPLICATION TO ANIMAL AND HUMAN CARCINOGEN BIOMONITORING OF GC-MS DETERMINATIONS OF HEMOGLOBIN ADDUCTS

6.1. Methylating Agents

The use of hemoglobin adducts as a monitor of exposure to methylating agents has been limited by the observation that there are relatively high levels of naturally occurring methylated amino acids in globin (see below). However, animal experiments with high doses (0.5–50 mg/kg) of methylmethanesulfonate showed a linear dose–response relationship for the production of S-methylcysteine (Bailey *et al.*, 1981). Nitrosodimethylamine gave a curved dose–response relationship for S-methylcysteine in rat hemoglobin with less than linear amounts produced at low doses. More recently, GC-MS has been used to demonstrate the production of S-methylcysteine in mouse hemoblogin following exposure to methyl bromide (Iwasaki, 1988).

Exposure to lower levels of methylating agents may be determined (in experimental situations) by using stable isotope-labeled analogues of the carcinogen, as there will be essentially no naturally occurring labeled methylated amino acids in hemoglobin. This procedure was illustrated by Farmer *et al.* (1986b) for exposure of rats to d_3-methylmethanesulfonate, d_3-nitrosomethylurea, and d_6-aminopyrine (administered in the presence of nitrite, which causes production of d_6-nitrosodimethylamine). All of these compounds produced d_3-S-methylcysteine.

No human studies on methylating agent exposure using S-methylcysteine as a dose monitor have been carried out, owing to the background level of this alkylated amino acid, which was found to be 16.4 ± S.D. 1.8 nmol/g globin (Bailey *et al.*, 1981). Studies on N^T-methylhistidine in human globin have shown that it is present in the globin of nonsmokers in the range 12–42 nmol/g (n = 4) and of smokers in the range 3.5–20 nmol/g (n = 5) (Törnqvist *et al.*, 1988a). N-Methylvaline is present in much smaller amounts (mean 0.50 nmol/g globin, n = 13) and there was no significant effect of smoking (mean 0.54 nmol/g globin, n = 7, plus one outlier 1.7 nmol/g globin) (Törnqvist *et al.*, 1988a).

6.2. Acrylamide

Exposure to acrylamide causes formation of S-3-amino-3-oxopropycysteine in hemoglobin, which may be determined following its hydrolysis to S-(2-carboxyethyl) cysteine. The amount of this adduct increases with increasing doses of acrylamide in rats, with increasing slope as the dose is raised (Bailey *et al.*, 1986). Human studies have been limited as again large background levels of the adduct were found in supposedly unexposed individuals.

6.3. Ethylene and Ethylene Oxide

The hemoglobin adducts that have been used for exposure monitoring to ethylene oxide and to its metabolic precursor ethylene are N^T-(2-hydroxyethyl) histidine and N-(2-hydroxyethyl)valine. The analysis of S-(2-hydroxyethyl)cysteine by GC-MS has only received limited study in experiments with mice exposed to ethylene (Ehrenberg *et al.*, 1977). Rats exposed to ethylene oxide by inhalation were shown to contain N^T-(2-hydroxyethyl)histidine, with the amount of the adduct being approximately proportional to the dose over the range 0–100 ppm (30 h/week) (Osterman-Golkar *et al.*, 1983).

The first GC-MS human monitoring study for ethylene oxide was that of Calleman *et al.* (1978), who determined N^T-(2-hydroxyethyl)histidine in hemoglobin from occupationally exposed and control individuals. The range of values for exposed samples was 0.5–13.5 nmol/g globin ($n = 7$) and 2 control samples had less than 0.05 nmol/g. Later, more extensive studies by Van Sittert *et al.* (1985) revealed no difference between exposed (2.08 ± S.E.M. 0.24 nmol/g, $n = 32$) and control (1.59 ± S.E.M. 0.18 nmol/g, $n = 31$) workers in an ethylene oxide plant. The source of the adduct in this control population is unknown. In a further study by Farmer *et al.* (1986a) of 10 workers in an ethylene oxide gas bottling plant, the N^T-(2-hydroxyethyl)histidine levels in exposed workers were notably higher (up to 8.0 nmol/g globin) than those in the control workers. This result was confirmed by N^T-(2-hydroxyethyl)valine measurements, which correlated well with N^T-(2-hydroxyethyl)histidine results.

The level of N^T-(2-hydroxyethyl)valine in hemoglobin was also increased by cigarette smoking. Fourteen nonsmokers were found to have an average level of this adduct of 58 pmol/g globin (± S. D. 25) compared to 389 pmol/g globin (± S.D. 138) in 11 smokers (Törnqvist, *et al.*, 1986b). In a separate study by Bailey *et al.* (1988) the observed values were 49.9 pmol/g globin in nonsmokers ($n = 24$) with an increase of 71 pmol/g globin per 10 cigarettes smoked per day. Thus N^T-(2-hydroxyethyl)valine levels appear to be a good biomonitor of hydroxyethylating agents produced as a result of inhalation of cigarette smoke. Another source of N^T-(2-hydroxyethyl)valine in hemoglobin is exposure to exhaust fumes. Exposure of hamsters to gasoline or diesel exhaust significantly increased levels of the adduct from control levels of ca. 100 pmol/g globin to ca. 700 pmol/g globin (Törnqvist *et al.*, 1988b). This increase was eliminated when a catalytic converter was used with gasoline exhaust.

6.4. Propylene and Propylene Oxide

Following an initial demonstration that exposure of rats to propylene oxide by inhalation led to a dose-related increase in the amount of N^T-(2-hydroxypropyl)histidine in their hemoglobin (Farmer *et al.*, 1982), the analysis of this adduct was carried out in a group of occupationally exposed workers. Exposure led to concentrations in the range 0.2–10.2 nmol/g globin ($n = 8$) of N^T-(2-hydroxypropyl)histidine compared to less than 0.1 nmol/g ($n = 9$) in a control population (Osterman-Golkar *et al.*, 1984). Exposure to exhaust fumes has been shown in animal experiments to be a potential source of the hydroxypropyl adduct at the N-terminal valine of hemoglobin, similar to the situation for N^T-(2-hydroxyethyl)valine (see above) (Törnqvist *et al.*, 1988b).

6.5. 4-Aminobiphenyl

4-Aminobiphenyl is a constituent of tobacco smoke and a known human bladder carcinogen. The feasibility of monitoring exposure to 4-aminobiphenyl by GC-MS determinations of the amine, following its hydrolytic release from its cysteine conjugate in hemoglobin, was first established by Green *et al.* (1984). The exposure of smokers to this amine has been determined by Bryant *et al.* (1987) with this technique.

The hemoglobin content of 4-aminobiphenyl in smokers was 154 ± 47 pg/g compared to a control level of 28 ± 13 pg/g. Furthermore, the level of 4-aminobiphenyl in the hemoglobin decreased in smokers who had stopped smoking, reaching the control level in 6–8 weeks. The association of adduct levels of 4-aminobiphenyl with smoking status was confirmed by the same group in later studies with more populations (Skipper *et al.*, 1988) and by Perera *et al.* (1987). The latter authors showed in one set of samples 154.5 ± 49.3 pg 4-aminobiphenyl/g hemoglobin in smokers compared to 32.2 ± 12.3 pg/g in nonsmokers, and found that the adduct level was significantly correlated with indices of active smoking (e.g., average packs/day, tar level, etc.).

6.6. Other Aromatic Amines

Analysis has also been carried out to determine the content in human hemoglobin of cysteine adducts with a variety of other aromatic amines (e.g., aniline, *o*-, *m*- and *p*-toluidine, 2-aminonaphthalene, 3-aminobiphenyl, 2-, 3-, and 4-ethylaniline, and 2,3-, 2,4-, 2,5-, and 2,6-dimethylaniline) (Stillwell *et al.*, 1987). The levels of the toluidine *o*- and *p*-isomers and of 2-aminonaphthalene were increased in smokers' blood. 3-Aminobiphenyl adducts were highly associated with cigarette smoking (Skipper *et al.*, 1988).

It is worth also pointing out that extensive studies on aniline adducts with hemoglobin cysteine have been carried out by Albrecht and Neumann (1985) and by Lewalter and Korallus (1985). For the latter study the detection of the amine was by gas chromatography using a nitrogen-sensitive detector, and mass spectrometry was not used. In workers exposed to aniline there were two distinguishable groups, "slow" and "fast" acetylators. All but one of the slow acetylators gave detectable levels of aniline adduct, whereas none of the fast acetylators gave detectable adducts. This emphasizes the importance of incorporating into aromatic amine biomonitoring studies an estimate of the acetylation polymorphic status of the exposed individuals.

7. BACKGROUND LEVELS OF MODIFIED AMINO ACIDS IN HEMOGLOBIN

A significant discovery in the above-mentioned studies of carcinogen adducts with hemoglobin was the detection of these adducts in control populations. The amounts of these "background" adducts range from 16.4 nmol/g globin for *S*-methylcysteine (Bailey *et al.*, 1981) to 0.04 pmol/g for 2-aminonaphthalene sulfinamide (Stillwell *et al.*, 1987) (Table 2). Their source is in general unknown.

S-Methylcysteine levels are species-dependent (Bailey *et al.*, 1981), and both this amino acid and N^T-methylhistidine may partially arise from misincorporation of these compounds (found, for example, in foods) into hemoglobin during its biosynthesis. On the other hand, *N*-methylvaline is not species-dependent and may be produced in hemoglobin by administration of formaldehyde, suggesting this (or possibly *S*-adenosylmethionine) as a source for this alkylated amino acid (Törnqvist *et al.*, 1988a). For the hydroxyethylated amino acids similar remarks apply. Both the cysteine and histidine adduct may be misincorporated during hemoglobin biosynthesis, whereas *N*-(2-hydroxyethyl)valine may not (Kautiainen *et al.*, 1986). For the latter compound the possibility of exposure to environmental ethylene or to hydroxyethylating agents produced endogenously (e.g., via lipid peroxidation) has been suggested (Kautiainen *et al.*, 1986; Törnqvist, *et al.*, 1986a).

No endogenous sources can be perceived for the adducts with the more complex electrophilic compounds, e.g., those derived from aromatic amines, and the production of these must be assumed to be due to low-level exogenous exposure of humans to these amines.

8. ANALYTICAL SHORTCOMINGS

Although, as described above, many successful human carcinogen biomonitoring studies have been achieved, there are still many practical problems associated with the assays. First, they require a high degree of technical skill for the successful and *reproducible* isolation of picomolar quantities of material in high purity. The procedures are slow and not adaptable at present for large-scale routine monitoring. As a result of this the reproducibility of assays on human samples has been hard to determine. Furthermore, the techniques involve the use of extremely expensive equipment, a fact that will further limit their routine use. However, the data acquired by mass spectrometry, however hard to obtain, have the great advantage that they are analytically correct to a high level of sensitivity with almost absolute selectivity for the carcinogen being studied. These data might thus be used as reference data in the development of less rigorous (and less expensive) analytical methods, which may be more suitable for routine monitoring.

TABLE 11–2. Background Levels of Modified Amino Acids in Hemoglobin

Amino Acid	Human nmol/g	Rat nmol/g	Reference
S-Methylcysteine	16.4	102	Bailey *et al.*, 1981
N^TMethylhistidine	12–42	...	Törnqvist *et al.*, 1988a
N-Methylvaline	0.5	0.5	Törnqvist *et al.*, 1988a
S-(2-Carboxyethyl)cysteine	8.5	21.4	Bailey *et al.*, 1986
N^T-(2-Hydroxyethyl)histidine	1.6	...	Van Sittert *et al.*, 1985
	...	1.3, 2.8	Osterman-Golkar *et al.*, 1983
N-(2-Hydroxyethyl)valine	0.05	...	Bailey *et al.*, 1988
	...	ca. 0.1	Törnqvist *et al.*, 1986a
N^T-(2-Hydroxypropyl)histidine	< 0.1	...	Osterman-Golkar *et al.*, 1984
Sulfinamides:			
4-Aminobiphenyl	0.00017	0.003–0.0178	Bryant *et al.*, 1987
Aniline	0.041	...	Stillwell *et al.*, 1987
o-Toluidine	0.00032	...	Stillwell *et al.*, 1987
m-Toluidine	0.0064	...	Stillwell *et al.*, 1987
p-Toluidine	0.00065	...	Stillwell *et al.*, 1987
2-Aminonaphthalene	0.00004	...	Stillwell *et al.*, 1987

Further problems with these biomonitoring assays relate to the continual requirement for further increases in sensitivity and selectivity. To some extent these objectives contradict each other. Thus, for example, the selectivity of analyses for *S*-(2-carboxyethyl)cysteine in globin may be increased by changing the mass spectrometer resolution from the normal value of 1000 to 10 000 (Farmer *et al.*, 1988). This procedure eliminates contaminating background peaks from the MS selected-ion recording. However, it also causes a dramatic reduction in sensitivity. Thus there is a continuing dependence upon the manufacturers of mass spectrometers to keep increasing instrumental sensitivity to meet our more challenging analytical needs, and upon research workers to develop more sensitive and selective isolation procedures for the adducts.

9. ANALYTICAL DEVELOPMENTS AND PROSPECTS FOR MONITORING PROTEIN–CARCINOGEN ADDUCTS

Further developments in isolation procedures for the adducts may be expected shortly, both in the extraction of adducts from the protein (e.g., by mild hydrolysis of carboxylic acid esters) and in their further purification. For the latter the use of immunoaffinity columns would appear to be a promising approach. These have successfully been used, for example, for DNA adducts with aflatoxin B_1 (Groopman *et al.*, 1985) and benzo[a]pyrene (Manchester *et al.*, 1988).

For the analysis of the adduct, immunoassay may also have potential. For example, the value of this technique was demonstrated by Wraith *et al.* (1988), who developed antibodies to the ethylene oxide adduct with *N*-terminal valine in hemoglobin. The method was validated against the established GC-MS approach and has been used in a study of ethylene oxide–exposed workers.

The development of mass spectrometry techniques depends largely on alterations in modes of ionization of the molecule. The extra sensitivity for some compounds generated by using negative chemical ionization has already been mentioned. Other prospects include the use of liquid chromatograph–mass spectrometer interfaces for polar and/or thermally labile adducts, and supercritical fluid chromatography for high-molecular-weight moderately polar compounds. The increasing sensitivity of benchtop mass spectrometers, including the ion trap, will also allow a greater variety of adducts to be determined by GC selective-ion recording on relatively cheap instruments. The capability of GC will also be increased by the current development of high-temperature stationary phases, which will allow chromatography of larger, more complex molecules.

One of the major gaps in the currently available MS techniques for biomonitoring is the ability to study mixed exposures to carcinogens. There are possibilities for using the modified Edman degradation technique for generalized screening for low-molecular-weight alkylating agents and epoxides, and for applying the mild alkaline hydrolysis technique for aromatic amine screening. However, a more general approach to detect all alkylating agents to which an individual is exposed is still required.

A requirement for each of the assays described above is a knowledge of the chemical involved and hence of the ion which should be monitored. Humans are always exposed to a variety of compounds, many of which are unknown. The use of tandem mass spectrometry should have considerable advantages in studying mixtures of adducts. If one selects a common fragmentation ion for a class of adducts (e.g., *N*-alkylvalines), it may be possible to identify the molecular weight, and from this the chemical nature, of all of these adducts present in a mixture, by carrying out a "parent ion scan." For this procedure the second mass spectrometer in the tandem system is focused continually on the fragment ion of interest. The first mass spectrometer is scanned over the entire mass range, and ions generated are focused into a collision cell between the two MS sectors. The parent ion scan consists of those ions that yield the ion of interest following collision-induced dissociation. The potential of this technique has recently been illustrated for *N*-7-alkylguanine derivatives (Farmer, 1988).

10. CONCLUSION

Considerable progress has been made over the past 10 years in the development of MS methods for carcinogen-protein monitoring, and one may now expect the generation of substantial amounts of data from humans. One must address oneself increasingly now to the evaluation of the meaning of these data to human carcinogenic risk, and to the necessity for collaboration between the chemical analysts and epidemiologists.

REFERENCES

W. Albrecht and H.-G. Neumann, Biomonitoring of aniline and nitrobenzene. Hemoglobin binding in rats and analysis of adducts, *Arch. Toxicol.* **57**, 1–5 (1985).

E. Bailey, T. A. Connors, P. B. Farmer, S. M. Gorf, and J. Rickard, Methylation of cysteine in hemoglobin following exposure to methylating agents, *Cancer Res.* **41**, 2514–2517 (1981).

E. Bailey, P. B. Farmer, I. Bird, J. H. Lamb, and J. A. Peal, Monitoring exposure to acrylamide by the determination of S-(2-carboxyethyl)cysteine in hydrolyzed hemoglobin by gas chromatography–mass spectrometry, *Anal. Biochem.* **157**, 241–248 (1986).

E. Bailey, A. G. F. Brooks, C. T. Dollery, P. B. Farmer, B. J. Passingham, M. A. Sleightholm, and D. W. Yates, Hydroxyethylvaline adduct formation in hemoglobin as a biological monitor of cigarette smoke intake, *Arch. Toxicol.* **62**, 247–253 (1988).

G. Birner and H.-G. Neumann, Biomonitoring of aromatic amines II. Hemoglobin binding of some monocyclic aromatic amines, *Arch. Toxicol.* **62**, 110–115 (1988).

M. S. Bryant, P. L. Skipper, S. R. Tannenbaum, and M. Maclure, Hemoglobin adducts of 4-aminobiphenyl in smokers and non-smokers, *Cancer Res.* **47**, 602–608 (1987).

C. J. Calleman, L. Ehrenberg, B. Jansson, S. Osterman-Golkar, D. Segerbäck, K. Svensson and C. A. Wachtmeister, Monitoring and risk assessment by means of alkyl groups in hemoglobin in persons occupationally exposed to ethylene oxide, *J. Environ. Pathol. Toxicol.* **2**, 427–442 (1978).

L. Ehrenberg, S. Osterman-Golkar, D. Segerback, K. Svensson, and C. J. Calleman, Evaluation of genetic risks of alkylating agents. III Alkylation of haemoglobin after metabolic conversion of ethene to ethene oxide *in vivo, Mutat. Res.* **45**, 175–184 (1977).

L. Ehrenberg and S. Osterman-Golkar, Alkylation of macromolecules for detecting mutagenic agents, *Teratogen. Carcinogen. Mutagen.* **1**, 105–127 (1980).

P. B. Farmer, Tandem mass spectrometry study of urinary alkylated purines, *Biomed. Environ. Mass Spectrom.* **17**, 143–145 (1988).

P. B. Farmer and E. Bailey, Protein-carcinogen adducts in human dosimetry, *Arch. Toxicol.*, suppl. 13, 83–90 (1989).

P. B. Farmer, E. Bailey, J. H. Lamb, and T. A. Connors, Approach to the quantitation of alkylated amino acids in haemoglobin by gas chromatography mass spectrometry, *Biomed. Mass Spectrom.* **7**, 41–46 (1980).

P. B. Farmer, The occurrence of S-methylcysteine in the hemoglobin of normal untreated animals. In: *Indicators of Genotoxic Exposure,* Banbury Report 13, Cold Spring Harbor Laboratory, Cold Spring Harbor, NY, pp. 169–175 (1982).

P. B. Farmer, S. M. Gorf, and E. Bailey, Determination of hydroxypropylhistidine in haemoglobin as a measure of exposure to propylene oxide using high resolution gas chromatography mass spectrometry, *Biomed. Mass Spectrom.* **9**, 69–71 (1982).

P. B. Farmer, E. Bailey, S. M. Gorf, M. Törnqvist, S. Osterman-Golkar, A. Kautiainen, and D. P. Lewis-Enright, Monitoring human exposure to ethylene oxide by the determination of haemoglobin adducts using gas chromatography–mass spectrometry, *Carcino.* **7**, 637–640 (1986a).

P. B. Farmer, D. E. G. Shuker, and I. Bird, DNA and protein adducts as indicators of *in vivo* methylation by nitrosatable drugs, *Carcinog.* **7**, 49–52 (1986b).

P. B. Farmer, H.-G. Neumann and D. Henschler, Estimation of exposure of man to substances reacting covalently with macromolecules, *Arch. Toxicol.* **60**, 251–260 (1987).

P. B. Farmer, J. Lamb, and P. D. Lawley, Novel uses of mass spectrometry in studies of adducts of alkylating agents with nucleic acids and proteins. In: *Methods for Detecting DNA Damaging Agents in Humans: Applications in Cancer Epidemiology and Prevention,* IARC Scientific Publication No. 89, eds. H. Bartsch, K. Hemminki, and I. K. O'Neill, IARC, Lyon, France, pp. 347–355 (1988).

P. G. Foiles, L. M. Miglietta, S. A. Akerkar, R. B. Everson, and S. S. Hecht, Detection of O^6-methyl deoxyguanosine in human placental DNA, *Cancer Res.* **48**, 4184–4188 (1988).

L.-S. Gan, J. S. Wishnok, J. G. Fox, and S. R. Tannenbaum, Quantitation of methylated hemoglobin via hydrolysis of methyl esters to yield methanol, *Anal. Biochem.* **179**, 326–331 (1989).

L. C. Green, P. L. Skipper, R. J. Turesky, M. S. Bryant, and S. R. Tannenbaum, *In vivo* dosimetry of 4-aminobiphenyl in rats via a cysteine adduct in hemoglobin, *Cancer Res.* **44**, 4254–4259 (1984).

J. D. Groopman, P. R. Donahue, J.-Zhu, J. Chen, and G. N. Wogan, Aflatoxin metabolism in humans: detection of metabolites and nucleic acid adducts in urine by affinity chromatography, *Proc Natl. Acad. Sci. USA* **82**, 6492–6496 (1985).

K. Hashimoto and W. N. Aldridge, Biochemical studies on acrylamide, a neurotoxic agent, *Biochem. Pharmacol.* **19**, 2591–2604 (1970).

K. Iwasaki, Determination of S-methylcysteine in mouse hemoglobin following exposure to methyl bromide, *Indust. Health* **26**, 187–190 (1988).

A. Kautiainen, S. Osterman-Golkar, and L. Ehrenberg, Mis-incorporation of alkylated amino acids into hemoglobin—a possible source of background alkylations, *Acta Chem. Scand.* **B40**, 453–456 (1986).

M. Kiese, *Methemoglobinemia: A Comprehensive Treatise*, CRC Press, Cleveland Ohio (1974).

S. Kim, P. D. Lotlikar, W. Chin, and P. N. Magee, Protein bound carboxyl-methyl ester as a precursor of methanol formation during oxidation of dimethylnitrosamine *in vitro*, *Cancer Lett.* **2**, 279–284 (1977).

J. Lewalter and U. Korallus, Blood protein conjugates and acetylation of aromatic amines. New findings on biological monitoring, *Int. Arch. Occup. Environ. Health* **56**, 179–196 (1985).

H. R. Mahler and E. H. Cordes, *Biological Chemistry*, 2nd ed., Harper and Row, New York, p. 187 (1971).

D. K. Manchester, A. Weston, J.-S. Choi, G. E. Trivers, P. V. Fennessey, E. Quintana, P. B. Farmer, D. l. Mann, and C. C. Harris, Detection of benzo[a]pyrene diol-epoxide-DNA adducts in human placenta, *Proc. Natl. Acad. Sci. USA,* **85**, 9243–9247 (1988).

M. B. Nördqvist, A. Löf, S. Osterman-Golkar, and S. A. S. Walles, Covalent binding of styrene and styrene 7,8-oxide to plasma proteins, hemoglobin and DNA in the mouse, *Chem. Biol. Interact.* **55**, 63–73 (1985).

H.-G. Neumann, Biomonitoring of aromatic amines and alkylating agents by measuring hemoglobin adducts, *Arch. Toxicol.* **60**, 151–155 (1988).

S. Osterman-Golkar, P. B. Farmer, D. Segerbäck, E. Bailey, C. J. Calleman, K. Svensson, and L. Ehrenberg, Dosimetry of ethylene oxide in the rat by quantitation of alkylated histidine in hemoglobin, *Teratogen. Carcinogen. Mutagen.* **3**, 395–405 (1983).

S. Osterman-Golkar, E. Bailey, P. B. Farmer, S. M. Gorf, and J. H. Lamb, Monitoring exposure to propylene oxide through the determination of hemoglobin alkylation, *Scand. J. Work Environ. Health* **10**, 99–102 (1984).

F. P. Perera, R. M. Santella, D. Brenner, M. C. Poirier, A. A. Munshi, H. K. Fischman and J. Van Ryzin, DNA adducts, protein adducts and sister chromatid exchange in cigarette smokers and non-smokers, *J. Natl. Cancer Inst.* **79**, 449–456 (1987).

D. H. Phillips, Chemical Carcinogenesis. In: *The Molecular Basis of Cancer*, eds. P. B. Farmer and J. M. Walker, Croom Helm, Beckenham, U. K., pp. 133–179 (1985).

R. M. Santella, C. D. Lin and N. Dharmaraja, Monoclonal antibodies to a benzo[a]pyrene diolepoxide modified protein, *Carcinog.* **7**, 441–444 (1986).

L. Shugart, Quantifying adductive modification of hemoglobin from mice exposed to benzo[a]pyrene, *Anal. Biochem.* **152**, 365–369 (1986a).

L. Shugart, Quantitating exposure to chemical carcinogens: *in vivo* alkylation of hemoglobin by benzo[a]pyrene, *Toxicol.* **34**, 211–220 (1986b).

P. L. Skipper, M. S. Bryant, and S. R. Tannenbaum, Determination of human exposure to carcinogenic aromatic amines from hemoglobin adducts in selected population groups. In: *Carcinogenic and Mutagenic Responses to Aromatic Amines and Nitroarenes*, eds. C. M. King, L. J. Romano, and D. Schuetzle, Elsevier, New York, pp. 65–71 (1988).

W. G. Stillwell, M. S. Bryant, and J. S. Wishnok, GC/MS analysis of biologically important aromatic amines. Application to human dosimetry, *Biomed. Environ. Mass Spectrom.* **14**, 221–227 (1987).

M. Törnqvist, J. Mowrer, S. Jensen, and L. Ehrenberg, Monitoring of environmental cancer initiators through hemoglobin adducts by a modified Edman degradation method, *Anal. Biochem.* **154**, 255–266 (1986a).

M. Törnqvist, S. Osterman-Golkar, A. Kautiainen, S. Jensen, P. B. Farmer, and L. Ehrenberg, Tissue doses of ethylene oxide in cigarette smokers determined from adduct levels in hemoglobin, *Carcinog.* **7**, 1519–1521 (1986b).

M. Törnqvist, S. Osterman-Golkar, A. Kautiainen, M. Näslund, C. J. Calleman, and L. Ehrenberg, Methylations in human hemoglobin, *Mutat. Res.* **204**, 521–529 (1988a).

M. Törnqvist, A. Kautiainen, R. N. Gatz, and L. Ehrenberg, Hemoglobin adducts in animals exposed to gasoline and diesel exhausts. I. Alkenes, *J. Appl. Toxicol.* **8**, 159–170 (1988b).

N. J. Van Sittert, G. De Jong, M. G. Clare, R. Davis, B. J. Dean, L. J. Wren, and A. S. Wright, Cytogenetic, immunological and haematological effects in workers in an ethylene oxide manufacturing plant, *Br. J. Ind. Med.* **42**, 19–46 (1985).

C. P. Wild, S. H. Lu and R. Montesano, Radioimmunoassay used to detect DNA alkylation in tissues from populations at high risk for oesophageal and stomach cancer. in: *The Relevance of N-Nitroso Compounds to Human Cancer: Exposures and Mechanisms*, IARC Scientific Publication No. 84, eds. H. Bartsch, I. K. O'Neill, and R. Schulte-Hermann, IARC, Lyon, France, pp. 534–537 (1987).

C. P. Wild, G. Smart, R. Saffhill, and J. M. Boyle, Radioimmunoassay of O^6-methyldeoxyguanosine in DNA of cells alkylated *in vitro* and *in vivo*, *Carcinog.* **4**, 1605–1609 (1983).

M. J. Wraith, W. P. Watson, C. V. Eadsforth, N. J. Van Sittert, M. Törnqvist, and A. S. Wright, An immunoassay for monitoring human exposure to ethylene oxide. In: *Methods for Detecting DNA Damaging Agents in Humans: Applications in Cancer Epidemiology and Prevention*, eds. H. Bartsch, K. Hemminki, and I. K. O'Neill, IARC Scientific Publication No. 89, IARC, Lyon, France, pp. 271–274 (1988).

Development of Immunoassays for the Detection of Carcinogen-DNA Adducts

Miriam C. Poirier
National Cancer Institute
National Institutes of Health
Bethesda, Maryland

1. INTRODUCTION

Since the mid-1970s a number of laboratories have been engaged in the quantitation and localization of DNA adducts using antisera that recognize DNA adducts and carcinogen-modified DNAs (Poirier, 1981, 1984). These antisera are sensitive probes for the presence of DNA-bound carcinogens and have proven useful in investigating mechanisms of chemical carcinogenesis in model systems (Müller and Rajewsky, 1981; Leng, 1985; Strickland and Boyle, 1984; Phillips, 1989) as well as detection of human exposure (Santella, 1988; Weston *et al.*, 1989b). The methodology that has made these studies possible evolved by adaptation of techniques previously developed for eliciting antisera against nucleosides, nucleotides, oligonucleotides, and nucleic acids. The immunogenicity of natural or unusual nucleosides covalently coupled to proteins and of nucleic acids coupled electrostatically through the negatively charged phosphates to methylated proteins had been clearly demonstrated (Stollar, 1980). The quantitative immunoassays developed with these antisera were highly sensitive and able to detect, for example, unusual bases in nucleic acids constituting a minute fraction of the total nucleotides

(Munns and Liszewski, 1980; Sawicki *et al.*, 1976). The adaptation to detection of chemical carcinogen-DNA adducts occurring at frequencies of 1 adduct in 10^5 bases or lower required primarily refinements and adaptations of the immunoassays above (Poirier, 1984; Müller and Rajewsky, 1981; Leng, 1985; Strickland and Boyle, 1984).

The focus of this chapter will be on the methodologies required to conjugate immunogens, immunize rabbits, establish immunoassays, characterize the antisera, and assay biological samples for the presence of carcinogen-DNA adducts. Included is a short discussion concerning problems that relate to standardization of immunoassays among different laboratories and a consideration of studies combining the use of adduct antisera with methods providing independent chemical identification.

2. PREPARATION OF IMMUNOGENS AND IMMUNIZATION

In order to achieve a sufficiently strong immunogenic stimulus, whether imunizing rabbits for polyclonal antisera or rodents for monoclonal antisera, it is necessary to couple the adduct hapten or the modified DNA to a protein carrier. Adduct haptens are generally bonded covalently (Erlanger, 1980), while modified DNAs are mixed with a methylated protein creating an electrostatic coupling between the nucleic acid phosphates and the protein methyl groups (Stollar, 1980; Plescia and Braun, 1967). Because of the polymeric nature of nucleic acids, this type of association is sufficiently stable to induce antibody formation. Standard covalent coupling procedures take advantage of available reactive moieties on the protein such as α-amino groups, sulfhydryl groups, phenolic hydroxyl groups, the imidazole group of histidine or the ϵ-amino group of lysine (Erlanger, 1980). The most common protein carriers are the serum albumins, hemocyanin, immunoglobulins, and ovalbumin. The serum albumins are soluble, while hemocyanins have a high molecular weight and are considered quite insoluble (Erlanger, 1980).

When coupling a nucleoside or nucleotide adduct the desired recognition site (the hapten), which is usually the chemical carcinogen bound to the base, should be distal from the site of protein attachment (Stollar, 1980). Frequent coupling sites are the 2' and 3' hydroxyls of ribose. Two common procedures are periodate oxidation of both 2' and 3' hydroxyls followed by condensation of the dialdehyde on lysine (Erlanger, 1980), and succinylation of a ribose 2' hydroxyl group with succinic anhydride and coupling of the resulting ester to protein with carbodiimide (Stollar, 1980). Coupling can also be achieved through a 5' phosphate by formation of phosphoramide conjugates through protein amino groups with water-soluble carbodiimides.

A deoxyribonucleoside lacking a 5' phosphate can be oxidized on the 5' hydroxyl to a carboxyl group and linked to a protein by water-soluble carbodiimide (Stollar, 1980).

For the production of polyclonal antisera in rabbits the ratio of hapten to protein carrier was first investigated by Landsteiner, who concluded that with serum albumin as carrier, 10 haptenic groups per protein molecule gave the optimum antigenic stimulus (Erlanger, 1980). Concentrations between 8 and 25 haptens per protein molecule have been reported by others to produce good antisera (Erlanger, 1980). Carcinogen-DNA adduct immunogens are no exception, and high-titer antiserum was first obtained against guanosin-(8-yl)-acetylaminofluorene conjugated to bovine serum albumin that contained 17–22 adduct haptens per molecule of protein (Poirier *et al.*, 1977). In general, 0.2–0.5 mg of adduct conjugated to bovine serum albumin can be injected once a week for 3–4 weeks either as bolus intramuscular injections and/or as a series of smaller intradermal injections (Stollar, 1980). Bleedings should begin one week after the last injection, and animals can be bled once a week for at least three months. If the titer starts to decline, intramuscular or intravenous boosting should restore it. Modified-DNA immunogens have produced a good antigenic stimulus with DNAs modified between 1% and 5% (1–5 adducts per 10^2 bases) mixed with an equal weight of methylated protein (Poirier, 1981). Each rabbit can be injected weekly with approximately 2 mg of the mixture (Poirier, 1984; Poirier *et al.*, 1980). The weekly injections should be stopped after 3 weeks because additional injections may elicit antisera specific for DNA (Poirier, unpublished). Similarly, boosting of animals injected with a modified-DNA immunogen may not be successful because it produces anti-DNA antibodies. A native DNA immunogen is less likely to produce anti-DNA antisera than denatured DNA. However, antiserum elicited against modified native DNA may still preferentially recognize modified denatured DNA (Fig. 3, radioimmunoassay; Poirier *et al.*, 1980).

For the production of monoclonal antibodies similar principles apply as in the case of polyclonals, but the details vary. Mice have been successfully immunized with 0.1–0.2 mg of adduct on days 1 and 21 and sacrificed on day 26 (Stollar, 1980). Usually injections are subcutaneous or intraperitoneal (Hertzog *et al.*, 1983). If only small quantities of immunogen are available, 20 μg of immunogen can be absorbed onto aluminum hydroxide and injected intradermally in Freunds adjuvant (Strickland and Boyle, 1984). If the titer is insufficient, boosters can be administered at 3–4 week intervals (Stollar, 1980; Hertzog *et al.*, 1983). Particular success in obtaining highly avid rat monoclonals has been achieved using keyhole limpet hemocyanin as protein carrier and absorbing it onto aluminum hydroxide before injection at several intradermal sites (Müller and Rajewsky, 1981). Selecting and screening of clones can be accomplished according to established procedures, as described

1) RIA

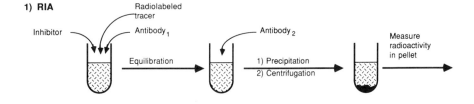

2) ELISA

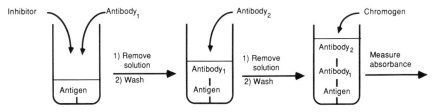

Figure 1. DNA-adduct determination by competitive immunoassay. For RIA, the radiolabeled tracer is more likely to be an adduct than a modified DNA; the inhibitor can be a standard, chemically synthesized adduct for DNA, as well as digested or undigested biological sample; antibody 1 is adduct- or modified-DNA-specific and antibody 2 can be anti-immunolglobulin or any other reagent that will precipitate the antigen-antibody complex. For ELISA, the antigen coated on the microtiter well can be either adduct or modified DNA; the inhibitor and antibody 1 are the same as for RIA; antibody 2 is an antiimmunoglobulin conjugated to an enzyme; the chromogen is a substrate for conjugated enzyme which is cleaved to a product measurable by spectrophotometry, fluorometry, radioactivity, or some other method of detection.

in Strickland and Boyle (1984). In practical terms, both polyclonals and monoclonals can be obtained with sufficient avidity to measure carcinogen and modification in biological samples. Polyclonals require less work, but monoclonals can be elicited with smaller amounts of immunogen, and once a clone is established the supply of antibody is virtually unlimited.

3. COMPETITIVE IMMUNOASSAYS

This section will describe radioimmunoassay (RIA) and enzyme-linked immunosorbent assay (ELISA) as prototypes for many immunoassay variations (Fig. 1). For clarity only one version of each type of assay will be discussed; however, assays that can be performed vary with the end point detected or the method of execution. The basic principles and problems remain the same in spite of the different variations. Both assays have in common the isolation of an antigen–antibody complex, where the antibody is bound either to an antigen on the bottom of a microtiter well (ELISA) or to a highly radioactively labeled antigen (RIA). Direct binding of antigen to antibody is useful in titering antiserum, but for the most consistent and sensitive results it is necessary to use these assays in a competititve mode (Fig. 1). That is, increasing concentrations of unlabeled (RIA) or unbound (ELISA) competitor are used to compete for antibody binding to the radiolabeled tracer (RIA) or immunogen binding to the bottom of the microtiter well (ELISA). In both assays the competitor is either a standard immunogen, used in increasing concentrations to construct a standard curve, or an unknown sample, which is quantitated by comparison with the known standard. The following sections will describe each assay separately.

Competitive RIA (Mayer and Walker, 1980; Butler, 1980; Zettner, 1973; Zettner and Duly, 1974) involves a concentration-dependent competition for an antibody haptenic site by two chemically identical immunogens (Fig. 1), one of which is highly radiolabeled and designated the ''tracer.'' The tracer is a reporter molecule used in a constant quantity in every tube of the assay. The RIA sensitivity will depend, in part, on the specific activity of the tracer, which should be above 10 Ci/mmol for determination of carcinogen–DNA adducts in biological samples. The isotope can be ^{3}H, ^{125}I, ^{131}I or ^{32}P, but the unstable radionuclides are generally less convenient than ^{3}H. The nonradioactive hapten or competitor can be either a standard immunogen or biological sample and is designated the ''inhibitor.'' It is used in increasing concentrations (in different tubes) to compete for antibody (antibody 1, Fig. 1) and inhibit the antibody–tracer binding. The reaction of tracer and/ or inhibitor with antibody is usually allowed to reach equilibrium, for example, at 37°C for 90 min or overnight at 4°C. The antigen–antibody complex is then separated from the whole mixture, and the radioactivity bound to antibody is measured. Separation is usually achieved with second antibody (anti-IgG, Fig. 1, antibody 2), high salt (NH_4SO_4), polyethylene glycol, protein A, or filters (Mayer and Walker, 1980; Butler, 1980). In the absence of inhibitor, a maximum amount of tracer will be found in the antigen–antibody complex, and this is termed B_0. As increasing concentrations of inhibitor are added, less and less radioactive tracer will become bound to the antibody because of inhibitor competition for the haptenic site. The degree of competition for antibody–tracer binding exhibited by each concentration of inhibitor can be expressed as a percentage (%Inhibition) of the ratio of the reduction in radioactivity at a particular inhibitor concentration and the radioactivity with no inhibitor (B_0):

$$\%\text{Inhibition} = \frac{B_0 - X_c}{B_0} \times 100$$

where X_c = radioactivity bound at inhibitor concentration c.

If the %Inhibition is plotted on a linear scale as a function of the inhibitor concentration on a logarithmic scale, the resulting curve will be linear in the range of 20–80% Inhibition. Each assay should contain a standard curve that is constructed with increasing concentrations of immunogen and consistent ($\pm$ 5–7%) for assays performed on different days. Unknown sample quantitation is achieved by comparison of the %Inhibition found from a biological sample with the inhibitor concentration at the same %Inhibition on a linear part of the standard curve. The adduct quantity is assumed to be the same in both at a given %Inhibition. This is a simple procedure for calculating RIA data but many other approaches can be used (Mayer and Walker, 1980; Zettner, 1973).

In general, RIA is a reliable and consistent assay that can be performed easily and inexpensively. There are several aspects of the assay that influence sensitivity (Müller and Rajewsky, 1981; Mayer and Walker, 1980; Butler, 1980; Zettner, 1973), and these include the antibody affinity, the specific activity of the tracer, the antibody dilution, the use of non-equilibrium incubation conditions (Zettner and Duly, 1974), the physical state of the biological sample DNA (native, denatured, or digested), and the amount of the sample that can be assayed without altering the standard curve. Sufficient sensitivity to assay for human DNA adducts generally requires antisera with affinity constants in the range of 10^8–10^9 L/mol and RIA tracers with specific activity $\geq$ 10 Ci/mmol (Müller and Rajewsky, 1981; Munns and Liszewski, 1980). The greater the dilution of the antiserum, the more sensitive the assay is likely to be. A common dilution range for rabbit antiserum is 1:500 to 1:2500. When titering the antiserum, B_0 will decrease linearly almost to background levels across a series of dilutions. The dilution chosen should be on the linear portion of the dilution curve and should produce tracer–antibody binding levels that are at least 4-fold above background, that is, a signal-to-noise ratio of 4:1. As described in the previous paragraph, the RIA is usually allowed to reach equilibrium for 90 min or longer before separation of the antigen–antibody complex (Zettner, 1973). However, an increase in sensitivity can be obtained using sequential saturation or non-equilibrium conditions (Zettner and Duly, 1974). Sequential saturation involves a brief tracer incubation immediately prior to separation of the antigen–antibody complex. In measuring DNA samples from biological sources, the assay sensitivity will also depend on the amount of DNA that can be introduced into the assay without alteration of the standard curve. An antiserum elicited against a DNA adduct recognizes adducts better than modified DNAs, therefore the DNA must be digested in order to measure all of the adduct present (Fig. 2; Poirier, 1981; Poirier and Connor, 1982). RIA tubes can often contain as much as 30 µg hydrolyzed DNA and accompanying hydrolytic enzymes without significantly affecting the assay. It is possible to remove the enzymes by boiling and centrifugation, although some adducts may be thermally labile.

The basic principles of competitive ELISA (Mayer and Walker, 1980;

Butler, 1980; Wisdom, 1976; Engvall, 1980; Klotz, 1982; Trivers *et al.*, 1983; Harris *et al.*, 1982) are similar to those of RIA, but the constant component, the tracer, has been replaced by an immunogen bound to the bottom of microtiter plate wells (Fig. 1, antigen). The immunogen coated on the wells can be either a modified DNA or an adduct coupled to a protein (albeit a different protein than that used for immunization) (Trivers *et al.*, 1983; Harris *et al.*, 1982). Control wells are coated either with unmodified DNA alone or the carrier protein alone. Before adding serum, coated microtiter plates are blocked with protein (animal serum or casein) to cover any unoccupied sites on the plastic that might bind the antibody nonspecifically (Trivers *et al.*, 1983; Harris *et al.*, 1982). Then, in the competition step, adduct- or modified DNA-specific antiserum (diluted 1:10,000–1:200,000) is incubated with increasing amounts of inhibitor in the microtiter wells, and

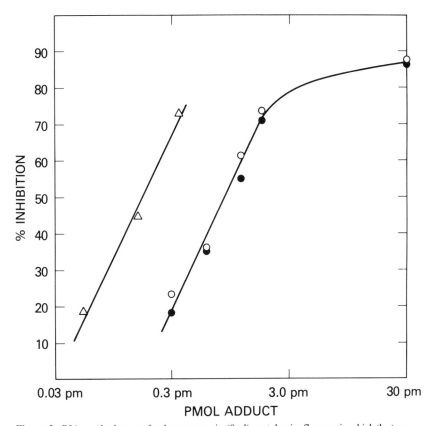

Figure 2. RIA standard curves for deoxyguanosin-(8-yl)-acetylaminofluorene in which the tracer is tritiated adduct, the antiserum was elicited against the protein-bound adduct (Poirier *et al.*, 1977), and the inhibitors are native (●), and denatured (○), and S₁ nuclease-hydrolyzed (△) claf thymus DNA modified to 1% (30 pmol/μg DNA) with *N*-acetoxy-acetylaminofluorene.

TABLE 12–1. BPDE–DNA Adduct (fmol Adduct/μg DNA), Determined by Radiolabeling and Immunoassay, in DNA from Primary Mouse Keratinocytes Exposed to ³H-BP*

Radiolabeling	*p*-nitro-phenyl phosphate ELISA	*m*-umbelliferyl-phosphate ELISA	USERIA
5.6	5.12 ± 1.33	6.2 ± 2.2	5.54 ± 1.38

*Immunoassay quantitation was accomplished by comparison with a standard BPDE–DNA modified to 4.5 fmol adduct/μg DNA.
Source: Santella *et al.*, 1988a.

at equilibrium the unbound mixture is removed by washing. The next step is incubation with an enzyme-conjugated second antibody that binds to the first antibody attached inside the wells (Fig. 1, antibody 2). An enzyme commonly used is alkaline-phosphatase (as in Table 1), which can cleave phosphorylated substrates to colorimetric (e.g., *p*-nitrophenol; Harris *et al.*, 1982) or fluorometric (e.g., methyl-umbelliferone; Kriek, *et al.*, 1984) end points (Fig. 1, chromogen). The product will be present in a quantity directly proportional to the original amount of antibody bound to the microtiter well. Variations on this assay include the use of different enzyme-substrate combinations (Kabakoff, 1980), a biotin-avidin step, which amplifies the signal (Roberts *et al.*, 1986), or the use of a radiolabeled substrate, as in the ultrasensitive enzyme RIA (USERIA; Trivers *et al.*, 1983; Harris *et al.*, 1982). Like the RIA, ELISA is an inverse assay; the wells with the lowest concentration of inhibitor have the highest amount of antibody bound and the most intense end point. Data can be processed in a fashion similar to that described for the RIA.

Automatic procedures for washing and reading solutions in microtiter plate wells and the lack of necessity for a radiolabeled tracer have made ELISA somewhat more convenient than RIA. In addition, ELISA can be more sensitive than RIA (Fig. 3). However, these advantages are often offset by greater intra- and interassay variability. In addition, this highly sensitive assay is executed in several steps; therefore troubleshooting can be difficult. Like RIA, ELISA sensitivity for the measurement of DNA adducts or modified-DNA samples is limited by the quantity of DNA that can be assayed in an individual microtiter well. In many cases 50 μg of native or denatured DNA can be added to each well without altering the standard curve, but this quantity may affect the intensity of color development (Perera *et al.*, 1982). Therefore it is important to use the same amount of DNA in all wells of an assay, including the standard curve, to eliminate this source of variability. The amount of carrier DNA that can be tolerated in the assay should be determined for each new antiserum.

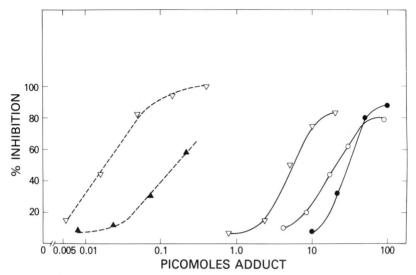

Figure 3. RIA (continuous line) (Poirier *et al.*, 1980) and ELISA (dashed line) (Santella *et al.*, 1988a) curves obtained with anti-BPDE-DNA. For RIA the tracer was tritiated BPdG[2], and for ELISA, wells were coated with denatured BPDE-DNA (the original immunogen) modified to 1% (30 pmol/μg). Competition curves for RIA are the immunogen DNA as native (○) and denatured (▽), and the individual BPdG adduct (●). Competition curves for ELISA are the denatured immunogen ▽) and a denatured, standard BPDE–DNA modified at 4.5 fmol/μg DNA (▲) (Santella *et al.*, 1988a).

4. CHARACTERIZATION OF ANTISERA

Extensive characterization of antisera is extremely important for all immunoassays. An antibody is a tool, and it is essential to know what that tool recognizes in order to interpret the experimental results. When first titering a new serum, it is customary to use a direct assay in which varying concentrations of antiserum from each bleed are incubated with excess antigen. Rabbit antisera titering should start at 1:100 for RIA and 1:1000 for ELISA. Over a range of dilutions a linear decrease in bound tracer (RIA) or end-point signal (ELISA) will be observed with decreasing antibody. A dilution in the middle of this linear range should be chosen for further competitive assays (Zettner, 1973; Wisdom, 1976). The original immunogen can then be competed over a wide range of concentrations to determine how much is required to inhibit either tracer–antibody binding (RIA) or microtiter well–antibody binding (ELISA). The most sensitive assay conditions can be achieved by decreasing the antiserum concentration and the constant immunogen (either the tracer for RIA or amount coated on microtiter plate wells for ELISA) to give the best competition with unlabeled (RIA) or unbound (ELISA) standard immunogen. Once a consistent standard curve is established, antibody specificity for other adducts or other modified-DNA samples can be determined by competitive inhibition in the same assay.

Cross-reactivity for unmodified DNA, normal DNA bases, and the carcinogen alone should be thoroughly investigated. In the case of a chemical–DNA adduct immunogen there will often be cross-reactivity for other related chemical–DNA adducts, including different adducts of the same carcinogen (Rajewsky *et al.*, 1980; Groopman *et al.*, 1984) and carcinogen–DNA adducts formed by compounds of the same chemical class (Poirier, 1981; Santella, 1988). Antisera elicited against a specific adduct often recognize the same adduct in modified DNA to a lesser extent, and that recognition may change if the DNA is native or denatured (Poirier, 1981; Poirier and Connor, 1982). In the case of a modified-DNA immunogen, cross-reactivity with individual adducts may be (Fig. 3, ●) but is not always seen (Reed *et al.*, 1991). Recognition of DNA samples modified with compounds of the same chemical class is frequent (Fig. 4; Poirier, 1982; Weston *et al.*, 1989a; Lippard *et al.*, 1983). In addition, the extent of modification may influence the antibody recognition (Fig. 3, ELISA). To be sufficiently immunogenic, a modified DNA must have adducts in the range of 1 adduct in 10^2 bases, and this type of DNA sample can be expected to be conformationally different than a biological sample modified in the range of 1 adduct in 10^6 bases. An antiserum elicited against a highly modified immunogen may not recognize all of the adducts in a DNA modified at a significantly lower level (van Schooten *et al.*, 1987; Santella *et al.*, 1988a), and the resulting underestimation will yield inaccurate adduct values, as discussed below.

5. QUANTITATION OF DNA ADDUCTS IN BIOLOGICAL SAMPLES

Whatever the source of biological sample (cultured cells, animal tissues, or human material), isolation of DNA is a necessary prerequisite for DNA-adduct analysis. The three types of DNA preparation used most frequently for these studies are CsCl buoyant density gradients (Flamm *et al.*, 1967), chloroform and phenol extraction with RNAse incubation (Bohr *et al.*, 1985), and proteinase K digestion followed by high salt extraction and ethanol precipitation (Miller *et al.*, 1988). In theory, any DNA extraction procedure that does not chemically alter the adducts and that gives RNA-free DNA should be appropriate. Since many adduct antisera recognize RNA adducts it is important to remove the RNA for exclusive measurement of chemical-DNA adducts (Poirier, 1981). The quantities of DNA required for immunoassays depend on the degree of modification of the DNA and the sensitivity of the specific assay. For cultured cells and animal tissues 50–100 μg should be enough to assay in duplicate or triplicate on two or more occasions. This amount of DNA can often be obtained from 10^8 cells or 100–200 mg of tissue. For human samples a minimum of 450 μg would be required to assay 50 μg of DNA per well in ELISAs on two separate occasions. This can readily be obtained from 40 ml of blood or 500 mg of human tissue.

Most antisera elicited against individual adducts do not recognize modified-DNA samples equally well (Poirier, 1981; Fig. 2). Therefore biological DNA samples generally require digestion prior to assay (Poirier and Connor, 1982). A chromatography step, inserted before the immunoassay, may remove the hydrolytic enzymes and at the same time separate the adducts for individual determination by immunoassay (Rajewsky *et al.*, 1980; Plooy *et al.*, 1985). If a biological DNA sample is likely to contain many adducts that cross-react with the same antibody, this may be the only way to obtain adduct specificity. It has the added advantage of increasing sensitivity since the amount of DNA that can be measured in one sample is virtually unlimited.

When using an antiserum elicited against a modified DNA, individual adducts are recognized less well than undigested DNA (Fig. 3, RIA), and therefore enzymatic hydrolysis of the biological samples is not necessary (Poirier, 1981; Poirier *et al.*, 1980; Santella *et al.*, 1988a). Other aspects of the immunoassay that require attention pertain to antibody recognition of native and denatured DNA (Fig. 3, RIA; Poirier, 1981; Santella, 1988; Poirier *et al.*, 1980) and the degree to which the antiserum recognizes DNA samples modified at levels lower than the original immunogen (Fig. 3, ELISA; van Schooten *et al.*, 1987; Santella *et al.*, 1988a). Competition with native and denatured immunogen DNAs should indicate whether or not the antibody recognizes one or the other preferentially. This appears to be a property of the antibody rather than the original immunogen. For example, some antisera elicited against native BPDE-DNA only recognize all of the adducts in denatured BPDE-DNA samples (Poirier *et al.*, 1980; Santella *et al.*, 1988a), while an antiserum against cisplatin-DNA recognizes native and denatured DNA samples equally (Fichtinger-Schepman *et al.*, 1985; Poirier, unpublished). BPDE-DNA is DNA modified with BPDE I [7β,8α-dihydroxy-9α,10α-epoxy-7,8,9,10-tetrahydrobenzo(a)pyrene-(anti-isomer)] such that the only adduct is *trans*-(7R)-N^2-{10(7β,8α,9α-trihydroxy-7,8,9,10-tetrahydrobenzo[a]pyrene)-yl}-deoxyguanosine. To reiterate, a modified-DNA antiserum elicited against a highly modified immunogen DNA may not efficiently recognize adducts in DNA modified to a lower extent (Fig. 3, ELISA). This can be investigated by assaying a series of DNA samples modified to varying extents in a competitive immunoassay. In addition, the total adducts in a biological sample DNA can often be determined by an independent method, for example radiolabeling, and the results compared with those obtained by immunoassay (Table 1; Santella, 1988; van Schooten *et al.*, 1987; Santella *et al.*, 1988a, 1988b). If the difference in antibody recognition between high and low modified DNAs is in the range of 100–

1000-fold, the assay may not be useful for the quantitative monitoring of biological samples. On the other hand, a discrepancy of several-fold might be overcome by comparison of the unknown biological samples with a DNA standard modified to a similar extent (Table 1; van Schooten *et al.*, 1987; Santella *et al.*, 1988a).

An antiserum that cross-reacts with adducts or adducted DNAs formed by compounds of the same chemical class as the original immunogen may be used to estimate adducts in biological samples provided that the limitations of the data are acknowledged. For example, the BPDE-DNA antiserum (Poirier *et al.*, 1980) has been shown to cross-react with DNAs modified by a variety of polycyclic aromatic hydrocarbons (PAHs) (Weston *et al.*, 1989a). Therefore, it is likely that inhibition exhibited by human samples in the BPDE–DNA ELISA is due to the presence of a variety of adducts. Since it is not feasible to determine these adducts individually, a provisional adduct estimation has been achieved by comparison of sample %Inhibition values with a BPDE–DNA standard curve. The results have been termed ''BP–DNA antigenicity'' or ''PAH–DNA adducts'' (Liou *et al.*, 1989).

To ensure adequate reproducibility in the ELISA, triplicate experimental wells and one control well should be analyzed simultaneously on each microtiter plate. In contrast, duplicate RIA tubes generally yield consistent results. Each sample should be assayed on two separate occasions (different RIAs or different microtiter plates), and can be considered positive if the %Inhibition observed with sample DNA is on the linear portion of the standard curve.

6. INTERLABORATORY COMPARISON OF ASSAYS

Immunoassays have been used with increasing frequency for the determination of DNA adducts in human samples (Bartsch *et al.*, 1988), and it has become apparent that the results vary from laboratory to laboratory. In a collaborative study a standard assay was established in which the same reagents were used by each laboratory in order to eliminate these discrepancies (Santella *et al.*, 1988a; Fig. 3, ELISA). In retrospect, it may be unnecessary to use exactly the same reagents in order to obtain comparable results, but certain sources of variability are worthy of note.

In establishing an immunoassay, the amount of standard adduct either alone or present in a given DNA sample is generally determined by UV absorbance (Jeffrey *et al.*, 1977; Poirier, 1980), radioactivity (Santella, 1988), or atomic absorbance spectroscopy (Reed *et al.*, 1988). These determinations may vary for the same type of adduct or modified DNA, and may result in large interlaboratory discrepancies in the standard curve. At the very least, the details of each method should be documented so that reasonable comparisons can be made.

Once an assay is developed and a standard curve is established, it is important to determine whether or not all of the adducts in a biological sample are being measured. One of the easiest ways to investigate this is often to expose cultured cells to a radioactive form of the chemical in question and determine adduct formation by radiochemical analysis (Table 1; Santella *et al.*, 1988a, 1988b; Poirier, 1980). Large differences between adducts measured by immunoassay and by radioactivity may be due to several factors, which will be discussed in the following paragraphs.

As mentioned previously, if the immunogen was a highly modified DNA (e.g., 1 adduct in 10^2 bases) the antiserum may not recognize all of the adducts in a biologically-modified-DNA sample with adduct levels in the range of 1 adduct in 10^6 bases (Fig. 3, ELISA). Complete recognition has been demonstrated for DNA samples modified to high and low extents with an antiserum elicited against 8-methoxypsoralen–DNA (Santella *et al.*, 1988b) and aflatoxin–DNA (Hsieh *et al.*, 1988). However, it is possible for the discrepancy to be several-hundred-fold (Poirier *et al.*, 1988 and unpublished), in which case the antiserum may not be useful for quantitative adduct determination. If such a discrepancy is only several-fold, as in the case of anti-BPDE–DNA, accurate determination of adducts in a biological sample may be achieved by comparison of biological samples with a standard DNA modified in the same range as the samples (van Schooten *et al.*, 1987; Santella *et al.*, 1988a).

It is also possible that the problem of adduct underestimation by an immunoassay may have a completely different source, for example, the assay end point. The ELISA for measurement of BPDE–DNA has been performed with an alkaline-phosphatase-conjugated-IgG and either *p*-nitrophenyl-phosphate or methyl-umbelliferyl-phosphate as the substrate (Santella *et al.*, 1988a). The latter end point, which uses fluorescence detection, has a greater discrepancy between the measurement of the high modified standard DNA and the biological samples than the colorimetric assay, which measures *p*-nitrophenol. Accurate sample determination can be achieved, however, by using a standard modified in the range of the biological samples (Table 1). Based on these experiences, the recommendation is to test anti-modified-DNA ELISAs against a radioactively labeled biological sample to determine whether or not adduct determination is complete or underestimated with a given assay.

Underestimation of adducts may also be related to antigen presentation in the assay. For example, an anti-modified-DNA antiserum may recognize adducts on denatured DNA better than those on native DNA (Fig. 3, RIA; Poirier *et al.*, 1980), and an anti-adduct antiserum may recognize individual adducts better than denatured modified DNA and native modified DNA (Fig. 2; Poirier and Connor, 1982; Poirier, 1980).

2. COMBINATION OF IMMUNOLOGICAL AND CHEMICAL TECHNIQUES FOR DETERMINING DNA ADDUCTS IN HUMAN SAMPLES

By comparison with other techniques that measure carcinogen-DNA adducts, immunoassays have their own particular strengths and weaknesses that are based primarily on the antisera specificities. Since the antisera often recognize adducts formed by many compounds within the same chemical class (Fig. 4), and since humans are exposed to many different chemicals, it is not possible to determine specific adducts individually without further purification of the sample. In the case of the hydrocarbons, many studies have been performed in which samples have been compared to a BPDE–DNA standard, and the results termed BP–DNA antigenicity [BP denotes benzo(a)pyrene] or PAH–DNA adducts, to reflect the presence of a mixture of hydrocarbon adducts. Ethyl and methyl adducts of deoxyguanosine and deoxyadenosine have been determined by immunoassay in human DNA samples after hydrolysis and high-pressure liquid chromatography (HPLC) separation of the adducts (Umbenhauer *et al.*, 1985). Besides allowing the determination of individual adducts, a prior HPLC separation increases the assay sensitivity since adducts from virtually unlimited (mg) quantities of DNA can be assayed.

The combination of HPLC adduct purification and immunoassay is most useful for simple adduct mixtures. When mixtures of adducts, for example from PAHs (Tierney *et al.*, 1986) or aflatoxins (Groopman *et al.*, 1985),

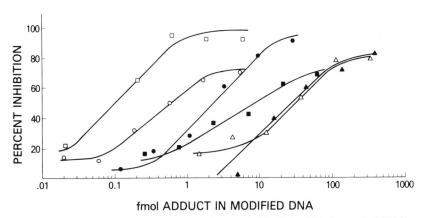

Figure 4. ELISA for BPDE–DNA (same conditions as in Fig. 3) in which the standard BPDE–DNA (1% modified) inhibitor is competed as denatured (●). The other curves are denatured DNAs, modified in the range of 0.1%–1.0%, with the diol-epoxides of chrysene (□), benzo[k]fluoroanthene (△), dibenz[a,c]anthracene (▲) and the bay region (○) and nonbay region (■) benz[a,c]anthracenes. This ELISA demonstrated that the anti-BPDE–DNA has specificity for all of the above modified DNAs.

are likely to be present in a human sample, determination of a specific adduct can be achieved by combining the antibody class specificity with other preparative and analytical techniques. Immunocolumns can be made by binding carcinogen–DNA adduct antisera to a solid matrix (Tierney *et al.*, 1986; Manchester *et al.*, 1988; Weston *et al.*, 1989a). Enzyme-hydrolyzed adducted DNA will attach to the column while unmodified DNA passes through and can later be quantitatively released (Manchester *et al.*, 1988; Weston and Poirier, unpublished). This method will serve to concentrate a specific class of adducts. The DNA samples collected in the column eluates can be subjected to further degradation, HPLC, and final identification. In order to combine immunoaffinity chromatography with highly sensitive adduct detection by ^{32}P-postlabeling (Manchester *et al.*, 1989), further method development is required for analytical specificity. Analytical techniques that could be employed at this point include fluorescence spectrophotometry (Weston *et al.*, 1989a; Manchester *et al.*, 1988), gas chromatography–mass spectrometry (Weston *et al.*, 1989b; Manchester *et al.*, 1988), or electrochemical detection (Adams *et al.*, 1986). For example, a combination of immunoaffinity chromatography and HPLC have been used to measure aflatoxin-bound guanine in the urine of Chinese individuals exposed to dietary aflatoxin (Sun *et al.*, 1988; Groopman *et al.*, 1985). In addition, a series of studies combining these techniques has been performed to characterize chemically the deoxyguanosine adduct of benzo(a)pyrene in DNA from human placenta containing a mixture of hydrocarbon adducts. Immunoassays were combined with HPLC and second-derivative synchronous scanning fluorescent spectrophotometry to obtain a chemical signal identical with the synthetic standard (Weston *et al.*, 1989a). The adducts obtained were further analyzed by ^{32}P postlabeling (Manchester *et al.*, 1989) and identified by gas chromatography–mass spectrometry (Manchester *et al.*, 1988). Although all of these steps could not feasibly be employed for routine sample testing, this type of approach is useful to assist validation of the different techniques available for human biomonitoring, and may provide essential chemical characterization of adducts found in human tissues.

8. CONCLUSIONS

Antisera specific for chemical–DNA adducts and chemically modified DNA have proven to be useful tools in the investigation of mechanisms of chemical carcinogenesis. However, the assays involved are complicated and require care; the data they yield will only be as good as the validation upon which they are established. As discussed here, the process of characterizing a new antiserum should carefully document many things, including cross-reactivity and the extent to which adducts are recognized in a biological sample. Extension of the technology described here into the realm of human biomonitoring may provide valuable information on human exposure, human cancer epidemiology, and risk assessment.

REFERENCES

H. Bartsch, K. Hemminki, and I. K. O'Neill (eds.), *Methods for Detecting DNA Damaging Agents in Humans: Applications in Cancer Epidemiology and Prevention,* International Agency for Research on Cancer Scientific Publication No. 89, IARC, Lyon France (1988).

V. A. Bohr, C. A. Smith, D. S. Okumoto, and P. C. Hanawalt, DNA repair in an active gene: Removal of pyrimidine dimers from the DHFR gene of CHO cells is much more efficient than in the genome overall, *Cell* **40,** 359–369 (1985).

J. E. Butler, Antibody-antigen and antibody-hapten reactions. In: *Enzyme-Immunoassay,* ed. E. T. Maggio, CRC Press, Boca Raton, FL pp. 6–52 (1980).

E. Engvall, Enzyme immunoassay ELISA and EMIT, *Methods Enzymol.* **70,** 419–439 (1980).

B. F. Erlanger, The preparation of antigenic hapten-carrier conjugates: A survey, *Methods Enzymol.* **70,** 85–104 (1980).

A. M. J. Fichtinger-Schepman, R. A. Baan, A. Luiten-Schuite, M. van Kijk, and P. H. M. Lohman, Immunochemical quantitation of adducts induced in DNA by *cis*-diamminedichloroplatinum (II) and analysis of adduct-related DNA-unwinding, *Chem. Biol. Interact.* **55,** 275–288 (1985).

W. G. Flamm, M. L. Birnstiel, and P. M. B. Walker, Preparation and fractionation, and isolation of single strands of DNA by isopycnic ultracentrifugation in fixed-angle rotors. In: *Subcellular Compounds: Preparation and Fractionation,* eds. G. Birnie and S. M. Fox, Butterworth, London, pp. 129–155 (1967).

J. D. Groopman, P. R. Donahyue, J. Q. Zhu, J. S. Chen, and G. N. Wogan, Aflatoxin metabolism in humans: detection of metabolites and nucleic acid adducts in urine by affinity chromatography, *Proc. Natl. Acad. Sci USA* **82,** 6492–6496 (1985).

J. D. Groopman, L. J. Trudel, P. R. Donahue, A. Marshak-Rothstein, and G. N. Wogan, High-affinity monoclonal antibodies for aflatoxins and their application to solid-phase immunoassays, *Proc. Natl. Acad. Sci. USA* **81,** 7728–7731 (1984).

C. C. Harris, R. H. Yolken, and I.-C. Hsu, Enzyme immunoassays: Applications in cancer research, *Methods Cancer Res.* **20,** 213–243 (1982).

P. J. Hertzog, A. Shaw, J. R. Lindsay Smith, and R. C. Garner, Improved conditions for the production of monoclonal antibodies to carcinogen-modified DNA, for use in enzyme-linked immunosorbent assays (ELISA), *J. Immunol. Methods* **62,** 49–58 (1983).

L.-L. Hsieh, S.-W. Hsu, D.-S. Chen, and R. M. Santella, Immunological detection of aflatoxin B$_1$-DNA adducts formed *in vivo, Cancer Res.* **48,** 6328–6331 (1988).

A. M. Jeffrey, I. B. Weinstein, K. W. Jennette, K. Grzeskowiak, K. Nakanishi, R. G. Harvey, H. Autrup, and C. Harris, Structures of benzo(a)pyrene-nucleic acid adducts formed in human and bovine bronchial explants, *Nature* **269,** 348–349 (1977).

D. S. Kabakoff, Chemical aspects of enzyme-immunoassay. In: *Enzyme-Immunoassay,* ed. E. T. Maggio, CRC Press, Boca Raton, FL, pp. 71–104 (1980).

J. L. Klotz, Enzyme-linked immunosorbent assay for antibodies to native and denatured DNA, *Methods Enzymol.* **84,** 194–201 (1982).

E. Kriek, L. Den Engelse, E. Scherer, and J. G. Westra, Formation of DNA modifications by chemical carcinogens identification, localization and quantification, *Biochim. Biophys. Acta* **738,** 181–201 (1984).

M. Leng, Immunological detection of lesions in DNA, *Biochim.* **67,** 309–315 (1985).

S.-H. Liou, D. Jacobson-Kram, M. C. Poirier, D. Nguyen, P. T. Strickland, and M. S. Tockman, Biological monitoring of firefighters: Sister chromatid exchange and polycyclic aromatic hydrocarbon-DNA adducts in peripheral blood cells, *Cancer Res.,* **49,** 4929–4935 (1989).

S. J. Lippard, H. M. Ushay, C. M. Merkel, and M. C. Poirier, Use of antibodies to probe the stereochemistry of antitumor platinum drug binding to deoxyribonucleic acid, *Biochem.* **22,** 5165–5168 (1983).

D. K. Manchester, A. Weston, J.-S. Choi, G. E. Trivers, P. V. Fennessey, E. Quintana, P. B. Farmer, D. L. Mann, and C. C. Harris, Detection of benzo[a]pyrene diol epoxide-DNA adducts in human placenta, *Proc. Natl. Acad. Sci. USA* **85,** 9243–9247 (1988).

D. K. Manchester, V. L. Wilson, I. C. Hsu, J. S. Choi, N. B. Parker, D. L. Mann, A. Weston, and C. C. Harris, Synchronous fluorescent spectroscopic, immunoaffinity chromatographic and ^{32}P-postlabeling analysis of human placental DNA known to contain benzo[a]pyrene diol epoxide adducts, *Carcinogenesis,* **11,** 553–559, 1990.

R. J. Mayer and J. H. Walker, Uses of antisera. In: *Immunochemical Methods in the Biological Sciences: Enzymes and Proteins,* Academic Press, New York (1980).

S. A. Miller, D. D. Dykes, and H. F. Polesky, A simple salting out procedure for extracting DNA from human nucleated cells, *Nucleic Acids Res.* **16,** 1215 (1988).

R. Müller and M. F. Rajewsky, Antibodies specific for DNA components structurally modified by chemical carcinogens, *J. Cancer Res. Clin. Oncol.* **102,** 99–113 (1981).

T. W. Munns and M. K. Liszewski, Antibodies specific for modified nucleosides: An immunochemical approach for the isolation and characterization of nucleic acids, *Prog. Nucleic Acid Res. Mol. Biol.* **24,** 109–165 (1980).

J. Adams, M. David and R. W. Giese, Pentafluorobenzylation of O^4-Ethylthymidine and Analogues by Phase Transfer catalysis for determination by Gas Chromatography with electron capture detection, *Anal. Chem.,* **58,** 345–348, (1986).

F. P. Perera, M. C. Poirier, S. H. Yuspa, J. Nakayama, A. Jaretzki, M. M. Curnen, D. M. Knowles, and I. B. Weinstein, A pilot project in molecular cancer epidemiology: Determination of benzo[a]pyrene-DNA adducts in animal and human tissues by immunoassays, *Carcinog.* **3,** 1405–1410 (1982).

D. H. Phillips, Modern methods of DNA adduct determination. In: *Handbook of Experimental Pharmacology,* Vol. 94/I eds. P. L. Glover and C. S. Cooper, Springer-Verlag, Berlin (1989), pp. 503–546.

O. J. Plescia and W. Braun, Nucleic acids as antigens, *Adv. Immunol.* **6,** 231–252 (1967).

A. C. M. Plooy, A. M. J. Fichtinger-Schepman, H. H. Schutte, M. van Kijk, and P. H. M. Lohman, The quantitative detection of various Pt-DNA-adducts in Chinese hamster ovary cells treated with cisplatin: application of immunochemical techniques, *Carcinog.* **6**, 561–566 (1985).

M. C. Poirier, Measurement of formation and removal of adducts of N-acetoxy-2-acetylaminofluorene. In: *DNA Repair,* Vol. 1, Part A, eds. E. C. Friedberg and P. C. Hanawalt, Dekker, New York, pp. 143–153 (1980).

M. C. Poirier, Antibodies to carcinogen-DNA adducts, *J. Natl. Cancer Inst.* **67**, 515–519 (1981).

M. C. Poirier, The use of carcinogen-DNA adduct antisera for quantitation and localization of genomic damage in animal models and the human population, *Environ. Mutagen.* **6**, 879–887 (1982).

M. C. Poirier, and R. J. Connor, Radioimmunoassay for 2-acetylaminofluorene-DNA adducts, *Methods Enzymol.* **84**, 607–618 (1982).

M. C. Poirier, M. J. Egorin, A. M. J. Fichtinger-Schepman, and E. Reed, DNA adducts of cisplatin and carboplatin in tissues of human cancer patients. in: *Methods for Detecting Damaging Agents in Humans: Applications in Cancer Epidemiology and Prevention,* eds. H. Bartsch, K. Hemminki, and I. K. O'Neill, International Agency for Research on Cancer Scientific Publication No. 89, Lyon, France, pp. 313–320 (1988).

M. C. Poirier, R. Santella, I. B. Weinstein, D. Grunberger, and S. H. Yuspa, Quantitation of benzo(a)pyrene-deoxyguanosine adducts by radioimmunoassay, *Cancer Res.* **40**, 412–416 (1980).

M. C. Poirier, S. H. Yuspa, I. B. Weinstein, and S. Blobstein, Detection of carcinogen-DNA adducts by radioimmunoassay, *Nature* **270**, 186–188 (1977).

M. F. Rajewsky, R. Müller, J. Adamkiewicz, and W. Drosdzick, Immunological detection and quantification of DNA components structurally modified by alkylating carcinogens (ethylnitrosourea). In: *Carcinogenesis: Fundamental Mechanisms and Environmental Effects,* eds. B. Pullman, P. O. P. Ts'o, and H. Gelboin, Reidel, Dordrecht, pp. 207–218 (1980).

E. Reed, S. Sauerhoff, and M. C. Poirier, Quantitation of platinum-DNA binding after therapeutic levels of drug exposure—a novel use of graphite furnace spectrometry, *At. Spectrosc.* **9**, 93–95 (1988).

E. Reed, S. Gupta-Burt, C. L. Litterst and M. C. Poirier, Characterization of the DNA damage recognized by an antiserum elicited against *cis*-diammine-dichloro-platinum (II)-Modified DNA, *Carcinogenesis,* in press, 1991.

D. W. Roberts, R. W. Benson, T. J. Flammang, and F. F. Kadlubar, Development of an avidin-biotin amplified enzyme-linked immunoassay for detection of DNA adducts of the human bladder carcinogen 4-aminobiphenyl. In: *Mechanisms of DNA Damage and Repair: Implications for Carcinogenesis and Risk Assessment,* eds. M. G. Simic, L. Grossman, and A. D. Upton, Plenum Press, New York, pp. 479–488 (1986).

R. M. Santella, Application of new techniques for the detection of carcinogen adducts to human population monitoring, *Mutat. Res.* **205**, 271–282 (1988).

R. M. Santella, A. Watson, F. P. Perara, G. T. Trivers, C. C. Harris, T. L. Young, D. Nguyen, B. M. Lee, and M. C. Poirer, Interlaboratory comparison of antisera and immunoassays for benzor[a]pyrene-diol-epoxide-I-Modified DNA, *Carcinog.* **9**, 1265–1269 (1988a).

R. M. Santella, X. Y. Yang, V. A. DeLeo, and F. P. Gasparro, Detection and quantification of 8-methoxypsoralen-DNA adducts. In: *Methods for Detecting DNA Damaging Agents in Humans: Applications in Cancer Epidemiology and Prevention*, International Agency for Research on Cancer Scientific Publication No. 89, eds. H. Bartsch, K. Hemminki, and I. K. O'Neill, IARC, Lyon, France, pp. 333–340 (1988b).

D. L. Sawicki, S. M. Beiser, D. Srinivasan, and P. R. Srinivasan, Immunochemical detection of 7-methylguanine residues in nucleic acids, *Arch. Biochem. Biophys.* **176**, 457–464 (1976).

B. D. Stollar, The experimental induction of antibodies to nucleic acids, *Methods Enzymol.* **70**, 70–85 (1980).

P. T. Strickland and J. M. Boyle, Immunoassay of carcinogen-modified DNA, *Prog. Nucleic Acid Res. Mol. Biol.* **31**, 1–58 (1984).

B. Tierney, A. Benson, and R. C. Garner, Immunoaffinity chromatography of carcinogen DNA adducts with polyclonal antibodies directed against benzo[a]pyrene diol-epoxide-DNA, *J. Natl. Cancer Inst.* **77**, 261–267 (1986).

G. E. Trivers, C. C. Harris, C. Rougeot, and F. Dray, Development and use of ultrasensitive enzyme immunoassays, *Methods Enzymol.* **103**, 409–434 (1983).

D. Umbenhauer, C. P. Wild, R. Montesano, R. Saffhill, J. M. Boyle, N. Huh, U. Kirstein, J. Thomale, M. F. Rajewsky, and S. H. Lu, O^6-Methyldeoxyguanosine in oesophageal DNA among individuals at high risk of oesophageal cancer, *Int. J. Cancer* **36**, 661–665 (1985).

F. J. van Schooten, E. Kriek, M. S. T. Steenwinkel, H. P. J. M. Noteborn, M. J. X. Hillebrand, and F. E. V. Leeuwen, The binding efficiency of polyclonal and monoclonal antibodies to DNA modified with benzo[a]pyrene diol epoxide is dependent on the level of modification. Implications for quantitation of benzo[a]pyrene-DNA adducts *in vivo, Carcinog.* **8**, 1263–1269 (1987).

A. Weston, D. K. Manchester, M. C. Poirier, J.-S. Choi, G. E. Trivers, D. L. Mann, and C. C. Harris, Derivative fluorescence spectral analysis of polycyclic aromatic hydrocarbon-DNA adducts in human placenta, *Chem. Res. Toxicol.*, **2**, 104–108, (1989).

A. Weston, D. K. Manchester, A. Povey, and C. C. Harris, Detection of carcinogen-macromolecular adducts in humans, *J. Am. Coll. Physicians*, **8**, 913–932, (1989).

G. B. Wisdom, Enzyme-immunoassay, *Clin. Chem.* **22**, 1243–1255 (1976).

A. Zettner, Principles of competitive binding assays (saturation analyses). I. Equilibrium techniques, *Clin. Chem.* **19**, 699–705 (1973).

A. Zettner and P. E. Duly, Principles of competitive binding assays (saturation analyses). II. Sequential saturation, *Clin. Chem.* **20**, 5–14 (1974).

Immunoassays for the Clinically Used DNA-Damaging Agents 8-Methoxypsoralen and *cis*-Diamminedichloroplatinum(II)

Regina M. Santella

Comprehensive Cancer Center and Division of
Environmental Sciences
School of Public Health
Columbia University
New York, New York

1. INTRODUCTION

A major difficulty in the development and validation of immunologic methods to monitor human exposure to chemical carcinogens has been the identification of unexposed, control populations. For most ubiquitous environmental carcinogens, such as benzo(a)pyrene, control, unexposed populations do not exist and "background" levels of adducts are found in the general population. In contrast to this situation, studies on patient populations treated with genotoxic agents provide a unique situation for the validation of different methodologies. Patients are exposed to high, well-defined doses, and control, unexposed populations are easily identified. Thus adduct levels in patients treated with 8-methoxypsoralen plus ultraviolet A light or *cis*-diamminedichloroplatinum(II) have been analyzed both for the purpose of validation of immunologic methods for DNA-adduct detection and to gain insight into the mechanisms of chemotheraputic effect.

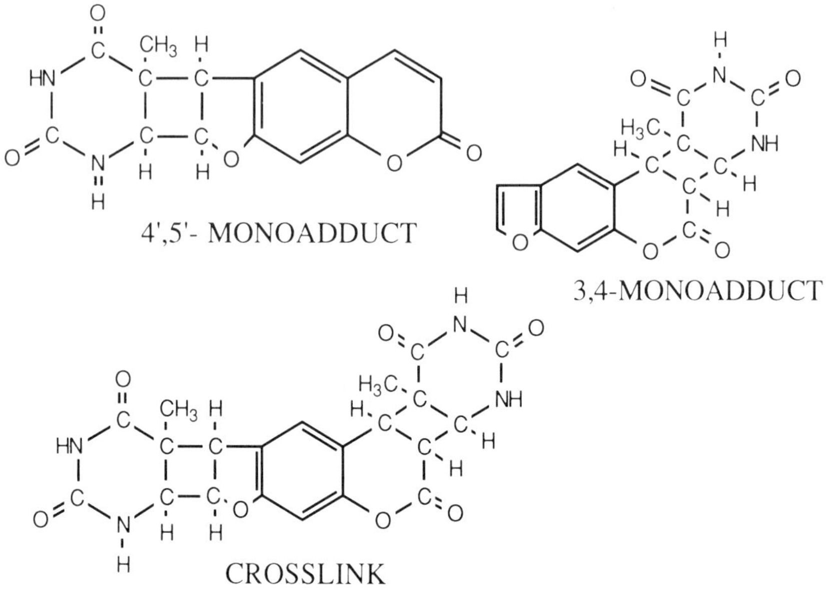

4',5'- MONOADDUCT

3,4-MONOADDUCT

CROSSLINK

Figure 1. Structure of 8-MOP–DNA adducts.

2. 8-METHOXYPSORALEN

8-Methoxypsoralen (8-MOP) plus ultraviolet A light (UVA, 320–400 nm), termed PUVA, is used clinically in the treatment of psoriasis and vitiligo (Parrish *et al.*, 1974) and extracorporeally as a cytoreductive treatment in the leukemic phase of cutaneous T-cell lymphoma (Edelson *et al.*, 1987). When photoactivated by UVA, 8-MOP reacts primarily with thymine bases in DNA, forming both monoadducts and cross-links (Fig. 1) (Ben-hur and Song, 1984; Song and Tapley, 1979). The 4',5' monoadduct is the predominant photoadduct (Gasparro *et al.*, 1984). PUVA is a known mutagen (Papadopoulo and Averbeck, 1985) and animal carcinogen-producing squamous and basal cell carcinomas in the skin (Young *et al.*, 1983). Several case reports and epidemiologic studies have suggested a higher risk of cutaneous carcinoma in treated patients (R. S. Stern, 1986; R. S. Stern *et al.*, 1984; W. J. Stern *et al.*, 1979). In contrast, two European studies have shown an increased risk only in patients also exposed to ionizing radiation and arsenic (Honigsmann *et al.*, 1980; Roenigk and Caro, 1981).

2.1. Detection of 8-Methoxypsoralen–DNA Adducts

We developed a panel of monoclonal antibodies that specifically recognize 8-MOP–DNA and do not react with free 8-MOP or nonmodified DNA (Santella *et al.*, 1985). The most sensitive antibody, 8G1, was characterized

**TABLE 13-1. Competitive ELISA of Monoclonal Antibodies
Against 8-MOP–DNA**

	Antibody	
	8G1	9D8
Competitor	Quantity in fmol causing 50% inhibition	
8-MOP–DNA	17	51
8-MOP–poly (dA-dT) · poly (dA-dT)	77	98
8-MOP–poly (dA) · poly (dT)	13	13
DMA–DNA*	104	100
AMT–DNA[†]	580	870

*DMA = dimethylangelicin.
[†]AMT = 4'-aminomethyl-4,5',8-trimethylpsoralen.

by enzyme-linked immunosorbent assays (ELISA) (Table 1). 8-MOP–modi-fied poly(dA) · poly(dT) contains only monoadducts and gave the highest sensitivity for adduct detection (50% inhibition at 13 fmol). This can be contrasted with the 77 fmol required for 50% inhibition with 8-MOP–poly (dA-dT) · poly(dA-dT), which contains both monoadducts and crosslinks. These data, in combination with immunoassays on high-pressure liquid chro-matography (HPLC) fractions obtained after enzyme digestion of modified poly(dA-dT) · poly(dA-dT), suggested that antibody 8G1 has primary speci-ficity for the 4',5' monoadduct. Characterization of antibody 9D8, used in immunofluorescence studies (see below), is also shown in Table 1. For quantitation of adducts in biological samples, a competitive ELISA with the fluorescence endpoint detection gives the highest sensitivity. When assaying 50 μg of DNA per well, one adduct in 10^8 nucleotides can be detected.

8-MOP–DNA adduct levels have been quantitated in a number of bio-logical samples. Cultured keratinocytes were treated *in vitro* with [^{3}H]8-MOP and UVA and adduct levels quantitated by ELISA with both color and fluorescence end point detection (Yang *et al.,* 1987). These values were compared to those determined by radioactivity (Table 2) and indicate that the ELISA can accurately measure adduct levels. Adduct levels were also monitored in animals treated *in vivo* both by dermal application of 8-MOP or i.p. injection followed by skin irradiation. Blood samples have also been obtained from cutaneous T-cell lymphoma patients treated by extracorporeal photopheresis. These patients take 8-MOP orally and are then attached to the photopheresis apparatus for irradiation of their lymphocytes. White blood cell DNA adducts, measured by ELISA (Table 2), were in the range of 5–7 adducts/10^7 nucleotides.

2.2. Immunofluorescence Localization of Adducts

Antibodies to specific adducts can also be utilized in conjunction with fluorescein isothiocyanate–labeled (FITC) second antibodies in immunofluorescence studies to localize adduct formation to specific cell or tissue types. Our initial studies were carried out on keratinocytes treated in culture with 8-MOP and UVA (Yang *et al.*, 1987). Ethanol-fixed cells were treated with RNase A to eliminate potential cross-reactivity with RNA adducts and with proteinase K to release proteins from the DNA and make it more accessible for antibody binding. This was followed by treatment with 4N HCl to denature the DNA and also increase antibody binding. Monoclonal antibody 9D8 (Santella *et al.*, 1985) was utilized for the immunofluorescence studies since it has the best sensitivity. Specific nuclear staining could be seen with treatments as low as 0.25 μg 8-MOP/ml and 12 J/cm^2 (Fig. 2a, Table 2). Controls, cells treated with vehicle alone, cells stained with nonspecific antiserum, cells treated with DNase before staining, and cells stained with antiserum preabsorbed with 8-MOP–DNA, were all negative. DNA was also isolated from treated cells and adduct levels quantitated by competitive ELISA or radioactivity (Table 2). The lowest level of adducts that could be visualized by the immunofluorescence technique is about 3 adducts/10^6 nucleotides, indicating that this method is much less sensitive than the competitive ELISA, which can detect 1 adduct/10^8 nucleotides.

Immunofluorescence staining has also been applied to the detection of adducts in human skin. Initial studies were with a volunteer treated topically with 75 μg 8-MOP/cm^2. A punch biopsy was taken and the sample irradiated *in vitro* with 22J/cm^2. Specific nuclear staining was visible in the stratified squamous epithelium of the epidermis (Fig. 2b). No staining was visible in the cytoplasm of these cells or in cells in the dermis. Skin biopsies were also obtained from PUVA-treated psoriasis patients in which the drug is taken orally and the skin irradiated two hours later. Three of five biopsies showed specific epidermal nuclear staining near the limit of detection of the immunofluorescence method, indicating adducts were in the range of 1 adduct/10^6 nucleotides (Yang *et al.*, 1989). Elevated levels of sister chromatid exchange (SCEs) (Bredberg *et al.*, 1983) and mutant lymphocytes (Strauss *et al.*, 1979) have been reported for PUVA patients, suggesting that a significant fraction of UVA light reaches the peripheral lymphocytes. To determine whether adducts could be detected in these cells, blood samples were obtained from the patients at the time of biopsy and DNA was isolated. Adducts were not detectable in the white blood cell DNA and thus were below 1 adduct/10^8 nucleotides, the limit of sensitivity of the assay. Drug plasma levels were also measured in these individuals by solid-phase absorption of the drug followed by HPLC quantitation, but these values were not related to positive immunofluorescence staining. Other factors such as variation in DNA repair or skin pigmentation, which influences UVA ab-

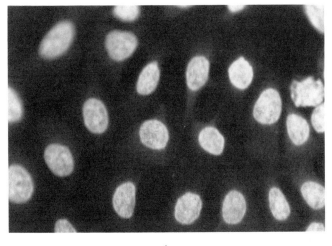

A

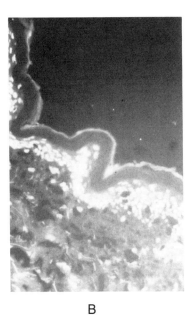

B

Figure 2. (a) Immunofluorescence staining of human keratinocytes treated with 0.25 μg/ml 8-MOP and 12J/cm² UVA. (b) Immunofluorescence staining of human skin biopsy treated with 75 μg/cm² 8-MOP and 22 J/cm² UVA.

TABLE 13–2. Immunofluorescence Staining and 8-MOP–DNA Adduct Levels

Sample	Dose 8-MOP	Dose UVA (J/cm²)	Relative immunofluorescence staining	8-MOP–DNA adduct level (μmol adduct/mol base) [³H]	ELISA Color	ELISA Fluorescent
In vitro						
Human keratinocyte	2.5 μg/m	12	+++	33	33	32
Human keratinocyte	0.5 μg/ml	12	++	27	12	14
Human keratinocyte	0.25 μg/ml	12	+		2.9	
Human keratinocyte	0	12	−			
In vivo						
Mouse skin i.d.	100 μg/cm²	12	++		18	
Mouse skin i.p.	30 μg/g	12	++		10	
Human skin topical	75 μg/cm²	22	+++			
Human skin oral	0.6 mg/kg	9	+			
Human lymphocytes Oral administration	0.6 mg/kg	2*			0.55–0.75	

*Lymphocytes are irradiated *ex vivo*.

sorption, may influence the extent of adduct formation. Prolonged treatment results in skin pigmentation and may decrease the effective UVA dose. It has been estimated that lymphocytes are exposed to approximately 1%–5% of the skin surface dose of UVA (Kraemer and Weinstein, 1977). It is therefore not surprising that lymphocytes have less than 1 adduct/10^8 nucleotides when skin adduct levels are approximately $1/10^6$ nucleotides.

Polyclonal antisera against low-molecular-weight 8-MOP–modified polynucleotides have been developed by another group. These antisera have been used to visualize 8-MOP–DNA adducts in treated lymphocytes (Zarebska *et al.*, 1984), in skin biopsies of animals treated topically, and in human skin treated *in vitro* (Pathak *et al.*, 1986). For these studies, sections were pretreated with ethanol and alkali or mild nuclease S_1 to enhance antibody binding. Repair after damage was also followed in animals treated *in vivo*.

We have developed flow cytometric methods for the analysis of 8-MOP–DNA adducts in PUVA-treated SV40 transformed keratinocytes (Yang *et al.*, 1988). Cells were stained with adduct-specific antibody 9D8 followed by fluorescein isothiocyanate–labeled second antibodies and counterstained with propidium iodide to simultaneously determine DNA content. The sensitivity of the method was determined by quantitation of adduct levels on isolated DNA by ELISA and indicated that 12 adducts/10^6 nucleotides could be detected, higher levels than the minimum needed for conventional immunofluorescence adduct detection. This technique was utilized to study adduct formation during various stages of the cell cycle. Two separate cell populations with a DNA content indicative of cells in G_1 but with different levels of FITC labeling were observed. When cells were pretreated with aphidicolin, which blocks cells in G_1, only one cell population with a G_1 DNA content was observed. These results suggest that those cells with a G_1 DNA content but with higher FITC staining may be in early S phase. Adduct formation may occur more readily in early S phase when unwinding of the DNA during replication may result in increased accessibility to the DNA. With further improvements in sensitivity, flow cytometry may be useful for studying adduct formation in specific cell types.

3. *cis*-DIAMMINEDICHLOROPLATINUM(II)

cis-Diamminedichloroplatinum(II), cisDDP, has been used successfully in the chemotherapy of several types of cancer, including testicular (Einhorn and Donohue, 1977) and ovarian (Young *et al.*, 1979). Antineoplastic activity is believed to be related to the formation of cisDDP–DNA damage. Four adducts have been identified in enzymatic digests of cisDDP–DNA (Fig. 3) (Eastman, 1987, Fichtinger-Schepman *et al.*, 1982). The two major adducts are cisPt(NH$_3$)$_2$d(pApG), derived from the bifunctional binding of Pt to

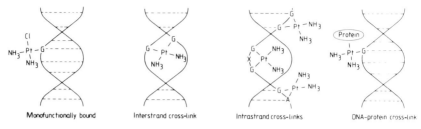

Figure 3. Structure of cisDDP–DNA adducts.

adenosine and guanosine in the sequence pApG, and $cisPt(NH_3)_2d(pGpG)$, resulting from intrastrand cross-links on two vicinal guanines in the sequence pGpG. These adducts account for approximately 30% and 60%, respectively, of the adducts formed on DNA *in vitro*. The two minor adducts are $Pt(NH_3)_3dGMP$, derived from the monofunctional binding of Pt to guanosine, and $cisPt(NH_3)_2d(GMP)_2$, resulting from intrastrand cross-links on the base sequence pG(pX)pG and from intrastrand cross-links on two guanines (Fichtinger-Schepman *et al.*, 1985). In all cases, reaction is at the N7 position of guanine or adenine.

3.1. Antibodies against *cis*DDP–DNA

A number of groups have developed polyclonal and monoclonal antibodies recognizing cisDDP–DNA adducts. The first polyclonal antibody developed against cisDDP–DNA (Malfoy *et al.*, 1981), while recognizing DNA modified *in vitro*, could not detect adducts produced *in vivo*. Subsequently, a polyclonal antiserum was produced against DNA modified *in vitro* to a high level (Pt/nucleotide = 0.044) complexed with methylated bovine serum albumin (Poirier *et al.*, 1982). By competitive ELISA, this antiserum recognized adducts formed in cultured mouse leukemia L1210 cells and in L1210 cells from the ascites fluid of tumor-bearing mice treated with cisDDP. It did not recognize cisDDP itself, DNA modified by *trans*-DDP, unmodified DNA, or individual cisDDP–DNA adducts. Studies with homopolymers and heteropolymers modified with cisDDP and calf thymus DNA modified with various platinum drug analogs indicated that the antibody specificity is directed towards DNA containing $cisPt(NH_3)_2dpGpG$ and pApG (Lippard *et al.*, 1983). This antiserum has been used to monitor adduct formation in white blood cell DNA of testicular and ovarian cancer patients receiving cisDDP or carboplatin. cisDDP chemotherapy is usually given in three to five cycles of five days of daily intravenous infusion. Most blood samples were collected on the morning of the sixth day. Initial studies on 120 samples indicated that 44 samples out of 102 taken from 45 treated patients were positive while all of 18 samples from the same number of controls did not have detectable adducts (Poirier *et al.*, 1985; Reed *et al.*,

1986b). Some patients did not form adducts even when given high doses. However, in approximately half of the patients, the quantity of measureable adducts increased as a function of cumulative dose of cisDDP, suggesting that adduct removal is slow. Disease response data, evaluated for 55 ovarian cancer patients, showed a positive correlation between the highest level of DNA adducts measured in each patient and response to drug therapy (Reed *et al.*, 1987). Values for median adduct levels in patients grouped by complete response, partial response, and no response were 212, 193, and 52 amol of adduct/μg of DNA (Figure 4a). Thus higher adduct levels correlate with disease response; eight patients without measurable adduct levels did not respond to therapy. A similar relationship between adduct level and response to therapy was seen in "poor prognosis" nonseminomatous testicular cancer patients. The median adduct level of individuals with complete response was 170 amol/μg DNA compared to 78 amol/μg in partial responders (Fig. 4b) (Reed *et al.*, 1988). Adducts were also detectable in kidney, spleen, or liver DNA isolated from a small number of patients (four)

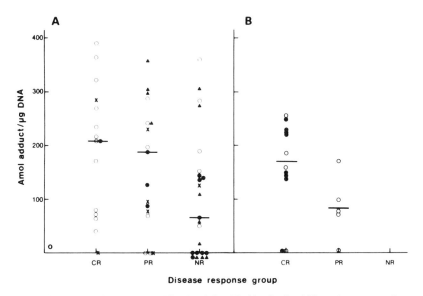

Figure 4. (a) Peak cisDDP–DNA adduct levels in white blood cells of 55 ovarian cancer patients treated with platinum-based chemotherapy. Treatment groups are CTX-DDP (open circles), HD-DDP (closed circles), diamminecyclobutane-dicarboxylatoplatinum (CBDCA, closed triangles), CHIP(s) (bold ×). For details of treatment regime see Reed *et al.*, 1987. Each data point represents the peak adduct level for a single patient. The median adduct levels are shown by the bold line. CR, complete response to therapy; PR, partial response and NR, no response (From Reed *et al.*, 1987, with permission). (b) Peak cisDDP–DNA adduct levels in white blood cells of 17 "poor prognosis" testicular cancer patients. PVeB, cisDDP, vinblastine, and bleomycin; PVeBV, cisDDP, vinblastine, bleomycin, and epidophyllotoxin. For details of the treatment regime see Reed *et al.*, 1988.

as much as 22 months after final treatment with cisDDP, indicating that the adduct may be removed slowly in human tissues (Poirier *et al.*, 1987; Reed *et al.*, 1986a, 1987).

Recently, adduct levels measured by this ELISA were compared to those determined by the G-Pt-GMP ELISA (see below) and atomic absorption spectroscopy (Poirier *et al.*, 1988). The cisDDP–DNA ELISA underestimates adducts since the highest values were 5 to 20 times lower than those obtained with the other methods. The original antigen for development of the polyclonal antisera was highly modified DNA, necessary for the production of an immune response. Studies with antibodies to benzo(a)pyrene diol epoxide–modified DNA have shown that antisera raised against highly modified DNAs have lower affinity for *in vivo*, low modified DNAs (Santella *et al.*, 1988; van Schooten *et al.*, 1987). Thus quantitation of adducts utilizing a standard curve of highly modified DNA will thus underestimate adduct levels. However, this is not always the case since antibodies to highly modified 8-methoxypsoralen–DNA, discussed above, have similar sensitivity for the adduct in low and high modified DNAs. These results highlight the importance of detailed characterization of antisera not only in terms of specificity but also in terms of sensitivity of adduct detection in DNA with different modification levels.

Monoclonal antibodies have also been developed against cisDDP–DNA (Pt/nucleotide = 0.06–0.08) as well as *trans*-DDP (Pt/nucleotide = 0.022) but these antibodies have not been utilized in studies on adduct formation in humans (Sundquist *et al.*, 1987). Another group has also developed both polyclonal and monoclonal antibodies against cisDDP–DNA (Mustonen *et al.*, 1987). The most sensitive antibody has 50% inhibition in a competitive ELISA at 50–100 fmol. While this antibody has not yet been applied to human samples, this group has utilized atomic absorption to measure white blood cell DNA and plasma protein adducts as well as free cisDDP levels in sera of 20 patients. DNA-adduct levels were in the range of 1 pg cisDDP/μg DNA ($\sim$4 fmol/μg).

Antibodies recognizing cisDDP–DNA adducts have also been generated by immunization of animals with the specific adducts coupled to protein (Fichtinger-Schepman *et al.*, 1985, 1988). cisPt(NH$_3$)$_2$dGuodGMP and its ribo analog were coupled to bovine serum albumin (Fichtinger-Schepman *et al.*, 1985). Polyclonal antisera detected these adducts in cisDDP–DNA but with picomole sensitivities. Modified DNA can be digested enzymatically to nucleotides and the adducts separated by anion-exchange chromatography (Fig. 5). Quantitation of the adducts by ELISA after HPLC separation had sensitivities in the femtomole range for detection of disPt(NH$_3$)$_2$d(pGpG) and cisPt(NH$_3$)$_2$d(GMP)$_2$. Subsequent development of antisera against Pt(NH$_3$)$_3$dGMP coupled to chicken gamma globulin and against cisPt(NH$_3$)$_2$d(pApG) coupled to bovine serum albumin has allowed the de-

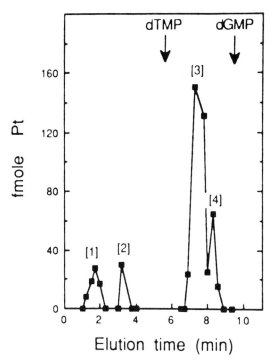

Figure 5. Anion-exchange chromatography (MonoQ column) of a digest of DNA isolated from SUSA cells, a line derived from a germ cell tumor of the testis, treated with cisDDP. Pt–DNA adducts were detected with specific antisera. Peak 1, cisPt(NH$_3$)$_3$dGMP; peak 2, cisPt(NH$_3$)$_2$d(pApG); peak 3, cisPt(NH$_3$)$_2$d(pGpG); and peak 4, cisPt(NH$_3$)$_2$d(GMP)$_2$. (From Bedford *et al.*, 1988, with permission).

tection of all adducts with sensitivity in the 4–9-fmol range with three antisera (Fichtinger-Schepman *et al.*, 1987a). All four adducts were detected in white blood cell DNA from treated patients with the predominant adduct (5.6 fmol/ μg DNA, ~65% of the total) resulting from intrastrand cross-link on pGpG. The pApG crosslink accounts for 22% of the adducts (1.9 fmol/μg) and the cisPt(NH$_3$)$_2$d(GMP)$_2$ about 13% (1.1 fmol/μg). Less than 1% of the adducts (0.06 fmol/μg) are Pt(NH$_3$)$_2$dGMP. This ratio of adducts is similar to those produced by modification of DNA *in vitro*. A 15-fold interindividual difference in levels of the major adduct, cisPt(NH$_3$)$_2$d(pGpG), was seen in four patients treated with 20 mg cisDDP/m^2. In addition, disappearance of this adduct was followed for 24 h after initial treatment in six male patients and indicated rapid removal (Fig. 6). Since cisDDP is usually given on five consecutive days, adduct levels were followed in a patient immediately after each infusion. Over the first three days the number of adducts increased, but no further increase was measured after the infusions on days 4 and 5 (Fichtinger-Schepman, *et al.* 1988). Substantial repair occurred by 24 h after the final infusion. Variations were also observed with regard to the day at which the highest adduct level was reached.

Study of the repair of adducts produced in cultured cells treated *in vitro* indicates that removal of all adducts proceeds with roughly the same kinetics (Dijt *et al.*, 1988). In addition, different cell lines contained different adduct levels after identical treatment. Studies were also carried out to determine if adduct levels in different tumor cell lines correlate to susceptibility to the cytotoxic effects of cisDDP treatment. Initial results suggest that repair capacities as well as initial adduct levels influence sensitivity to cisDDP (Fichtinger-Schepman *et al.*, 1988). Since clinical response to cisPt chemotherapy may vary with the individual, an *in vitro* assay to determine an individual's sensitivity to cisDDP therapy would be very useful to the clinician. Adduct levels formed *in vitro* by cisDDP in whole blood from patients prior to the start of chemotherapy were found to correlate (r = 0.91) with those formed *in vivo* during therapy (Fichtinger-Schepman *et al.*, 1987b).

In order to use immunocytochemical methods to follow adduct formation in intact cells, a polyclonal antiserum was raised in rabbits by immuniza-

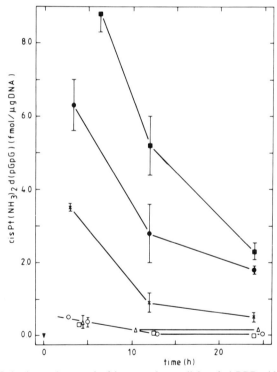

Figure 6. Induction and removal of intrastrand cross-links of cisDDP with two neighboring guanines [cisPt(NH₃)₂d(pGpG)] in DNAs isolated from the white blood cells of six patients who received their first treatment with cisDDP. Cross-links were determined by ELISA after enzymatic digestion of the DNA and separation of adducts by anion-exchange chromatography. (From Fichtinger-Schepman *et al.*, 1987, with permission).

tion of cisDDP–DNA complexed with methylated bovine serum albumin (Terheggen *et al.*, 1987). This antiserum was used in a double peroxidase–antiperoxidase staining procedure to follow adduct formation in animals treated with cisDDP. Specific nuclear staining was observed in all tissues examined, and the extent of staining was dose and time dependent. This antiserum was also used to follow adducts in buccal epithelial and urinary cells from cancer patients by immunoperoxidase staining (Terheggen *et al.*, 1988). Nuclear staining density was measured by microdensitometry with a reported detection limit of 2 adducts/10^6 nucleotides. Binding in buccal cells increased in proportion to the cumulative dose of cisDDP. Although the staining density of urinary cells was higher than that in buccal cells in the small number of samples studied, their morphology was poor and few cells could be examined.

4. SUMMARY

Studies on patients treated with genotoxic chemotheraputic agents have provided valuable information for the development and validation of immunologic methods to quantitate DNA adducts. Information on the formation of adducts as well as their repair should be useful to increase the understanding of the basic mechanism of action of these agents as well as to gain insight into the process of resistance.

ACKNOWLEDGMENTS

The work in the author's laboratory was supported by grants from the National Cancer Institute CA21111, and National Institute of Environmental Health Science ES03881. The important contributions to this work of Xiao Yan Yang, Francis Gasparro, Vincent DeLeo, Marina Stefanidis, and You Li are gratefully acknowledged.

REFERENCES

P. Bedford, A. M. Fichtinger-Schepman, S. A. Shellard, M. C. Walker, J. R. W. Masters, and B. T. Hill, Differential repair of platinum-DNA adducts in human bladder and testicular tumor continuous cell lines, *Cancer Res.* **48**, 3019–3024 (1988).

E. Ben-hur, P. S. Song, The photochemistry and photobiology of furocoumarines (psoralens), *Adv. Radiat. Biol.*, **11**, 131–171 (1984).

A. Bredberg, B. Lambert, A. Lindblad, G. Swanbeck, and G. Wennersten, Studies of DNA and chromosome damage in skin fibroblasts and blood lymphocytes from psoriasis patients treated with 8-methoxypsoralen and UVA irradiation, *J. Invest. Dermatol.* **81**, 93–97 (1983).

F. J. Dijt, A. M. J. Fichtinger-Schepman, F. Berends, J. Reedijk, Formation and repair of cisplatin-induced adducts to DNA in cultured normal and repair-deficient human fibroblasts, *Cancer Res.*, **48**, 6058–6062 (1988).

A. Eastman, The formation, isolation and characterization of DNA adducts produced by anticancer platinum complexes, *Pharmacol. Ther.* **34**, 155–166 (1987).

R. Edelson, C. Berger, F. Gasparro, B. Jegasothy, P. Heald, B. Wintroub, E. Vonderheid, and R. Knobler, Treatment of cutaneous T-cell lymphoma by extra-corporeal photochemotherapy, *N. Engl. J. Med.* **316**, 297–303 (1987).

L. H. Einhorn, J. Donohue, *cis*-diamminedichloroplatinum, vinblastine and bleo-mycin combination chemotherapy in disseminated testicular cancer, *Ann. Intern. Med.* **87**, 293–298 (1977).

A. M. Fichtinger-Schepman, P. H. M. Lohman, J. Reedijk, Detection and quanti-fication of adducts formed upon interaction of diamminedichloroplatinum(II) with DNA, by anion-exchange chromatography after enzymatic degradation, *Nucleic Acids Res.* **10**, 5345–5357 (1982).

A. Fichtinger-Schepman, R. Baan, A. Luiten-Schuite, M. VanDijk, and P. H. M. Lohman, Immunochemical quantitation of adducts induced in DNA by *cis*-diam-minedichloroplatinum(II) and analysis of adduct related DNA-unwinding, *Chem. Biol. Interact.* **55**, 275–288 (1985).

A. M. Fichtinger-Schepman, A. T. van Oosterom, P. H. M. Lohman, and F. Berends, *cis*-diamminedichloroplatinum(II)-induced DNA adducts in peripheral leukocytes from seven cancer patients: quantitative immunochemical detection of the adduct induction and removal after a single dose of *cis*-diamminedichloropla-tinum(II), *Cancer Res.* **47**, 3000–3004 (1987a).

A. M. J. Fichtinger-Schepman, A. T. van Oosterom, P. H. M. Lohman, and F. Berends, Interindividual human variation in cisplatinum sensitivity, predictable in an *in vitro* assay, *Mutat. Res.* **190**, 59–62 (1987b).

A. M. J. Fichtinger-Schepman, F. J. Dijt, P. Bedford, A. T. van Oosterom, B. T. Hill, and F. Berends, Induction and removal of cisplatin-DNA adducts in human cells *in vivo* and *in vitro* as measured by immunochemical techniques. In: *Methods for Detecting DNA Damaging Agents in Humans: Applications in Cancer Epide-miology and Prevention,* IARC Scientific Publication No. 89, eds. H. Barrsch, K. Hemminki, and I. K. O'Neill, IARC, Lyon, France (1988).

F. P. Gasparro, J. Bagel, and R. L. Edelson, HPLC analysis of 4′,5′-monoadduct formation in calf thymus DNA and synthetic polynucleotides treated with UVA and 8-methoxypsoralen, *Photochem. Photobiol.* **42**, 95–98 (1984).

H. Honigsmann, H. Wolff, F. Gschnait, W. Brenner, and E. Jaschke, Keratoses and non-melanoma skin tumors in long-term photochemotherapy (PUVA), *J. Am. Acad. Dermatol.* **3**, 406–414 (1980).

K. H. Kraemer and G. D. Weinstein, Decreased thymidine incorporation in circu-lating leucocytes after treatment of psoriasis with psoralen and long-wave ultra-violet light, *J. Invest. Dermatol.* **69**, 211–214 (1977).

S. J. Lippard, H. M. Ushay, C. M. Merkel, and M. C. Poirier, Use of antibodies to probe the stereochemistry of antitumor platinum drug binding to DNA, *Biochem.* **22**, 5165–5168 (1983).

B. Malfoy, B. Hartmann, J. P. Macquet, and M. Leng, Immunochemical studies of DNA modified by *cis*-diamminedichloroplatinum (II), *Cancer Res.* **41**, 4127–4131 (1981).

R. Mustonen, K. Hemminki, A. Alhonen, P. Heitanen, and M. Kiilunen, Determination of cisplatin in blood compartments of cancer patients. In: *Methods for Detecting DNA Damaging Agents in Humans: Applications in Cancer Epidemiology and Prevention,* eds. H. Bartsch K. Hemminki and I. K. O'Neill, IARC, Lyon, France, pp. 329–332, in press (1988).

D. Papadopoulo and D. Averbeck, Genotoxic effects and DNA photoadducts induced in Chinese hamster V79 cells by 5-methoxypsoralen and 8-methoxypsoralen, *Mutat. Res.* **151,** 281–291 (1985).

J. A. Parrish, T. B. Fitzpatrick, M. A. Pathak, L. Tannenbaum, Photochemotherapy of psoriasis with oral methoxysalen and long wave ultraviolet light, *N. Engl. J. Med.* **291,** 1207–1211 (1974).

M. A. Pathak, Z. Zarebska, M. C. Mihm, M. Jarzabek-Chorzelska, T. Chorzelski, and S. Jablonska, Detection of DNA psoralen photoadducts in mammalian skin, *J. Invest. Dermatol.* **86,** 308–315 (1986).

M. C. Poirier, M. J. Egorin, A. M. Fichtinger-Schepman, and E. Reed, DNA adduct of cisplatin and carboplatin in tissues of human cancer patients. In: *Methods for Detecting DNA Damaging Agents in Humans: Applications in Cancer Epidemiology and Prevention,* IARC Publication No. 89, eds. H. Bartsch, K. Hemminki, and I. K. O'Neill, IARC, Lyon, France (1988).

M. C. Poirier, S. Lippard, L. A. Zwelling, M. Ushay, D. Kerrigan, R. M. Santella, D. Grunberger, and S. H. Yuspa, Antibodies elicited against *cis*-diamminedichloroplatinum(II)-modified DNA are specific for *cis*-diamminedichloroplatinum(II)-DNA adducts formed in vivo and in vitro, *Proc. Natl. Acad. Sci. USA* **79,** 6443–6447 (1982).

M. C. Poirier, E. Reed, R. F. Ozols, T. Fasy, and S. H. Yuspa, DNA adducts of cisplatin in nucleated peripheral blood cells and tissues of cancer patients, *Prog. Expt. Tumor Res.* **31,** 104–113 (1987).

M. C. Poirier, E. Reed, L. A. Zwelling, R. F. Ozols, C. L. Litterst, and S. H. Yuspa, Polyclonal antibodies to quantitate *cis*-diamminedichloroplatinum(II)-DNA adducts in cancer patients and animal models, *Environ. Health Persp.* **62,** 89–94 (1985).

E. Reed, R. F. Ozols, T. Fasy, S. H. Yuspa, M. C. Poirier, C. Ramel, B. Lombert, and J. Magnusson, Biomonitoring of cisplatin-DNA adducts in cancer patients receiving cisplatin chemotherapy. In: *Genetic Toxicology of Environmental Chemicals, Part G: Genetic Effects and Applied Mutagenesis,* Alan R. Liss, New York, pp. 247–252 (1986a).

E. Reed, S. H. Yuspa, L. A. Zwelling, R. F. Ozols, and M. C. Poirier, Quantitation of *cis*-diamminedichloroplatinum II (cisplatin)-DNA-intrastrand adducts in testicular and ovarian cancer patients receiving cisplatin chemotherapy, *J. Clin. Invest.* **77,** 545–550 (1986b).

E. Reed, R. F. Ozols, R. Tarone, S. H. Yuspa, and M. C. Poirier, Platinum-DNA adducts in nucleated peripheral blood cell DNA correlate with disease response in ovarian cancer patients receiving platinum-based chemotherapy, *Proc. Natl. Acad. Sci. USA* **84,** 5024–5028 (1987).

E. Reed, R. F. Ozols, R. Tarone, S. H. Yuspa, and M. C. Poirier, The measurement of cisplatin-DNA adduct levels in testicular cancer patients, *Carcinog.* **9,** 1909–1911 (1988).

H. H. Roenigk, W. A. Caro, Skin cancer in the PUVA-48 cooperative study, *J. Am. Acad. Dermatol.* **4**, 319–324 (1981).

R. M. Santella, N. Dharmaraja, F. P. Gasparro, and R. L. Edelson, Monoclonal antibodies to DNA modified by 8-methoxypsoralen and ultraviolet A light, *Nucleic Acids Res.* **13**, 2533–2544 (1985).

R. M. Santella, A. Weston, F. P. Perera, G. T. Trivers, C. C. Harris, T. L. Young, D. Nguyen, B. M. Lee, and M. C. Poirier, Interlaboratory comparison of antisera and immunoassays for benzo(a)pyrene-diol-epoxide-I-modified DNA, *Carcinog.* **9**, 1265–1269 (1988).

P. S. Song and J. K. Tapley, Photochemistry and photobiology of psoralens, *Photochem. Photobiol.* **29**, 1177–1197 (1979).

R. S. Stern, PUVA carcinogenesis after ten years: Prospect and retrospectives, *Photodermatology* **3**, 257–260 (1986).

R. S. Stern, N. Laird, J. Melski, J. A. Parrish, T. B. Fitzpatrick, and H. I. Blech, Cutaneous squamous-cell carcinoma in patients treated with PUVA, *N. Engl. J. Med.* **310**, 1156–1161 (1984).

W. J. Stern, L. A. Thihodeau, R. A. Kleinerman, A. B. Parrish, T. B. Fitzpatrick, Risk of cutaneous carcinoma in patients treated with oral methoxypsoralen photochemotherapy for psoriasis, *N. Engl. J. Med.* **300**, 804–813 (1979).

G. H. Strauss, R. J. Albertini, P. A. Krusinski, and R. D. Baughman, 6-Thioguanine resistant peripheral blood lymphocytes in humans following psoralen, long wave ultraviolet light (PUVA) therapy, *J. Invest. Dermatol.* **73**, 211–216 (1979).

W. I. Sundquist, S. J. Lippard, and B. D. Stollar, Monoclonal antibodies to DNA modified with *cis*- or *trans*-diamminedichloroplatinum (II), *Biochem.* **84**, 8225–8229 (1987).

P. M. A. B. Terheggen, R. Dijikman, A. C. Begg, R. Dubbelman, B. G. J. Floot, A. A. M. Hart, and L. denEngelse, Monitoring of interaction products of *cis*-diamminedichloroplatinum(II) and *cis*-diammine(l,-cyclobutanedicarboxylato)-platinum(II) with DNA in cells from platinum-treated cancer patients, *Cancer Res.* **48**, 5597–5603 (1988).

P. M. A. B. Terheggen, B. G. J. Floot, E. Scherer, A. C. Begg, A. M. J. Fichtinger-Schepman, and L. denEngelse, Immunocytochemical detection of interaction products of *cis*-diamminedichloroplatinum(II) and *cis*-diammine-(1,1-cyclobutane dicarboxylato)platinum(II) with DNA in rodent tissue sections, *Cancer Res.* **47**, 6719–6725 (1987).

F. J. van Schooten, E. Kriek, M. S. T. Steenwinkel, H. P. J. M. Noteborn, M. J. X. Hillebrand, and F. E. vanLeeuwen, The binding efficiency of polyclonal and monoclonal antibodies to DNA modified with benzo[a]pyrene diol epoxide is dependent on the level of modification. Implications for quantitation of benzo[a]pyrene-DNA adducts in vivo, *Carcinog.* **8**, 1263–1269 (1987).

X. Y. Yang, V. DeLeo, and R. M. Santella, Immunological detection and visualization of 8-methoxypsoralen-DNA photoadducts, *Cancer Res.* **47**, 2451–2455 (1987).

X. Y. Yang, T. Delohery, and R. M. Santella, Flow cytometric analysis of 8-methoxypsoralen-DNA photoadducts in human keratinocytes, *Cancer Res.*, **48**, 7013–7017 (1988).

X. Y. Yang, F. P. Gasparro, V. A. DeLeo, and R. M. Santella, 8-Methoxypsoralen-DNA adducts in patients treated with 8-methoxypsoralen and ultraviolet A light, *J. Invest. Dermatol.*, **92**, 59–63 (1989).

A. R. Young, I. A. Magnus, A. C. Davies, and N. P. Smith, A comparison of the phototumorigenic potential of 8-MOP and 5-MOP in hairless albino mice exposed to solar simulated radiation, *Brit. J. Derm.* **108**, 507–518 (1983).

R. C. Young, D. D. VonHoff, P. Gormley, R. Makuch, J. Cassidy, D. Howser, and J. M. Bull, *cis*-Diamminedichloroplatinum(II) for the treatment of advanced ovarian cancer, *Cancer Treat. Rev.* **63**, 1539–1544 (1979).

Z. Zarebska, R. Jarzabek-Chorzelska, G. Zesa, W. Glinski, M. Pawinskia, T. Choozelski, and S. Jablonska, Detection of DNA-Psoralen Photo Adducts *In Situ*, *Photochem. Photobiol.* **39**, 307–312 (1984).

Chapter

14

Application of Fluorescence to Analysis of Genotoxicity

A. M. Jeffrey
Division of Environmental Science
Institute of Cancer Research and Department of Pharmacology
Columbia University
New York, New York

1. BACKGROUND TO FLUORESCENCE

Fluorescence, as well as the related pheonomenon phosphorescence, is the emission of light from molecules that have absorbed photons of higher energy. These two processes are related in that a molecule in its ground state absorbs a photon to produce an excited-state molecule. It is the nature of the excited state that distinguishes these two phenomena. Typically organic molecules are in a singlet state: they have an even number of electrons, the spins of which are paired and opposite ($+\frac{1}{2}$ and $-\frac{1}{2}$) so that the resultant spin is zero. The multiplicity *(M)* of the state is singlet because $M = 2S + 1$, where S is the total spin state. When a photon is absorbed an electron is moved from the uppermost filled orbital of the ground state S_0 to a vacant orbital of higher energy. If the spin of the excited state is unchanged, then the total spin and multiplicity of the excited states remain zero and singlet, respectively. If the spins are no longer opposed, then the spin becomes unity and the multiplicity triplet. Because the transition from the excited to ground state occurs with high probability only when the states are of the same multiplicity, the two relaxation processes have markedly different lifetimes. Relaxation from the first excited singlet state $S_1 \rightarrow S_0$ is called fluorescence and typically has lifetimes of 10^{-7} to 10^{-10} s but may be as long as 10^{-3} s.

Phosphorescence, from $T_1 \rightarrow S_0$, has lifetimes of 10^{-3} to 10 s. For most biological organic molecules in which all the atoms are of low atomic mass, e.g. H, C, N, and O, the transition between the ground-state singlet and excited triplet states is forbidden and therefore improbable, and the absorption is weak. The strongest absorption band in most organic molecules corresponds to the excitation from the lowest ground state S_0 to the lowest excited singlet state S_1. Since the lowest excited triplet state is usually of lower energy than S_1, phosphorescence is generally of longer wavelength than fluorescence and occurs by intersystem crossing from S_1 to T_1. Depending upon the wavelength of the photon absorbed, the transition may fall into one of several vibrational sublevels within S_1. Biological samples of solids or liquids undergo rapid relaxation to the lowest vibrational state of S_1 within 10^{-13} to 10^{-12} s. Returning to S_0 may occur by fluorescence or internal conversion. In the latter case the energy is transformed into thermal energy. Again the return to S_0 may occur at any one of several vibrational sublevels before reaching the ground vibrational state. For these reasons the emission spectra will often appear as the mirror image of the excitation spectra. Changes in the dipole orientation or hydrogen bonding between the S_0 and S_1 states may alter this relationship of the spectra. The lowest energy band in absorption and the highest energy band in emission should be identical. This occurs in gases but generally not in solids or liquids. After correcting for characteristics of the particular instrument on which they are measured, the excitation spectra should be equivalent to the absorption spectra for a given compound.

Beyond measuring the excitation and emission spectra, other important parameters from which information on the state of the molecule can be gleaned include their lifetimes, quantum yields, and polarization. The fluorescence intensity follows an exponential decay once the exciting light source is removed. The observed lifetime of a molecule may be shortened by collisions with other solute or solvent molecules. The quantum yield is defined as the number of fluorescence quanta emitted per quanta absorbed to a singlet excited state and therefore ranges from 0 to 1. This is simple to measure in a relative way but difficult to determine as an absolute value. When polarized light is used to excite molecules, those molecules with their transition moments in the same direction as the light will be preferentially excited. If the molecule does not move before it fluoresces, then the emitted photon will also be polarized. Decreased motion or decreases in the fluorescence lifetime of the molecules will increase the polarization of the emitted photons. As a complement to fluorescence polarization studies, information can be obtained by monitoring recovery of fluorescence after photobleaching. Samples are bleached by short intense pulses of light either parallel or perpendicular to the axis of the much weaker illumination used to measure reorientation and angular redistribution. From comparisons of these two recovery curves corrections can be made for nonrotational components in the recovery curves [1].

Specific applications of fluorescence to molecular dosimetry will now be discussed. An extensive collection of reviews on fluorescence microscopy have been published recently. Wang, Y.-L. and Taylor, D. L. Methods Cell Res. 29 and 30 (1989).

2. BROADBAND FLUORESCENCE SPECTRA

The pioneering work in the field of polycyclic aromatic hydrocarbon (PAH)–DNA adducts was by Vigny and Duquesne. They found differences between the fluorescence spectra of DNA extracted from cells treated *in vitro* with 7-methylbenz[a]anthracene and those obtained by modifying DNA *in vitro* with 7-methylbenz[a]anthrene 5,6-oxide, the metabolite thought at that time to be the most likely intermediate responsible for DNA-adduct formation. Experiments by Wang *et al.* showed that, of all the metabolites of benzo[a]pyrene (BP) tested, only the 7,8-dihydrodiol bound better to DNA than benzo[a]pyrene itself when activated by microsomes. This led Sims *et al.* to test further the 7,8-dihydrodiol and suggest that it, rather than the 4,5 K region epoxide, was a key intermediate metabolite in the formation of a diol epoxide as the ultimate carcinogen of BP. Fluorescence spectra supported these conclusions. Analysis of the fluorescence spectra of DNA adducts formed with 9-hydroxybenzo[a]pyrene in the presence of microsomes suggested that activation of the 4,5 positions generated the metabolites that bound to guanine and adenine bases. In mouse skin these appeared to be minor adducts.

The instrument used by Vigny and Duquesne for their experiments was a photon-counting device that provided good sensitivity by signal averaging. Ivonovic *et al.* used a conventional spectrofluormeter but increased sensitivity by cooling the samples to liquid-nitrogen temperatures. This can potentially increase the quantum yield by decreasing both intramolecular quenching, which can occur when the S_0 and S_1 vibrational energy levels are isoenergetic, allowing internal conversion, and intermolecular quenching, by collisions with other solute or solvent molecules such that the energy levels are perturbed or the barriers for crossing decreased. However, relatively long-lived complexes may also be formed that have longer lifetimes or lower quantum yields, potentially decreasing the sensitivity of the assay.

Many other studies have used this type of approach to investigate DNA-adduct formation. One limitation to this approach is the absorption of the DNA itself. The major adducts formed between DNA and dimethylbenzanthracene (DMBA), BP, or related compounds do not present this problem since the longest wavelength of these adducts' absorption is beyond that of

DNA. However, had DMBA, for example, been activated primarily at the 8,11 positions, the residual phenanthrene chromophore would have been much more difficult to detect. One strategy to circumvent this problem has been to digest the DNA and collect the adduct after high-performance liquid chromotography (HPLC) separation. Chrysene was shown to bind to the DNA of hamster embryo cells, yielding phenanthrene-like adducts. In cases where multiple adducts are formed, they may be resolved chromatographically and the individual spectra then measured. Such an approach has been used for the analysis of simple adducts such as ethenoadenine and O^6-alkylguanine adducts, which all absorb at relatively short wavelengths. Further hydrolysis of the modified base may improve the sharpness of the fluorescence spectra by eliminating interactions between the carcinogen and base moieties of the adduct. Chirally or structurally distinct adducts may yield the same product and so information will be lost. Additional reactions may occur, resulting in decomposition of the carcinogen moiety, which can further complicate the analysis. Thus to optimize the analysis of BP, tetraols released from DNA by acid hydrolysis requires careful control of hydrolysis conditions.

The sensitivity of detection may be further increased using more powerful light sources such as lasers and windowless cells to reduce scattering. Capillary zone electrophoresis is just starting to be applied to these types of compounds and, in conjunction with a variety of detection techniques, including fluorescence, has considerable potential in this field.

The combination of slight differences in spectra and lifetimes can be used to detect the presence of DNA adducts that are closely related structurally. Two distinct excitation spectra and lifetimes of tomamycin-DNA were obtained. The differences, although small, together with other data allowed the identification of two enantiomeric adducts.

3. SYNCHRONOUS SCANNING

Rather than varying either the excitation or emission monochromators, if both are scanned synchronously with a separation of their wavelength corresponding to the wavelength difference between peaks in the excitation and emission spectra, then the changes in fluorescence intensities will be maximized. With this approach, after release of BP dihydrodiol epoxide (DE) adducts as tetraol, 20 fmol/100 μg DNA have been detected, which corresponds to 1 adduct per 1.4×10^7 bases. Depending upon the wavelengths chosen, the resulting spectra may be as simple as a single peak. Thus while sensitivity is increased, the certainty that the peak results from the specific adduct for which the analysis is being performed is reduced. This approach has enabled the detection of fluorescence in samples of DNA from exposed epidermis and lymphocytes from coke-oven workers and in 1 of 30 samples of lymphocyte DNA from aluminum plant workers. It was also used as one step towards the identification of BPDE adducts in placental DNA. The problems associated with this approach have been clearly demonstrated using second-derivative analysis of the data. However, after HPLC and immunoaffinity chromatography, spectra characteristic of BP tetraol are obtained. More recently synchronous fluorescence, in conjunction with immunoaffinity chromatography and ^{32}P-postlabeling has been used to characterize PAH-DNA adducts in human placenta. Manchester *et al.*

4. FLUORESCENCE LINE NARROWING

In contrast to synchronous scanning, fluorescence line narrowing (FLN) provides a more characteristic fingerprint spectrum. If molecules are isolated from each other by a uniform matrix and cooled to near absolute zero OR to about 4 K to remove vibrational effects, then the absorption spectra of those molecules and their corresponding emission spectra may become very sharp. While these techniques have been widely applied to simple organic molecules, including, for example, the polycyclic aromatic hydrocarbons only more recently have spectra been obtained for PAH–DNA adducts. For these molecules an ideal isolation matrix is not available. PAH–DNA adducts have been analyzed in cooled ethylene glycol–water–ethanol glasses, but the chromophores are not in a uniform environment and still show broad spectra if excited with conventional light sources. Use of a laser, however, enables the few molecules absorbing at the narrowbandpass of the laser to be excited, and these may emit characteristically sharp fluorescence spectra. Such spectra allow the resolution of closely related adducts without prior physical separation of the adducts. Analysis of DNA adducts using this approach has now reached the femtomole level, which makes it similar to the enzyme-linked immunosorbent assay (ELISA) in sensitivity.

Nonphotochemical hole burning has been useful in the analysis of DNA adducts by FLN. As samples are irradiated the fluorescence intensity may decrease for several reasons. The adduct may photochemically decompose, the rate of which depends upon its structure and conformation. The minor adducts from BPDE, particularly those from the $(-)$ enantiomer that tend to intercalate, are most sensitive to photochemical dissociation. This chemistry has been used in the location of sites of modification of BPDE adducts. In a glass the excited molecules may relax into slightly different microenvironments with different excitation spectra. These molecules are not destroyed but are no longer excited by the laser. Recovery from nonphotochemical hole burning may be achieved by warming the sample, normally to 20–40 K, or irradiating with white light so the hole will be repopulated as excited species relax into it.

5. APPLICATIONS IN IMMUNOLOGY

The selectivity of immunoassays and the high speed and automation with which they can be performed makes them valuable in the analysis of large numbers of samples. They are applicable to the parent carcinogen, metabolites, or DNA adducts. The limit on the sensitivity of the assay often depends upon the combination of the affinity of the antibody for the antigen and the lack of cross-reactivity with other components in the mixture being tested. Biologically significant levels of DNA modification are typically less than 1 adduct per 10^6 bases. Unless the adducts are first purified, the cross-reactivity with unmodified DNA must be very low. For this reason and the increased difficulty associated with obtaining low backgrounds caused by the hydrolysis of the indicator substrate, often the ELISA use color rather than fluorescence as the end points, although the latter is in principle more sensitive. Analysis of DNA adducts in single cells by immunofluorescence microscopy or image analysis not only enhances the sensitivity of immunological assays based on quantity of DNA needed but also enables the identification of a subpopulation of modified cells. A comparison has been made of immunofluorescence and immunoautoradiography for the N-acetoxy-2-acetylaminofluorene adducts formed in DNA human fibroblasts exposed *in vitro*. The limits of detection occurred at 20% and 60% survival, respectively. Considerable progress is being made in data collection and image analysis; thus the sensitivity of immunofluorescence is increasing significantly. DNA adducts formed *in vivo* have also been detected although generally only as a result of high exposures to genotoxic compounds. Examples of compounds analyzed by these approaches include alkylating agents, psoralen, thymidine dimers, and benzo[a]pyrene adducts.

Time-resolved fluorescence spectroscopy has been used to circumvent some of the problems described above that are associated with background fluorescence. Chelates of some lanthanides, typically those of Eu^{3+} and Tb^{3+}, and certain β-diketone or dicarboxylic acids are highly fluorescent. In addition, their excitation and emission maxima, which vary with the chelating agent used, are well separated and sharp. Depending upon the ion used, their lifetimes are 10^3–10^6 times longer than those of typical fluorescent probes. Excitation by a brief flash of microsecond duration followed by a delay of several hundred μs allows most background fluorescence to decay before that of the lanthanide chelate is measured. The systems operate between 10 and 1000 Hz. In a recent paper, measurement of cortisol serum levels in the picomole range was reported, but, based upon the fluorescence intensity of the complex, analysis in the femtomole range should be possible.

Fluorescence polarization measurements, which depend upon the decreased rate of rotation of fluorescent probes when bound to antibodies, has been reviewed recently. There are two important limitations of this approach.

It is most suitable for the detection of low molecular weight compounds since the rate of rotation, and hence the extent of polarization of the fluorescent ligand, must be reduced upon binding to the antibody. In principle, the assay can detect a high-molecular-weight compound in a competitive assay. The second difficulty is the low sensitivity achieved, which is better suited to the analysis of drug metabolites than DNA adducts. Lifetimes can also be determined by phase-resolved fluorescence, and this approach has been applied to the measurement of phenobarbital.

6. FLOW CYTOMETRY

In this technique individual cells are separated and analyzed as they pass through the detector. At that time many properties, including fluorescence, can be measured. Flow cytometry has been widely used to identify small cell subpopulations as needed, for example, in the routine pathological detection of bladder cancer. These approaches may be complemented by fluorescence image analysis. β-Globin mutants at a frequency of 1 in 10^8 cells can be detected by flow or scanning cytofluorometry. Recently, cells differentially incorporating bromodeoxyuridine have been distinguished by subsequently labeling the cells with two dyes: one for total DNA and the other for bromo-deoxyuridine content.

7. HIGH-PERFORMANCE LIQUID CHROMATOGRAPHY

Alkylation of DNA produces many fluorescent derivatives that may be isolated and separated by HPLC using their fluorescence for selective detection. The sensitivity of this approach makes it most suitable for the analysis of *in vitro* modified samples or samples from animal exposed to high doses. The etheno derivative formed from chloroacetaldehyde or vinyl chloride epoxide and guanine residues has been analyzed in this way. Similar methods were used to detect the reaction products from malondialdehyde and DNA. Degradation products from aflatoxin–DNA adducts have been detected in human urine by HPLC analysis. Aflatoxin–lysine adducts in albumin have been found in those exposed to this fungal toxin through their diets. As little as 0.5 ml serum or plasma was used, and the adducts were purified by immunoaffinity chromatography prior to HPLC analysis (Wild *et al.* and Sabbioni).

8. DRUG METABOLISM

Fluorescent measurements in the study of drug metabolism are widely used. One of the early assays for benzo[a]pyrene metabolism was based upon the formation of predominantly 3-hydroxy BP, which could be separated by extraction from the parent hydrocarbon. This assay is now often replaced by a more detailed analysis of the metabolites formed by HPLC separation. Use of ethoxyresorufin as a P-450 substrate has the advantage that the assay is continuous but is complicated by the simultaneous reduction of the product by the cytochrome P-450 reductase. The specificity for different P-450 isoenzymes varies, depending upon the alkoxy group.

9. DNA REPAIR

Ethidium bromide is much more fluorescent in double- than single-stranded DNA. Strand separation, which occurs under alkaline conditions, initiates preferentially at the end of DNA strands, and therefore the rate is increased in nicked DNA. Strand breaks in DNA have been detected using the decrease in fluorescence resulting from increased strand separation. For many compounds this assay agrees well with that of alkaline elution, although exceptions have been reported.

10. APPLICATION IN DNA SEQUENCING

DNA sequence analysis, when first introduced, was a highly labor-intensive process. Major improvements have been made in the automation of this analysis by using fluorescent labels. Two distinct but related approaches have been developed that circumvent the use of conventional gels. These methods have been recently described. The location of specific sequences within DNA fragments or even chromosomes is possible by hybridization with fluorescently labeled probes. This in turn allows for easier automation of, for example, the measurements of sister chromatid exchanges.

11. SUMMARY AND FUTURE PROSPECTS

Fluorescence is a highly sensitive technique that has been used directly and interfaced with other methods not only to detect compounds but also to provide structural and conformational information. Photons can be measured with high accuracy, but for many samples, background fluorescence and scattered light from the excitation source often limit the sensitivity. Improved instrument design can help, but the application of lasers that can provide strong but brief pulses of highly monochromatic light, together with extensive data manipulation both during and after collection, has expanded the range of information that can be collected and interpreted. Thus, for routine assays, polarization studies are not feasible except with an automated instrument. As the sensitivity of fluorescence is increased to that of radiolabeled assays, there will be a tendency for the latter to be replaced. As new technologies develop, for example, those from our increased knowledge of the structure of the human genome, the applications of fluorescence will find an expanding role (Lichter *et al.* and Wilson *et al.*).

REFERENCES

H. Autrup, K. A. Bradley, A. K. M. Shamsuddin, J. Wakhisi and Q. Wasunna, Detection of putative adduct with fluorescence characteristics identical to 2,3-dihydro-2-(7′-guanyl)-3-hydroxyaflatoxin B_1 in human urine collected in Murang's district, Kenya, *Carcinog.* **4**, 1193–1195 (1983).

M. D. Barkley, S. Cheatham, D. E. Thurston, and L. H. Hurley, Pyrrolo[1,4]-benzodiazepine antitumor antibiotic: Evidence for two forms of Tomaymycin bound to DNA, *Biochem.* **25**, 3021–3031 (1986).

J. R. Barrio, J. A. Secrist, and N. J. Leonard, Fluorescent adenosine and cytosine derivatives, *Biochem. Biophys. Res. Commun.* **46**, 597–604 (1972).

R. A. Becker, L. R. Barrows, and R. C. Shark, Methylation of liver DNA guanine in hydrazine hepatoxicity: dose-response and kinetic characteristics of 7-methyl-guanine and O^6-methylguanine formation and persistence in rats, *Carcinog.* **2**, 1181–1188 (1981).

M. A. Bedell, M. C. Cyroff, G. Doerjer, and J. A. Swenberg, Quantitation of etheno adducts by fluorescence detection. In: *The Role of Cyclic Nucleic Acid Adducts in Carcinogenesis and Mutagenesis*, IARC Scientific Publication No. 70, eds. B. Singer and H. Bartsch, Oxford Univ. Press, London, pp. 425–436 (1986).

H. C. Birnboim and J. J. Jevcak, Fluorometric method for rapid detection of DNA strand breaks in human white blood cells produced by low doses of radiation, *Cancer Res.* **41**, 1889–1892 (1981).

T. C. Boles, and M. E. Hogan, High-resolution mapping of carcinogen binding sites on DNA, *Biochem.* **25**, 3039–3043 (1986).

F. V. Bright and L. B. McGown, Homogenous immunoassay of phenobarbital by phase-resolved fluorescence spectroscopy, *Talanta* **32**, 15–18 (1985).

S. Cheatham, A. Kook, L. H. Hurley, M. D. Barkley, and W. Remers, One- and two-dimensional ¹H NMR, fluorescence, and molecular modeling studies on the tomaymycin-d(ATGCAT)₂ adduct. Evidence for two covalent adducts with opposite orientations and stereo- chemistries at the covalent linkage site, *J. Med. Chem.* **31**,583–590 (1988).

R. S. Cooper, R. Jankowiak, J. M. Hayes, L. Pei-qi, and G. J. Small, Fluorescence line narrowing spectrometry of nucleoside-polycyclic aromatic hydrocarbon adducts on thin-layer chromatographic plates, *Anal. Chem* **60**, 2692–2694 (1988).

H. A. Crissman and J. A. Steinkamp, A new method for rapid and sensitive detection of bromodeoxyuridine in DNA-replicating cells, *Exp. Cell Res.* **173**, 256–261 (1987).

P. Daudel, M. Duquesne, P. Vigny, P. L. Grover, and P. Sims, Fluorescence spectral evidence that benzo[a]pyrene-DNA products in mouse skin arise from diol epoxides, *FEBS Lett.* **57**, 250–253 (1975).

N. J. Dovichi, J. C. Martin, and R. A. Keller, Attogram detection limit for aqueous dye samples by laser-induced fluorescence, *Science* **219**, 845–947 (1983).

E. P. Diamandis, V. Bhayana, K. Conway, E. Reichstein, and A. Papanastasiou-Diamandis, Time-resolved fluoroimmunoassay of cortisol in serum with a europium chelate as label, *Clin. Biochem.* **21**, 291–296 (1988).

D. R. Dutton, D. A. Reed, and A. Parkinson, Redox cycling of resorufin catalyzed by rat liver microsomal NADPH-cytochrome P450 reductase, *Arch. Biochem. Biophys.* **268**, 605–616 (1989).

H. V. Gelboin, Benzo[a]pyrene metabolism, activation and carcinogenesis: Role and regulation of mixed-function oxidases and related enzymes, *Physiol. Rev.* **60**, 1107–1166 (1980).

J. W. Gray, P. N. Dean, J. C. Fuscoe, D. C. Peters, B. J. Trask, G. J. van den Engh, and M. A. Van Dilla, High-speed chromosome sorting, *Science* **238**, 323–329 (1987).

T. M. Guenthner, B. Jernstrom, and S. Orrenius, On the effect of cellular nucleophiles on the binding of metabolites of 7,8-dihydroxy-7,8-dihydrobenzo[a]pyrene and 9-hydroxybenzo[a]pyrene to nuclear DNA, *Carcinog.* **1**, 407–418 (1980).

D. Haidukewych, Therapeutic drug monitoring by automated fluorescence polarization immunoassay, *Immunoassay Technol.* **2**, 71–103 (1987).

A. Haugen, G. Becher, C. Benestad, K. Vahakangas, G. E. Tivers, M. J. Newman, and C. C. Harris, Determination of polycyclic aromatic hydrocarbons in the urine, benzo[a]pyrene diol epoxide-DNA adducts in lymphocyte DNA, and antibodies to the adducts in sera from coke oven workers exposed to measured amounts of PAHs in the work atmosphere, *Cancer Res.* **46**, 4178–4183 (1986).

C. C. Harris, K. Vahakangas, M. J. Newman, G. E. Trivers, A. Shamsuddin, N. Sinopoli, D. L. Mann, and W. E. Wright, Detection of benzo[a]pyrene diol epoxide-DNA adducts in peripheral blood lymphocytes and antibodies to the adducts in serum from coke oven workers, *Proc. Natl. Acad. Sci. USA* **82**, 6672–6676 (1985).

V. Heisig, A. M. Jeffrey, M. J. McGlade, and G. J. Small, Fluorescence line narrowed spectra of polycyclic aromatic carcinogen-DNA adducts, *Science* **223**, 289–291 (1984).

K. Hemminki, Fluorescence studies of DNA alkylation by epoxides, *Chem. Biol. Interact.* **28**, 269–278 (1979).

L. Hood, Biotechnology and medicine of the future, *J. Am. Med. Assoc.* **259**, 1837–1844 (1988).

V. Ivanovic, N. Geacintov, and I. B. Weinstein, Cellular binding of benzo[a]pyrene to DNA characterized by low temperature fluorescence, *Biochem. Biophys. Res. Commun.* **70**, 1172–1179 (1976).

R. Jankowiak, R. S. Cooper, D. Zamzow, G. J. Small, G. Doskocil, and A. M. Jeffrey, Double spectral selection using fluorescence line narrowing spectrometry and nonphotochemical hole burning as applied to analytical studies of DNA and globin adducts, *J. Chem. Toxicol.* **1**, 60–68 (1988).

A. M. Jeffrey, DNA modification by chemical carcinogens, *J. Pharmacol. Ther.* **83**, 241–245 (1985).

A. M. Jeffrey, Alternative methods for/to the analysis of carcinogen-DNA adducts. In: *Primary Changes and Control Factors in Carcinogenesis*, eds. T. Friedberg and F. Oesch, Deutscher Fachschriften-Verlag, Wiesbaden, pp. 156–160 (1986).

B. Jolles, P. Demerseman, O. Chalvet, H. Strapelias, T. Herning, R. Royer, and M. Duquesne, Reactivity with DNA of three pyrenofuran analogues of benzo[a]pyrene and benzo(e)pyrene, *Nucl. Acids Res.* **15**, 9487–9497 (1987).

H. W. S. King, M. H. Thompson, and P. Brookes, The role of 9-hydroxy-benzo[a]pyrene in the microsome mediated binding of benzo[a]pyrene to DNA, *Int. J. Cancer* **18**, 339–344 (1976).

S. R. Koepke, M. B. Kroeger-Koepke, W. Bosan, B. J. Thomas, W. G. Alvorod, and C. J. Michejda, Alkylation of DNA in rats by N-nitrosomethyl-(2-hydroxy-ethyl)amine: dose response and persistence of the alkylated lesion in vivo, *Cancer Res.* **48**, 1537–1542 (1988).

S. A. Lesko, W. Li, G. Zheng, D. Callahan, D. S. Kaplan, W. R. Midden, and P. T. Strickland, Quantitative immunofluorescence assay for cyclobutyldithymidine dimers in individual mammalian cells, *Carcinog.* **10**, 641–646 (1989).

P. Lichter, C. J. Tang, K. Call, G. Hermanson, G. A. Evans, D. Housman, and D. C. Ward, High-resolution mapping of human chromosome 11 by *in situ* hybridization with cosmid clones. *Science* **247**, 64–69 (1990).

D. K. Manchester, A. Weston, J.-S. Choi, G. E. Trivers, P. V. Fennessey, E. Quintana, P. B. Farmer, D. L. Mann, and C. C. Harris, Detection of benzo[a]pyrene diol epoxide DNA adducts in human placenta, *Proc. Natl. Acad. Sci. USA* **85**, 9243–9246 (1988).

D. K. Manchester, V. L. Wilson, I. C. Hsu, J. S. Choi, N. B. Parker, D. L. Mann, A. Weston, C. C. Harris, *Carcinogenesis* **11**, 553–559 (1990).

M. L. Mendelsohn, W. L. Bigbee, E. W. Branscomb, and G. Stamatoyannopoulos, The detection and sorting of rare sickle-hemoglobin containing cells in normal human blood, *Flow Cytometry* **4**, 311–313 (1980).

R. C. Moschel and N. J. Leonard, Fluorescent modification of guanine. Reaction with substituted malondialdehydes, *J. Org. Chem.* **41**, 294 (1976).

R. C. Moschel, W. R. Hudgins, and A. Dipple, Fluorescence of hydrocarbon-deoxyribonucleoside adducts, *Chem. Biol. Interact.* **27**, 69–79 (1979).

M. A. Muysken-Schoen, R. A. Baan, and P. H. M. Lohman, Detection of DNA adducts in N-acetoxy-2-acetylaminofluorene-treated human fibroblasts by means of immunofluorescence microscopy and quantitative immunoautoradiology, *Carcinog.* **6**, 999–1004 (1985).

F. Oesch and G. Doerjer, Detection of $N^{2,3}$-ethenoguanine in DNA after treatment with chloroacetaldehyde in vitro, *Carcinog.* **3**, 663–665 (1982).

W. L. Parry and G. P. Hemstreet, Cancer detection by quantitative fluorescence image analysis, *J. Urol* **139**, 270–274 (1988).

R. S. Paules, M. C. Poirier, M. J. Mass, S. H. Yuspa, and D. G. Kaufman, Quantitation by electron microscopy of the binding of highly specific antibodies to benzo[a]pyrene-DNA adducts, *Carcinog.* **6**, 193–198 (1985).

D. Pinkel, T. Straume, and J. W. Gray, Cytogenetic analysis using quantative, high sensitivity, fluorescence hybridization, *Proc. Natl. Acad. Sci. USA* **83**, 2934–2938 (1986).

J. M. Prober, G. L. Trainor, R. J. Dam, F. W. Hobbs, C. W. Robertson, R. J. Zugursky, A. J. Cocuzza, M. A. Jensen, and K. Baumeister, A system for rapid DNA sequencing with fluorescent chain-terminating dideoxynucleotides, *Science* **238**, 336–341 (1987).

R. O. Rahn, S. S. Chang, J. M. Holland, and I. R. Shugart, A fluorometric-HPLC assay for quantitating the binding of benzo[a]pyrene metabolites to DNA, *Biochem. Biophys. Res. Commun.* **109**, 262–268 (1982).

R. O. Rahn, S. S. Chang, J. M. Holland, T. J. Stephens, and L. H. Smith, Binding of benzo[a]pyrene to epidermal DNA and RNA as detected by synchronous luminescence spectrometry at 77 K, *J. Biochem. Biophys. Methods* **3**, 285–291 (1986).

A. E. Rettie, F. M. Williams, M. D. Rawlins, R. T. Mayer, and M. D. Burke, Major differences between lung, skin and liver in the microsomal metabolism of homologous series of resorufin and coumarin ethers, *Biochem. Pharmacol.* **35**, 3495–3500 (1986).

L. Roberts, Genome mapping goal now in reach, *Science* **244**, 424–425 (1989).

G. Sabbioni, Chemical and physical properties of the major serum albumin adduct of aflatoxin B_1 and their implications for the quantification in biological samples. *Chem. Biol. Interact.* **75**, 1–15 (1990).

M. J. Sanders, R. S. Cooper, G. J. Small, V. Heisig, and A. M. Jeffrey, Identification of polycyclic aromatic hydrocarbon metabolites in mixtures using fluorescence line narrowing spectroscopy, *Anal. Chem.* **57**, 1148–1152 (1985).

P. D. Sattsangi, N. J. Leonard, and C. R. Frihart, 1,N^2-Ethenoguanosine and $N^{2,3}$-ethenoguanine. Synthesis and comparison of electronic and spectral properties of these linear and angular triheterocycles related to the Y base, *J. Org. Chem.* **42**, 3292–3296 (1977).

B. A. Scalettar, P. R. Selvin, D. Axelrod, J. E. Hearst, M. P. Klein, A fluorescence photochemical study of the microsecond reorientational motions of DNA, *Biophys. J.* **53**, 215–226 (1988).

L. Shugart, J. M. Holland, and R. O. Rahn, Dosimetry of PAH skin carcinogenesis: Covalent binding of benzo[a]pyrene to mouse epidermal DNA, *Carcinog.* **4**, 195–198 (1983).

P. Sims, P. L. Grover, A. Swaisland, K. Pal, and A. Hewer, Metabolic activation of benzo[a]pyrene proceeds by a diol epoxide, *Nature* **252**, 326–327 (1974).

R. C. Stroupe, P. Tokousbalides, R. B. Dickinson, E. L. Wehry, and G. Mamantov, Low-temperature fluorescence spectrometric determination of polycyclic aromatic hydrocarbons by matrix isolation, *Anal. Chem.* **49**, 701–705 (1977).

M. Taningher, R. Bordone, P. Russo, S. Grilli, and S. Parodi, Major discrepancies between results obtained with two different methods for evaluating DNA damage: Alkaline elution and alkaline unwinding. Possible explanations, *Anticancer Res.* **7**, 669–680 (1987).

A. D. Tates, L. F. Bernini, A. T. Natarajan, J. S. Ploem, N. P. Verwoerd, J. Cole, M. H. L. Green, and P. N. Norris, Detection of somatic mutants in man: HPRT mutations in lymphocytes and hemoglobin mutations in erythrocytes, *Mutat. Res.* **213**, 73–82 (1989).

K. Vahakangas, A. Haugen, and C. C. Harris, An applied synchronous fluorescence spectrophotometric assay to study benzo[a]pyrene-diolepoxide-DNA adducts, *Carcinog.* **6,** 1109–1116 (1985).

P. Vigny and M. Duquesne, *Photochem. Photobiol.* **20,** 15–25 (1974).

P. Vigny and M. Duquesne, Fluorometric detection of DNA-carcinogen chemical carcinogen complexes, *DNA* **1,** 85–110 (1979).

P. Vigny, Y. M. Ginot, M. Kindts, C. S. Cooper, P. L. Grover, and P. Sims, Fluorescence spectral evidence that benzo[a]pyrene is activated by metabolism in mouse skin to a diol-epoxide and a phenol-epoxide, *Carcinog.* **1,** 945–954 (1980).

P. Vigny, M. Spiro, R. M. Hudgson, P. L. Grover, and P. Sims, Fluorescence spectral studies on the metabolic activation of chrysene by hamster embryo cells, *Carcinog.* **3,** 1491–1493 (1982).

I. Y. Wang, E. Rasmussen, and T. T. Crocker, Isolation and characterization of an active DNA-binding metabolite of benzo[a]pyrene from hamster liver microsomal incubation systems, *Biochem. Biophys. Res. Commun.* **49,** 1142–1149 (1972).

A. Weston, M. Rowe, M. Poirier, G. Triivers, K. Vahakangas, M. Newman, A. Haugen, and D. Manchester, The application of immunoassays and fluorometry to the detection of polycyclic hydrocarbon-macromolecular adducts and anti-adduct antibodies in humans, *Int. Arch. Occup. Environ. Health* **60,** 157–162 (1988).

A. Weston, D. K. Manchester, M. C. Poirier, J.-S. Choi, G. E. Trivers, D. L. Mann, and C. C. Harris, Derivative fluorescence spectral analysis of polycyclic aromatic hydrocarbon-DNA adducts in human placenta, *Chem. Res. Toxicol.* **2,** 104–108 (1989).

C. P. Wild, B. Chapot, E. Scherer, L. Den-Engelse, and R. Montesano, Application of antibody methods to the detection of aflatoxin in human body fluids. In: IARC Sci. Publ., (ed.), pp. 67–74 (1988).

C. P. Wild, Y. Z. Jiang, G. Sabbioni, B. Chapot, and R. Montesano, Evaluation of methods for quantitation of aflatoxin-albumin adducts and their application to human exposure assessment. *Cancer Res.* **50,** 245–251 (1990).

R. K. Wilson, C. Chen, and L. Hood, Optimization of asymmetric polymerase chain reaction for rapid fluorescent DNA sequencing. *Biotechniques* **8,** 184–189 (1990).

X. Y. Yang, F. P. Gasparro, V. A. DeLeo, and R. M. Santella, 8-Methoxypsoralen-DNA adducts in patients treated with 8-methoxypsoralen and ultraviolet A light, *J. Invest. Dermatol.* **92,** 59–63 (1989).

D. Zinger, N. Geacintov, and R. G. Harvey, Conformations and selective photodissociation of heterogeneous benzo[a]pyrene diol epoxide enantiomer-DNA adducts, *Biophys. Chem.* **27,** 131–138 (1987).

Detection of Alkylated DNA Adducts in Human Tissues

Christopher P. Wild and Ruggero Montesano
Unit of Mechanisms of Carcinogenesis
International Agency for Research on Cancer
Lyon, France

1. INTRODUCTION

Exposure of humans to the alkylating agents N-nitrosamines (NNO) results from the presence of these carcinogens in the environment and/or from their formation *in vivo* from various precursors (e.g., nitrates, nitrites, secondary amines) (see Bartsch and Montesano, 1984). It has been estimated that the daily intake of NNO in the U.S. is approximately 17 μg/person from cigarette smoking, 0.32–20 μg from diet, and 10–120 μg from occupational exposure (National Research Council, 1981). Similar estimations have been made for other parts of the world (see Preussman and Eisenbrand, 1984).

These chemicals have been shown to be carcinogenic in a variety of animal species (Bogovski and Bogovski, 1981); however, there is at present no reliable evidence that any specific human cancers are causally associated with exposure to NNO. This is probably due to the poor sensitivity of the epidemiological studies performed and in particular, the difficulty in defining quantitatively an exposure that (i) might affect a large proportion of the population; (ii) occurs in general at a relatively low level, and (iii) presents a difficulty in defining other factors that contribute and modify the carcinogenic effect of these chemicals. These limitations of epidemiological studies are not confined to this particular situation but apply to various other environmental exposures. Consequently, in recent years laboratory scientists have attempted to develop methods that would permit a better assessment of exposure at an individual level to carcinogens and thus possibly increase the sensitivity of epidemiological studies (see Montesano *et al.*, 1987).

In fact, it is predictable from experimental studies that environmental exposure data (i.e., levels of carcinogens in food, tobacco, ambient air, etc.) would not necessarily provide a reliable indication of exposure or biologically effective dose for an individual or a population; for example, one would not expect to observe great differences in exposure among populations with different risks of a given cancer attributable to NNO, based on such environmental measurements, if the exposure at an individual level is predominantly determined by factors that affect their *in vivo* formation or their metabolism. These considerations explain some of the difficulties of epidemiological studies in ascertaining the degree of causal association between certain cancer(s) in man and exposure to NNO. This article describes the present methodology for the measurement of DNA alkylation adducts and reports the results so far obtained of their presence in human tissues. In general, three considerations are paramount: (i) which adduct to measure, (ii) the cells or tissue in which to make the measurement, and (iii) which method to employ in these measurements. As is discussed below, these considerations are interrelated.

2. ADDUCT FORMATION

The monofunctional alkylating NNO, following metabolism by P450 microsomal enzymes, results in alkylation at the N or O atoms of the various DNA bases. The proportion of alkylation at the different sites in DNA varies with the type of alkylating agent (e.g., dimethyl- versus diethylnitrosamine), and the N-7 position of guanine, O^6-alkylguanine, and phosphotriester are preferentially alkylated, whereas a lower alkylation level is observed, for example, at the O^4 position of thymine and at other sites (see Singer and Grunberger, 1983, Fig. 1). The initial level of these DNA adducts depends mainly on the nucleophilicity of the N or O atoms of the DNA bases with the NNO metabolites; however, their persistence and accumulation during continuous exposure is greatly affected by the efficacy of the tissue or cell to repair these adducts and by cell turnover (see Belinsky *et al.*, 1987; Planche-Martel, *et al.*, 1985). For example, while 7-methyldeoxyguanosine (7-medG) and phosphotriesters are slowly removed from DNA, O^6-methyldeoxyguanosine (O^6-medG) in most instances is actively removed by an efficient repair process. In addition, the biological importance of the adducts in, for example, inducing mutations is different, with O^6-alkyldG being promutagenic while 7-alkyldG is not directly mutagenic (Saffhill *et al.*, 1985).

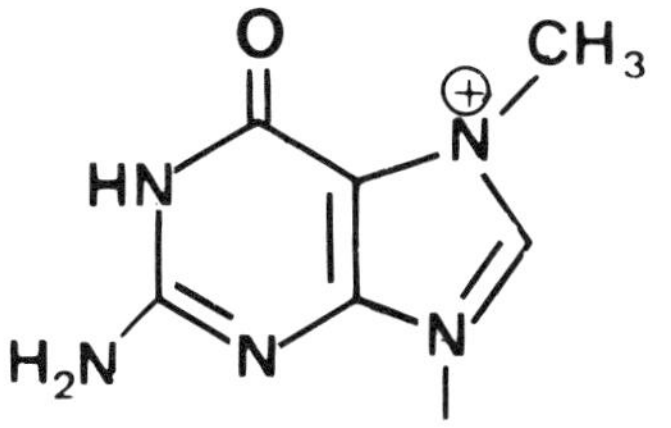

7-methylguanine

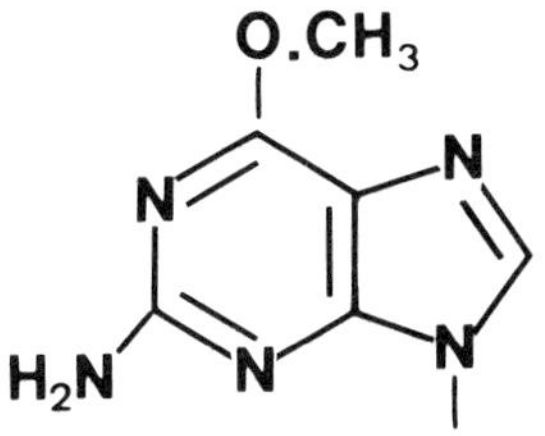

O^6-methylguanine

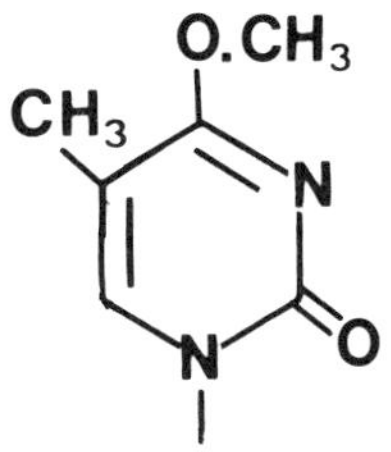

O^4-methylthymidine

Figure 1. Structure of some methylated nucleosides.

Thus the level of DNA adducts at a given time is an integrated value of a number of variables that include actual exposure, absorption and body distribution, tissue specific activation, chemical stability of the metabolites and DNA adducts, efficacy of DNA repair enzymes, and DNA replication. These variables can be affected by genetic and environmental factors and in the case of NNO by factors that could enhance or inhibit the *in vivo* formation of these carcinogens. The fact that the formation of these DNA adducts is at the end of such a complex chain of events makes this approach particularly informative in assessing differences in levels among various individuals or populations. However, particular awareness should be paid to the above-noted contributing variables in the interpretation of the results.

3. METHODS OF DETECTION

The methods that have been applied to the measurement of methylated DNA adducts in human tissues are described, and the development of these methods is outlined together with a discussion of other potential methodological approaches that can be envisaged. A critique is made of the suitability of these methods for human exposure assessment within the framework of epidemiological studies. The discussion is limited to DNA adducts in cellular DNA, as opposed to nucleic acid adducts excreted in the urine, and focuses mainly on the methods that have already been applied to human DNA.

**TABLE 15–1. Assay Requirements for
Measurements of DNA Alkylation Adducts in
Epidemiological Studies**

High sensitivity (low false-negatives)
High specificity (low false-positives)
Long-term exposure information
Quantitative measure of DNA damage in target organ
Large sample capacity
Noninvasive sampling

3.1. General Methodological Requirements

The characteristics of a laboratory assay of alkylated DNA that would
be suitable for epidemiological investigations are given in Table 1. No assay
to date satisfies all these demands, and most have been developed primarily
with sensitivity and specificity as their major criteria. These qualities alone
may not be sufficient to make the assay of use in an epidemiological study.

A summary of the methods used on human samples is presented in Table
2. It is important to stress that assay sensitivity is linked not only to the
theoretical detection limit in the assay but equally to the amount of DNA
available for assay or the amount of DNA that can be incorporated into the
assay. For example, the modification level that can be detected in a radioimmunoassay (RIA) high-pressure liquid chromatography (HPLC) method for
O^6-medG may be far lower than the ^{32}P-postlabeling HPLC method when
the quantity of DNA is unlimited, i.e., with surgical tissue samples of 10
mg DNA, where 10 fmol adduct per mg DNA can be measured by RIA
HPLC. However, if only 100 µg DNA were available, then the ^{32}P-postla-
beling method would be some three times more sensitive than RIA because
in this case only 1000 fmol O^6-medG per mg DNA can be measured by RIA
HPLC.

In this regard, the immuno-slot-blot technique reported originally
by Nehls *et al.* (1984) may provide a valuable alternative when very low
amounts of DNA are available. Using 3 µg DNA this assay can detect 0.1
fmol of O^4-ethylthymidine (O^4etT) (~33 fmol mg^{-1}), which is equivalent
to the sensitivity obtained with 20 mg DNA by HPLC immunoassay (Table
2). However, one disadvantage is that the specificity of the immuno-slot-
blot method relies completely upon the antibody properties, and this aspect
is considered below. In general, then, the method of choice for adduct
analysis (even when considering one single factor such as sensitivity) will
vary depending upon the nature and availability of the DNA sample.

Specificity, when measuring very low levels of adduct (less than 1 in
10^6 unmodified bases), is a particular challenge to the assay methodologies.
The antibody specificity or the chromatographic properties of an adduct alone

TABLE 15–2. Methods Applied to the Detection of DNA Alkylation Adducts in Human Tissues

Methods	Adduct*	Sensitivity			Specificity Determined by	Reference
		Detection limit (fmol)	Quantity DNA used (mg)	Modification level (fmol mg^{-1})		
RIA + HPLC	O^6-medG	105[†]	3[‡]	35	(i) Chromatography	Umbenhauer et al., 1985
	O^6-etdG	75	3	25	(ii) Antibody specificity	Umbenhauer et al., 1985
	O^4-etT	600	20	30		Huh et al., 1989
ELISA + HPLC	O^6-medG	720	2	360		Foiles et al., 1988
³²P-postlabeling + HPLC	O^6-medG	35	0.1	350	Chromatography by (i) HPLC; (ii) TLC[§]	Wilson et al., 1988, 1989
Fluorescence + HPLC[‖]	O^6-meG	500	1	500	(i) Chromatography	Belinsky et al., 1987
	7-meG	10,000	1	10,000	(ii) Fluorescence	(and references therein)

*O^6-medG, O^6-meG: O^6-methyldeoxyguanosine, O^6-methylguanine; O^6-etdG, O^6-ethyldeoxyguanosine; O^4-etT, O^4-ethylthymidine.

[†]Detection limit for immunoassays is based on 6 times the absolute limit in the assay, to allow repeat of assay in triplicate.

[‡]In practice the amount of DNA that may be used is not restricted to that presented in the table but is limited only by the capacity of the HPLC column. Consequently the lower level of modification detectable will vary. Amounts of DNA presented in the table are actual values from the published studies concerned.

[§]TLC = thin-layer chromatography.

[‖]Data are the lowest values reported by Swenberg and colleagues in experimental studies (Belinsky et al., 1987).

is insufficient to identify unequivocally a specific adduct, and a minimum requirement should be the presence of at least two determinants of spec- ificity, e.g., chromatography and immunoassay combined (see Table 2). An improved approach would be either to assay the same sample by two of the different methods listed in Table 2, e.g., HPLC + immunoassay and HPLC + fluorescence or one of those listed in the section on further developments (see below), or to subject one-half of the sample to a treatment that is known to specifically remove the adduct of interest, i.e., chemical depurination, for 7-methylguanine (7-meG), or exposure to a specific repair enzyme (for O^6- methylguanine) and to use this as a control for the untreated half of the sample.

Ideally it would be desirable to have sufficient alkylation adduct to carry out structural analysis of the adduct by gas chromatography–mass spectro- metry (GC-MS), as has been done for benzo[a]pyrene residues in DNA pooled from several human placentas (Manchester *et al.*, 1988). However, although a theoretical sensitivity for 7-medG of 1 pmol by HPLC–tandem MS has been reported (Chang *et al.*, 1986) it is difficult to envisage the routine application of this method in epidemiological studies where large sample capacity (Table 1) is a critical criteria and where the amount of DNA available is restricted.

In addition, the use of the multiple analyses suggested above requires more sample and may consequently reduce the sensitivity of the method because less DNA is available for any one type of assay. Also, while DNA from surgical tissue may allow multiple analyses, the application of such samples in epidemiological studies is limited because of the invasiveness of sampling. In contrast, peripheral blood cell DNA is more widely available, but the same rigorous set of analyses cannot be performed due to lower amounts of available DNA. The solution to optimizing the approach is one area in which the continuous interaction of epidemiologists and laboratory scientists is essential.

Linked to the question of method specificity is the background level of adducts present in human samples. For example, it is important to establish the level of adducts in populations at low exposure, where a low incidence of specific cancer occurs, in order to have a database with which to compare individuals and/or populations with a suspected high exposure. This has been of particular relevance to the question of methylation adducts and the choice of whether to measure DNA or protein adducts. For methylation the approach to measuring protein adducts has been limited because, for example, the background levels of S-methylcysteine in rat hemoglobin required a dose of 10 mg/kg dimethylnitrosamine to give a detectable increase in alkylation level (Bailey *et al.*, 1981). This limitation may be overcome for alkyl adducts other than methyl (Farmer *et al.*, 1988). The corresponding background level of methyl adducts in DNA, while much lower, is poorly defined at present due to the lack of data available (see below).

**TABLE 15–3. Antibodies Available to
DNA Alkylation Adducts**

Alkyl Group	Modification
Methyl	7dG; 7G; O^6dG; O^4T; O^2T; 3A
Ethyl	O^6dG; O^4T; O^2T; 3A
Propyl	O^6dG
Butyl	O^6dG; O^4T; O^2T; 3A

Sources: Strickland and Boyle, 1984; Adamkiewicz *et al.*, 1986; Degan *et al.*, 1988; Shuker and Farmer, 1988; Shuker, 1988; Rajewsky *et al.*, personal communication; and references in Table 2.

3.2. Immunoassay

3.2.1. Antibodies

Antibodies to many alkylation adducts are now available in a number of laboratories (Table 3). These monoclonal (rat or mouse) and polyclonal (rabbit) antibodies have in the majority of cases been raised against the corresponding ribose compound by periodate coupling to the carrier proteins bovine serum albumin or hemocyanin. Exceptions are the antibodies to the 7-meG (Shuker, 1988) and 3-alkyladenine (Shuker and Farmer, 1988; Rajewsky and Eberle personal communication) where analogues of the adducts were prepared to allow coupling in the absence of a sugar moiety. The detection limits in immunoassay for the antibodies tested in Table 3 are generally between 1 and 500 fmol with the most sensitivity normally being obtained with the larger butyl and ethyl adducts as compared with methyl for the same modified parent deoxynucleoside. A comparison of the properties of the monoclonal and polyclonal antibodies shows that in terms of the lower limit of detection in immunoassay and of specificity with regard to other modified or unmodified deoxynucleosides, they are very similar (e.g., Wild *et al.*, 1983). However, the availability of unlimited quantities of monoclonal antibodies is advantageous when supplying reagents for other laboratories and particularly with regard to their application in affinity purification procedures (see below).

3.2.2. Sample preparation

Initially it was hoped that specific antibodies would allow the quantitation of DNA methylation adducts in intact DNA. However, even the most specific monoclonal and polyclonal antibodies have a reaction with unmodified deoxynucleosides at levels 10^5- to 10^7-fold in excess of the modified deoxynucleoside (Müller and Rajewsky, 1978; 1980; Saffhill *et al.*, 1982; Wild *et al.*, 1983). In the case of human DNA modified by environmental levels of alkylating agents, the levels of modification observed to date have generally been lower than 1 adduct in 10^6 unmodified deoxynucleosides (see Fig. 2), and thus a purification step is required. In all the reports shown in Fig. 2 this has been achieved by HPLC, which improves the sensitivity and the specificity of the assay.

3.2.3. Assay

The approaches used to obtain the data in Fig. 2 were RIA except for the work of Foiles *et al.* (1988), where an enzyme-linked immunosorbent assay (ELISA) was used. The sensitivity of the RIA depends upon the availability of a stable tracer of high specific activity (> 10 Ci $\cdot$ mmol^{-1}). In ELISA where a protein–DNA adduct conjugate is used as antigen, sufficient specificity for the adduct as compared to the protein must be obtained. This can be achieved either by removing from polyclonal antisera through affinity chromatography the antibodies recognizing primarily the protein, or by using different proteins for immunization and ELISA antigen to minimize cross-reactivity with protein epitopes.

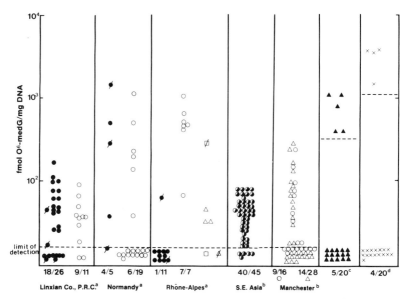

Figure 2. Level and prevalence of positive samples of O^6-methyldeoxyguanosine in human tissues. •, esophagus; ○, stomach; ◑, esophagus or stomach; △, colon; □, liver; ▲, placenta; ×, oral mucosa. The samples with a bar refer to tumor samples. The prevalence of positive samples are indicated at the bottom of each column. Data are from (a) Umbenhauer *et al.*, 1985 and Montesano *et al.*, unpublished data; (b) Saffhill *et al.*, 1988; (c) Foiles *et al.*, 1988; (d) Wild *et al.*, 1989.

3.3. Fluorescence Detection

In the only study (Cooper and Kimbrough, 1980; Herron and Shank, 1980) in which O^6-meG and 7-meG has been detected by HPLC fluorescence in DNA from human tissues, the limit of detection for O^6-meG was 8 μmol per mole guanine, while for the much less fluorescent 7-meG it was 222 μmol per mole guanine. These levels of detection have been reduced 5- to 10-fold (Table 2) with technological improvements but remain at present less sensitive than immunoassay. However, chemical derivatization of the poorly fluorescent 7-meG has been reported, giving a highly fluorescent product (Sabbioni *et al.*, 1987), and together with further improvements in instrumentation that can be envisaged, this method should attain still lower detection limits. A major potential advantage of this approach is the recovery of the adduct for further confirmatory analysis using, for example, immunological or mass spectral characterization.

3.4. ³²P-Postlabeling

In the original studies of Randerath *et al.* (1981, 1984) the highly sensitive ³²P-postlabeling technique was not applicable for detecting low levels of methylated adducts because of the similar elution of modified and unmodified 3′,5′ biphosphate deoxynucleotides. Wilson *et al.* (1988) combined HPLC purification of deoxyribonucleotide-3′-monophosphates, modified and unmodified, with subsequent ³²P-postlabeling and thin-layer chromatography. The limit of detection was 1 adduct in 10^7 parent deoxynucleosides in 100 μg DNA and was applied successfully to the measurement of O^6-methyl- and ethyldeoxyguanosine in human tissues (Wilson *et al.*, 1989; Fig. 2). Recently Shields *et al.* (1989) reported a similar approach to the quantitation of 7-medG with a detection limit, using marker compounds of 1 adduct in 10^6 parent deoxynucleotides. Den Engelse *et al.* (1989) used a ³²P-postlabeling assay for alkylphosphotriesters also following HPLC purification, which achieved a detection limit of 1 dTp(ethyl)dT adduct per 10^8 nucleotides in 1 mg DNA. To date the latter two assays have not been applied to human tissues.

3.5. Other Potentially Applicable Assays

The application of antibodies to human exposure monitoring may be further developed in two directions. First, as a purification step: Affinity chromatography has been used to purify aflatoxin and benzo(a)pyrene DNA adducts (Groopman *et al.*, 1985; Tierney *et al.*, 1986) as well as 3-methyladenine (D. Shuker, personal communication), O^6-medG, and 7-medG (D. P. Cooper, personal communication; C. P. Wild, unpublished data). Purified adducts could subsequently be quantitated by fluorescence, electrochemical detection (see below), ^{32}P-postlabeling, or GC-MS. Second, the application of antibodies to measure alkylation adducts at the single-cell level has been demonstrated in various experimental studies (e.g., Menkveld *et al.*, 1985; Van Benthem *et al.*, 1988). One potential advantage of this method is to have information on adduct levels at a single-cell level, since differences in biologically effective dose to target cell populations within a tissue may exist between individuals but be masked by analyses of DNA extracted from the whole tissue. This has been illustrated in the lung in rats with 4-(*N*-methyl-*N*-nitrosamino)-1-(3-pyridyl)-1-butanone (NNK), where a high level of O^6-medG was observed specifically in the Clara cells as compared to other types of cells within this organ (Belinsky *et al.*, 1987). Certainly an awareness of this cellular heterogeneity is required when interpreting the presence of alkylation adducts in human DNA. It has been argued that the present limits of sensitivity of these methods (5,000 to 10,000 adducts per cell) are insufficient for human DNA-adduct analysis. However, if adducts are concentrated in subpopulations of slowly dividing, repair-deficient cells, then levels of alkylation more than 30-fold higher than average levels in whole-tissue DNA can be observed (Belinsky *et al.*, 1987). Another potential advantage, apart from more specific information regarding intercellular adduct distribution, is that a few hundred cells may provide sufficient material for assay, e.g., lymphocytes obtained from a finger-prick blood sample, buccal cells, cells from a cervical smear, etc.

Other potential advances include the more routine application of mass spectrometry. Sensitive detection of 7-medG (Chang *et al.*, 1986) and O^4-etT after derivatization (Saha *et al.*, 1988) has been reported. Many technological advances have led to the sensitive detection of several carcinogen–macromolecular adducts (see Farmer *et al.*, 1988). As mentioned above, this approach has the major advantage of identifying the chemical nature of the adduct but is limited by the cost and small number of samples that can be analyzed. In addition, a sensitive electrochemical detection of 7-meG and O^6-meG with HPLC has been reported (Park *et al.*, 1989). Both these methods could be used after affinity purification of adducts to increase sample throughput and/or specificity of the approach.

A recent assay development has taken advantage of the specificity and suicide repair activity of the O^6-alkylguanine DNA alkytransferase (AGT) protein (Souliotis and Kyrtopoulos, 1989). DNA is incubated with a known quantity of AGT, and a certain fraction of AGT will be inactivated corresponding directly to the amount of O^6-medG present in the sample. The remaining AGT can then be quantitated by reaction with standard O^6-medG containing oligonucleotide (^{35}S-labeled) to allow an indirect measure of O^6-medG levels in the initial DNA sample. With use of AGT from *E. coli* and with the assumption that it is inactivated only by O^6-methylguanine adducts, then this method should be specific and would be applicable to human studies as it requires only up to 10 μg DNA and can detect 0.5 fmol of O^6-medG.

4. DNA ALKYLATION ADDUCTS IN HUMAN TISSUES

Data from various studies are now available that sought determination of the level of DNA alkylation adducts in human tissues. The main objectives of these studies have been (a) to determine whether these abnormal DNA alkylation adducts were present in human tissues and at what levels; (b) to ascertain if the levels found were attributable to known or suspected environmental exposure(s); (c) to examine if variation in DNA adducts could be observed among individuals at a population level from geographic areas at different incidence of a given cancer; and (d) to provide information on the degree of variability that may exist in the level of these adducts at an individual level. In respect to the above-noted points, the studies completed so far (see Fig. 2) have been informative but at the same time have posed questions that demand further examination and clarification prior to the application and integration of these exposure measurements in epidemiological studies.

In the studies described, the level of DNA alkylation is expressed as femtomole of modified nucleoside per milligram of DNA. The majority of the data refers to O^6-medG and O^4-etT; the information on 7-medG is more limited. The tissues examined are tumorous or nontumorous samples either derived from patients who underwent surgery or from autopsy; the presence of DNA alkylation adducts in peripheral blood cells has been determined in a few instances.

The presence of O^6-medG and 7-medG in human liver DNA was described for the first time in two persons who were poisoned by the intentional addition of dimethylnitrosamine to lemonade and milk that were consumed by the victims; the levels of O^6-medG and 7-medG were, respectively, approximately 2×10^5 and 8×10^5 fmol/mg DNA, which corresponds to an estimated exposure to dimethylnitrosamine greater than 20 mg/kg (Herron and Shank, 1980). More recently, with use of immunoassay and monoclonal antibodies against O^6-medG, the presence of this DNA adduct was detected (Umbenhauer *et al.*, 1985; our unpublished data) in surgical specimens of esophageal and stomach mucosa from patients originating from populations at different risk of cancer at these sites. In Lin-xian County (People's Republic of China), a high-risk area for esophageal and stomach cancer where high exposure to NNO has been suspected, 18 out of 26 samples of esophageal tissue and 9 out of 11 samples of stomach tissue showed levels of O^6-medG above the detection limit (approximately 1 adduct in 10^8 molecules of guanine) with levels ranging from less than 50 and up to 161 fmol/mg DNA. In Normandy, France, a second high-risk area for esophageal cancer, the prevalence of positive samples was also high for the esophagus (4/5), and 6 stomach mucosa samples out of 19 tested had levels of O^6-medG up to approximately 1000 fmol/mg DNA. In the region of Rhône-Alpes (France), a low-risk area for esophageal cancer, only 1 sample of esophageal tissue out of 11 was positive, whereas all 7 samples of stomach mucosa were positive with levels of O^6-medG, similar to that observed in Normandy. Another study (Saffhill *et al.*, 1988) also reported a consistent detection of O^6-medG in esophageal and/or stomach tissues from Southeast Asia (40/45) as compared to stomach (9/16) and colon (14/28) tissues from Manchester (U.K.), although in a few instances a high level of O^6-medG of > 200 fmol/mg DNA was seen (see Fig. 2).

Two studies examined the presence of O^6-medG in placenta (Foiles *et al.*, 1988) and peripheral lung DNA (Wilson *et al.*, 1989) of smoking and nonsmoking individuals. Five out of 20 placentas were shown to contain O^6-medG (360 to 960 fmol/mg DNA), however, no relationship was observed with smoking habits; in the lung tissue this DNA adduct was clearly present in one sample (3100 fmol/mg) and at <1500 fmol/mg DNA in all others while O^6-ethyldG was also present in various lung samples. The presence of O^6-medG was detected in 4 samples out of 20 of oral mucosa DNA from cigarette smokers but not in nonsmokers, and the level of adduct was between 1,300 and 4,200 fmol/mg DNA (Wild *et al.*, 1989).

One study (Huh *et al.*, 1989) searched for the presence of O^4-etdT, an adduct that is repaired slowly and can accumulate upon repeated exposure (Swenberg *et al.*, 1984), and among 33 human liver autopsy specimens from Japan, 30 showed levels of this adduct in their DNA and higher levels (> 30 to 200 adducts/10^{-8} dT) were observed in the DNA of patients with liver cancer or other types of cancers compared with noncancerous diseases (Fig. 3).

5. DISCUSSION

The methods described and the studies so far available reporting the presence of these DNA adducts in human tissues indicate that this approach is sufficiently sensitive and specific to assess human exposure to alkylating agents like NNO, to which it was estimated that human exposures can occur at levels of 1 to 10 μg per person per day and with higher exposures in certain occupations (Preussmann and Stewart, 1984; National Research Council, 1981). However, most of the studies completed so far (see Fig. 2) examined the presence of DNA alkylation adducts in internal organs obtained through surgery or autopsy. While these studies have been significant in demonstrating the presence of promutagenic adducts in the target organ for induction of the tumor, they are limited in their application. Specifically, it is not possible to compare in a random fashion the differences in the level and prevalence of the DNA adducts in various individuals and/or populations suspected of being exposed to different levels of carcinogenic alkylating agents. In this respect, the detemination of these adducts in peripheral blood

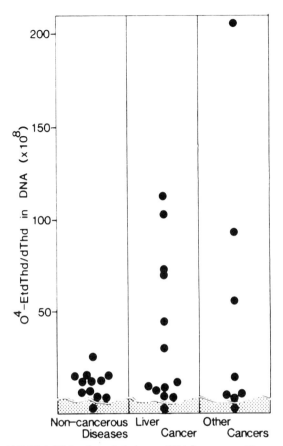

Figure 3. O^4-EtdThd:dThd molar ratio in human DNA from cases with noncancerous diseases, primary liver cancer, or cancers other than liver. One case in each group exhibited a O^4-EtdThd:dThd molar ratio below the detection limit of O^4-EtdTh:dThd $= 3 \times 10^{-4}$ (shaded area) (from Huh *et al.,* 1989).

cell DNA would be valuable. A clear distinction between the use of peripheral blood cells simply as a marker of exposure and their use as a marker of biologically effective dose to the target organ is necessary. Initially, to address the latter of these two points, proper pharmacokinetic data should be measured in experimental animals to determine the relationship between the levels observed in peripheral blood cell DNA and those observed in internal organs with respect to different exposure regimens. These studies, which are critical to the evaluation of the observations in humans, have been remarkably few to date. Recently, antibodies have been produced to allow measurements of 7-medG (Degan *et al.,* 1988), a DNA adduct that, as mentioned previously, occurs at higher levels than other adducts and that may be more persistent than rapidly repaired adducts such as O^6-medG. It has been shown that this DNA adduct can be detected in peripheral blood cell DNA of rats systemically treated with dimethylnitrosamine at doses less than 1 mg/kg and that adduct levels were similar to those seen in liver DNA (Degan *et al.*, 1988). In addition, this adduct has been detected in patients treated with methylating chemotherapeutic drugs (Wild *et al.*, 1988). Preliminary studies (Wild, Likhachev and Montesano, unpublished data) also showed that 7-medG can be detected in peripheral blood cell DNA from heavy cigarette smokers.

As already mentioned, humans are exposed to alkylating carcinogenic NNO (including methylating agents) through various types of exposures including tobacco, and it would be informative to assess and distinguish the relative contributions of these exposures to the formation of these adducts. Tobacco has been shown to contain a specific nitrosamine NNK, which is a DNA alkylating agent, that produced methylation and nonmethylation adducts such as oxobutylated DNA and hemoglobin adducts (Hecht *et al.*, 1988; Carmella and Hecht, 1987). The measurement of these different adducts could differentiate the contribution of tobacco from other sources of exposure in the formation of DNA methylation adducts.

The present view has been restricted to a discussion of the development and validation of methods used to detect DNA alkylation adducts and the problems inherent in their application of assessing human exposure to carcinogenic alkylating agents. The biological relevance of these DNA adducts to mutagenesis and carcinogenesis has been discussed elsewhere (Montesano, 1981; Preussmann and Stewart, 1984; Saffhill *et al.*, 1985; Singer and Grunberger, 1983).

ACKNOWLEDGMENTS

These studies were partly supported by U.S. NIEHS Grant No. 5 UO1 ESO 4281-02 and CEC Contract No. EV4V 0040-F (CD). The authors thank Mrs. Collard-Bianchi for typing the manuscript.

REFERENCES

J. Adamkiewicz, O. Ahrens, G. Eberle, P. Nehls, and M. F. Rajewsky, Monoclonal antibody-based immunoanalytical methods for detection of carcinogen-modified DNA components. In: *The Role of Cyclic Nucleic Acid Adducts in Carcinogenesis and Mutagenesis,* IARC Scientific Publication No. 70, eds. B. Singer and H. Bartsch, IARC, Lyon, France, pp. 403–411 (1986).

E. Bailey, T. A. Connors, P. B. Farmer, S. M. Gorf, and J. Rickard, Methylation of cysteine in hemoglobin following exposure to methylating agents, *Cancer Res.* **41**, 2514–2517 (1981).

H. Bartsch and R. Montesano, Relevance of nitrosamines to human cancer, *Carcinog.* **5**, 1381–1393 (1984).

S. A. Belinsky, C. M. White, T. R. Devereux, J. A. Swenberg, and M.W. Anderson, Cell selective alkylation of DNA in rat lung following low dose exposure to the tobacco specific carcinogen 4-(N-methyl-N-nitrosamino)-1-(3-pyridyl)-1-butanone, *Cancer Res.* **47**, 1143–1148 (1987).

P. Bogovski and S. Bogovski, Animal species in which N-nitroso compounds induce cancer, *Int. J. Cancer* **27**, 471–474 (1981).

S. G. Carmella and S. S. Hecht, Formation of hemoglobin adducts upon treatment of F344 rats with the tobacco-specific nitrosamines 4-(methylnitrosamino)-1-(3-pyridyl)-1-butanone and N'-nitrosonornicotine, *Cancer Res.* **47**, 2626–2630 (1987).

C.-J. Chang, D. J. Ashworth, I. Isern-Flecha, X.-Y. Jiang, and R. G. Cooks, Modification of calf thymus DNA by methylmethane-sulphonate. Quantitative determination of 7-methyldeoxyguanosine by mass spectrometry, *Chem. Biol. Interact.* **57**, 295–300 (1986).

S. W. Cooper and R. D. Kimbrough, Acute dimethylnitrosamine poisoning outbreak, *J. Forensic Sci,* **25**, 874–882 (1980).

P. Degan, R. Montesano, and C. P. Wild, Antibodies against 7-methyldeoxyguanosine: its detection in rat peripheral blood lymphocyte DNA and potential applications to molecular epidemiology, *Cancer Res.* **48**, 5065–5070 (1988).

L. Den Engelse, R. J. de Brij, M. P. M. N. Vissers, and J. T. Lutgerink, Analysis of alkylphosphotriesters by HPLC and ^{32}P-postlabeling, formation and stability of enantiomers, *Proc. Am. Assoc. Cancer Res.* **30**, 132 (1989).

P. B. Farmer, J. Lamb, and P. D. Lawley, Novel uses of mass spectrometry in studies of adducts of alkylating agents with nucleic acids and proteins. In: *Methods for Detecting DNA Damaging Agents in Humans: Applications in Cancer Epidemiology and Prevention,* IARC Scientific Publication No. 89, eds. H. Bartsch, K. Hemminki, and I.K. O'Neill, IARC, Lyon, France, pp. 347–355 (1988).

P. G. Foiles, L. M. Miglietta, S. A. Akerkar, R. B. Everson, and S. S. Hecht, Detection of O^6-methyldeoxyguanosine in human placental DNA, *Cancer Res.* **48**, 4184–4188 (1988).

J. D. Groopman, P. R. Donahue, J. Zhu, J. Chen, and G. N. Wogan, Aflatoxin metabolism in humans: detection of metabolites and nucleic acid adducts in urine by affinity chromatography, *Proc. Natl. Acad. Sci. USA* **82**, 6492–6496 (1985).

S. S. Hecht, T. E. Spratt, and N. Trushin, Evidence for 4-(3-pyridyl)-4-oxobutylation of DNA in F344 rats treated with the tobacco-specific nitrosamines 4-(methylnitrosamino)-1-(3-pyridyl)-1-butanone and N'-nitrosonornicotine, *Carcinog.* **9**, 161–165 (1988).

D. C. Herron and R. C. Shank, Methylated purines in human liver DNA after probable dimethylnitrosamine poisoning, *Cancer Res.* **40**, 3116–3117 (1980).

N.-H. Huh, M. S. Satoh, J. Shiga, M. F. Rajewsky, and T. Kuroki, Immunoanalytical detection of O^4-ethylthymine in liver DNA of individuals with or without malignant tumors, *Cancer Res.* **49**, 93–97 (1989).

D. K. Manchester, A. Weston, J.-S. Choi, G. E. Trivers, P. V. Fennessey, E. Quintana, P. B. Farmer, D. L. Mann, and C. C. Harris, Detection of benzo[a]pyrene diol epoxide-DNA adducts in human placenta, *Proc. Natl. Acad. Sci. USA* **85**, 9243–9247 (1988).

G. J. Menkveld, C. J. Van Der Laken, T. Hermsen, E. Kriek, E. Scherer, and L. Den Engelse, Immunohistochemical localisation of O^6-ethyldeoxyguanosine and deoxyguanosin-8-yl-(acetul)aminofluorene in liver sections of rats treated with diethylnitrosamine, ethylnitrosourea or N-acetylaminofluorene, *Cancer Res.* **6,** 263–270 (1985).

R. Montesano, Alkylation of DNA and tissue specificity in nitrosamine carcinogenesis, *J. Supramol. Struct. Cell. Biochem.* **17,** 259–273 (1981).

R. Montesano, D. M. Parkin, and L. Tomatis, Environmental causes of cancer in man. In: *Cancer Biology and Medicine. Biology of Carcinogenesis,* eds. M. J. Waring and B. A. J. Ponder, MTP Press, Lancaster, pp. 141–163 (1987).

R. Müller and M. F. Rajewsky, Sensitive radioimmunoassay for detection of O^6-ethyldeoxyguanosine in DNA exposed to the carcinogen ethylnitrosourea, *Z. Naturforsch.* **33,** 897–901 (1978).

R. Müller and M. F. Rajewsky, Immunological quantification by high-affinity antibodies of O^6-ethyldeoxyguanosine in DNA exposed to N-ethyl-N-nitrosourea, *Cancer Res.* **40,** 887–896 (1980).

National Research Council, *The Health Effects of Nitrate, Nitrite and N-Nitroso Compounds,* National Academy Press, Washington, D. C. (1981).

P. Nehls, J. Adamkiewicz, and M. F. Rajewsky, Immuno-slot-blot: a highly sensitive immunoassay for the quantitation of carcinogen-modified nucleosides in DNA, *J. Cancer Res. Clin. Oncol.* **108,** 23–29 (1984).

J.-W. Park, K. C. Cundy, and B. N. Ames, Detection of DNA adducts by high-performance liquid chromatography with electrochemical detection, *Carcinog.* **10,** 827–832 (1989).

G. Planche-Martel, A. Likhachev, C. P. Wild, and R. Montesano, Modulation of repair of O^6-methylguanine in parenchymal and non-parenchymal liver cells of rats treated with dimethylnitrosamine, *Cancer Res.* **45,** 4768–4773 (1985).

R. Preussmann and G. Eisenbrand, N-nitroso carcinogens in the environment. In: *ACS Monograph 182, Chemical Carcinogens,* Vol. 1, pp. 829–868 (1984).

R. Preussmann and B. W. Stewart, N-Nitroso carcinogens. In: *ACS Monograph 182, Chemical Carcinogens,* Vol. 2, pp. 643–828 (1984).

K. Randerath, M. V. Reddy and R. C. Gupta, ^{32}P-labeling test for DNA damage, *Proc. Natl. Acad. Sci. USA* **78,** 6126–6129 (1981).

K. Randerath, E. Randerath, H. P. Agrawal and M. V. Reddy, Biochemical (postlabeling) methods for analysis of carcinogen-DNA adducts. In: *Monitoring Human Exposure to Carcinogenic and Mutagenic Agents,* IARC Scientific Publication No. 59, eds. A. Berlin, M. Draper, K. Hemminki, and H. Vainio, IARC, Lyon, France, pp. 217–231 (1984).

G. Sabbioni, P. L. Skipper, G. Büchi, and S. R. Tannenbaum, Isolation and characterization of the major serum albumin adduct formed by aflatoxin B_1 *in vivo* in rats, *Carcinog.* **8,** 819–824 (1987).

R. Saffhill, A.F. Badawi, and C. N. Hall, Detection of O^6-methylguanine in human DNA. In: *Methods for Detecting DNA Damaging Agents in Humans: Applications in Cancer Epidemiology and Prevention,* IARC Scientific Publication No. 89, eds. H. Bartsch, K. Hemminki, and I. K. O'Neill, IARC, Lyon, France, pp. 301–305 (1988).

R. Saffhill, G. P. Margison and P. J. O'Connor, Mechanisms of carcinogenesis induced by alkylating agents, *Biochim. Biophys. Acta* **823**, 111–145 (1985).

R. Saffhill, P. T. Strickland, and J. M. Boyle, Sensitive radioimmunoassays for O^6-n-butyldeoxyguanosine, O^2-n-butylthymidine and O^4-n-butylthymidine, *Carcinog.* **3**, 547–552 (1982).

M. Saha, O. Minnetian, D. Fisher, E. Rogers, R. Annan, G. Kresbach, P. Vouros, and R. Giese, Pentafluorobenzylation of alkyl and related DNA base adducts facilitates their determination by electrophore detection. In: *Methods for Detecting DNA Damaging Agents in Humans: Applications in Cancer Epidemiology and Prevention,* IARC Scientific Publication No. 89, eds. H. Bartsch, K. Hemminki, and I. K. O'Neill, IARC, Lyon, France, pp. 356–360 (1980).

P. G. Shields, A. C. Povey, V. L. Wilson, and C. C. Harris, ^{32}P-postlabeling of N^7-methyldeoxyguanosine (N^7-medG), *Proc. Am. Assoc. Cancer Res.* **30**, 316 (1989).

D. E. G. Shuker, Determination of 7-methylguanine by immunoassay. In: *Methods for Detecting DNA Damaging Agents in Humans: Applications in Cancer Epidemiloogy and Prevention,* IARC Scientific Publication No. 89, eds. H. Bartsch, K. Hemminki, and I. K. O'Neill, IARC, Lyon, France, pp. 296–300 (1988).

D. E. G. Shuker and P. B. Farmer, Urinary excretion of 3-methyladenine in humans as a marker of nucleic acid methylation. In: *Methods for Detecting DNA Damaging Agents in Humans: Applications in Cancer Epidemiology and Prevention,* IARC Scientific Publication No. 89, eds. H. Bartsch, K. Hemminki, and I. K. O'Neill, IARC, Lyon, France, pp. 92–96 (1988).

B. Singer and D. Grunberger, *Molecular Biology of Mutagens and Carcinogens,* Plenum Press, New York (1983).

V. L. Souliotis and S. A. Kyrtopoulos, (1989) A novel, sensitive assay for O^6-methyl- O^6-ethylguanine in DNA, based on repair by the suicide enzyme O^6-alkylguanine–DNA-alkyltransferase in competition with an oligonucleotide continuing O^6-methylguanine. *Cancer Res.* **49**, 6997–7001.

P. T. Strickland and J. M. Boyle, Immunoassay of carcinogen-modified DNA, *Prog. Nucl. Acid Res.* **31**, 1–58 (1984).

J. A. Swenberg, M. C. Dyroff, M. A. Bedell, J. A. Popp, N. Huh, U. Kirstein, and M. F. Rajewsky, O^4-Ethyldeoxythymidine, but not O^6-ethyldeoxyguanosine, accumulates in hepatocyte DNA of rats exposed continuously to diethylnitrosamine, *Proc. Natl. Acad. Sci. USA* **81**, 1692–1695 (1984).

B. Tierney, A. Berson, and R. C. Garner, Immunoaffinity chromatography of carcinogen-DNA adducts with polyclonal antibodies directed against benzo(a)pyrene diol-epoxide-DNA, *J. Natl. Cancer Inst.* **77**, 261–267 (1986).

D. Umbenhauer, C. P. Wild, R. Montesano, R. Saffhill, J. M. Boyle, N. Huh, U. Kirstein, J. Thomale, M. F. Rajewsky, and S. H. Lu, O^6-Methydeoxyguanosine in oesophageal DNA in populations at high risk of oesophageal cancer, *Int. J. Cancer* **36**, 661–665 (1985).

J. Van Benthem, C. P. Wild, E. Vermeulen, H. H. K. Winterwerp, L. Den Engelse, and E. Scherer, Immunocytochemical localization of DNA adducts in rat tissues following treatment with N-nitrosomethylbenzylamine. In: *Methods for Detecting DNA Damaging Agents in Humans; Applications in Cancer Epidemiology and Prevention,* IARC Scientific Publication No. 89, eds. H. Bartsch, K. Hemminki, and I. K. O'Neill, IARC, Lyon, France, pp. 102–106 (1988).

C. P. Wild, P. Degan, H. Brésil, M. Serres, R. Montesano, M. Gershanovitch, and A. Likhachev, Quantitation of 7-methyldeoxyguanosine (7-medG) in peripheral blood cell DNA after exposure to methylating agents, *Proc. Am. Assoc. Cancer Res.* **29**, 260 (1988).

C. P. Wild, G. Smart, R. Saffhill, and J. M. Boyle, Radioimmunoassay of O^6-methyldeoxyguanosine in DNA of cells alkylated *in vitro* and *in vivo, Carcinog.* **4**, 1605–1609 (1983).

C. P. Wild, H. F. Stich, and R. Montesano, Presence of alkylated DNA in oral mucosal cells from cigarette smokers, *Proc. Am. Assoc. Cancer Res.* **30**, 318 (1989).

V. L. Wilson, A. K. Basu, J. M. Essigmann, R. A. Smith, and C. C. Harris, O^6-Alkyldeoxyguanosine detection by ^{32}P-postlabeling and nucleotide chromatography analysis, *Cancer Res.* **48**, 2156–2161 (1988).

V. L. Wilson, A. Weston, D. K. Manchester, G. E. Trivers, D. W. Roberts, F. F. Kadlubar, C. P. Wild, R. Montesano, J. C. Willey, D. L. Mann, and C. C. Harris, Alkyl and aryl carcinogen adducts detected in human peripheral lung, *Carcinog.* **10**, 2149–2153 (1989).

Chapter

16

Noninvasive Methods for Measuring Exposure to Alkylating Agents: Recent Studies on Human Subjects

Helmut Bartsch, Hiroshi Ohshima,
and David E. G. Shuker
Unit of Environmental Carcinogens and Host Factors
International Agency for Research on Cancer
Lyon, France

1. INTRODUCTION

Humans are exposed to a wide range of nitrogen-containing compounds that can react with nitrosating agents to form *N*-nitroso compounds (NOC), a versatile class of carcinogens (National Research Council, 1981; Shephard *et al.*, 1987; Druckrey *et al.*, 1967; Preussmann and Stewart, 1984). In addition, nitrosation of certain polyaromatic hydrocarbons and phenolic compounds results in the formation of *C*-nitroso and *C*-nitro arenes or diazonium compounds, some of which have been reported to be mutagenic and carcinogenic [International Agency for Research on Cancer (IARC), 1984; Ohshima *et al.*, 1989]. Humans are also exposed to various types of nitrosating agents in the diet, tobacco smoke, air, and water. Nitrite, nitrate, and nitrosating agents can also be synthesized endogenously in reactions mediated by bacteria and activated macrophages (Marletta, 1988; Stuehr and Marletta, 1985; Calmels *et al.*, 1987a, 1987b; Miwa *et al.*, 1987). In this way, endogenous formation of NOC can occur in various ways at various sites in the body, including nitrosation in the oral cavity, stomach, and intestine, reaction of nitrogen oxides (NO_x) in the lung, and the reactions mediated by bacteria and macrophages in infected or inflamed organs.

Exposure to endogenously formed NOC has been associated with an increased risk of cancer of the stomach, esophagus, and urinary bladder, but convincing epidemiological evidence is lacking (Bartsch and Montesano, 1984). One of the reasons has been the absence of reliable methods for estimating the extent of formation of NOC *in vivo*. In addition, any nitrosation reaction occurring *in vivo* is influenced by many factors. For all these reasons, we developed a simple and sensitive method, called the *N*-nitrosoproline (NPRO) test, for the quantitative estimation of endogenous nitrosation in humans (Ohshima and Bartsch, 1981, 1988; Ohshima *et al.*, 1982b). It is based on our findings and reports in the literature that certain nitrosamino acids (NAA) such as NPRO are excreted unchanged almost quantitatively in urine (Ohshima *et al.*, 1982a; Chu and Magee, 1981; Dailey *et al.*, 1975). This NPRO test has for the first time allowed study of the kinetics and factors affecting nitrosation *in vivo* in human subjects, as well as in animals (Ohshima *et al.*, 1982a, 1983).

In order to evaluate the possible role of NOC in human cancer at specific sites, we have embarked on the following projects: (i) to measure endogenous nitrosation rates in healthy human subjects in order to collect data on geographic and interindividual variation; (ii) to study dietary and host factors that affect nitrosation, in particular to elucidate the molecular mechanism by which bacteria are involved in this process; (iii) to compare NOC exposure in subjects with precancerous conditions of the stomach with subjects without such lesions; (iv) to compare NOC exposure in asymptomatic subjects from high- and low-risk areas for esophageal and stomach cancer and subjects with different exposure to NOC precursors (e.g., tobacco users). Data on NOC exposure are then evaluated in relation to epidemiological and up-to-date clinical observations.

In addition, immunoassay methods have been developed to quantitate 3-alkyladenines in urine, indicative of DNA alkylation, and these assays are being applied to monitor human exposure to alkylating nitroso compounds from the environment and from cancer therapy.

A brief account of the results obtained to date after applying these noninvasive methods to clinical and epidemiological studies (Table 1) is presented below.

TABLE 16–1. Summary of Clinical and Epidemiological Studies using Urinary *N*-Nitrosamino Acids (NAA) as Indices of Exposure to NOC

Observations	References
Studies in Tobacco Users	
Increased NAA formation in cigarette smokers	Hoffmann and Brunnemann, 1983; Ladd *et al.*, 1984; Bartsch *et al.*, 1984; Tsuda *et al.*, 1986; Nair *et al.*, 1986
Nitrosation in the oral cavity of betel-quid chewers	Nair *et al.*, 1986, 1987
Clinical Studies	
Studies on intragastric nitrosation and pre-cancerous lesions of the stomach: no increase in NPRO in subjects with chronic atrophic gastritis or after cimetidine dose	Bartsch *et al.*, 1984
No increase in NPRO formation in subjects with pernicious anaemia and gastrectomy	Hall *et al.*, 1987a, 1987b
No increase in NPRO formation in subjects with high bacteria count and high gastric pH	Crespi *et al.*, 1987
Significant increase in NPRO level (after a proline dose without nitrate) in urine of gastrectomy and post vagotomy patients; moderate increase in pernicious anemia patients	Houghton *et al.*, 1989
Increased background NPRO and other NAA in cirrhosis patients	N. Habib, H. Ohshima, H. Bartsch*
Epidemiological Studies	
Field study in high- and low-risk areas for esophageal cancer in China; higher levels of NAA in a high-risk population; endogenous nitrosation of proline and its inhibition by vitamin C; NAA levels correlated positively with mortality rate of esophageal cancer in eight areas of China	Lu *et al.*, 1986, 1987
Field study in 26 counties in China: moderate correlation between esophageal cancer mortality rates and NPRO formation after intake of proline	Chen *et al.*, 1987
Field study in high- and low-risk areas for stomach cancer in Japan: increased nitrosation potential in a high-risk population	Kamiyama *et al.*, 1987
Higher urinary excretion of NPRO in subjects with liver fluke, a risk factor for cholangiocarcinoma in Thailand	Srianujata *et al.*, 1987; Srivanatakul *et al.*, 1990

*In IARC, 1985, Annual Report p. 36

2. METHODS OF URINE ANALYSIS

2.1. Procedures for the NPRO test (Oshima and Bartsch, 1981)

Several forms of the method (see below) are being applied in clinical and epidemiological studies. L-Proline is utilized as a probe for nitrosatable amines, and NPRO excreted in urine is determined as a marker for endogenous nitrosation. The rationale for applying this test in human studies is based on the following: (i) NPRO has been reported to be neither carcinogenic nor mutagenic (IARC, 1978; Mirvish *et al.*, 1980); (ii) after gavage of rats with ^{14}C-NPRO, ^{14}CO$_2$ production and DNA alkylation were negligible (Chu and Magee, 1981), but urinary excretion of NPRO (as the unchanged compound) was rapid and almost complete (Dailey *et al.*, 1975; Chu and Magee, 1981; Ohshima *et al.*, 1982a); (iii) in humans, preformed NPRO ingested in food extracts was also eliminated rapidly and almost quantitatively in the urine within 24 h after ingestion (Ohshima *et al.*, 1982b). The difference between the amount of NPRO excreted in the 24-h urine and that ingested in foods can therefore be used as an indicator of daily endogenous nitrosation (Ohshima and Bartsch, 1981). Application of the NPRO test does not entail risk to the health of study subjects, and it was cleared by the IARC ethical committee.

2.1.1. Procedure A

(i) Human subjects are given 200 ml beetroot juice (260 mg nitrate); (ii) 30 min later they are given L-proline (500 mg); (iii) the subjects then fast for 2 h; (iv) 24-h urine samples are collected in the presence of NaOH (during urine collection, foods rich in nitrate or nitrite are avoided); (v) 100-ml urine aliquots are stored at $-20°C$ prior to analysis.

2.1.2. Procedure B

Alternatively, three 24-h urine samples are collected from each subject according to the following protocols: (i) before dosing, (ii) after intake of 100 mg L-proline three times a day after each meal, and (iii) after intake of 100 mg L-proline and 100 mg vitamin C three times a day.

2.1.3. Procedure C

Proline is added to betel-quid and/or tobacco samples used by chewers or dippers, and saliva is subsequently analyzed for NPRO content as an index of nitrosation reactions occurring within the oral cavity.

Whenever possible, information is collected from each study subject on demographic data, smoking, drinking, and dietary habits, and clinical findings.

2.2. Sample Collection and Analysis of NAA

Samples of 24-h urine are collected in plastic bottles containing sodium hydroxide or ammonium sulfamate in dilute sulfuric acid to avoid artifactual formation of NPRO during collection and storage of the sample. Samples of 12-h urine collected without preservatives are divided into two aliquots: either sodium hydroxide or ammonium sulfamate is added and used for analyses of nitrite, nitrate, and NAA; the other is stored without alkali or acid to be used for analyses of creatinine and other compounds.

Urine samples are spiked with *N*-nitrosopipecolic acid as an internal standard and analyzed for NPRO and other NAA after conversion to their methyl esters by diazomethane in a gas chromatograph with a thermal energy analyzer, a nitrosamine-specific detector. Other derivatizing agents, such as boron trifluoride–methanol (Ladd *et al.*, 1984; Leaf *et al.*, 1987) and penta-fluorobenzyl bromide (Garland *et al.*, 1986), have been used; however, the sulfur-containing NAA are acid-labile and decompose during derivatization with boron trifluoride or hydrochloric acid in methanol (Ohshima *et al.*, 1984a).

2.3. Urinary Alkylpurines

N-methyl-*N*-nitroso compounds, such as dimethylnitrosamine (DMN), are metabolically converted into methylating agents that react at many sites in DNA. The major sites of modification are *N*-7- and O^6-deoxyguanosine and *N*-3-deoxyadenosine. O^6-methyldeoxyguanosine is considered to be of major importance as a mutagenic lesion; however, an efficient repair system exists, which acts by transferring the alkyl group to a receptor protein (Karran and Lindahl, 1985). In contrast, repair of *N*-alkyl adducts is mediated by glycosylases through excision of the intact alkylpurines, which ultimately are replaced by normal purines. Subsequently, the alkylpurines, for example, *N*-7-methylguanine (7-MeGua) and *N*-3-methyladenine (3-MeAde), are excreted, intact, via urine (Craddock and Magee, 1967; Hanski and Lawley, 1985; Shuker *et al.*, 1987a). For practical purposes the detection of carcinogen-derived 7-MeGua in urine is rendered impossible due to the presence of large amounts of RNA-drived 7-MeGua, although in experimental animals, and in some cases in humans, this problem can be overcome by use of isotopically labeled precursors (Farmer *et al.*, 1986). For 3-MeAde, however, no such source of "background" appeared to exist, and its use as a marker for nucleic acid, particularly DNA, methylation has been proposed (Lawley, 1976). Recent advances in analytical methodology have permitted the ready determination of urinary 3-MeAde, and these methods are summarized below. Some of the applications of this monitoring method are described later in this chapter.

3-MeAde and also d_3-3-MeAde can be separated and quantified by gas chromatography–mass spectrometry (GC-MS), after derivatization, following the series of clean-up steps involving XAD-2 polystyrene resin column chromatography and high-pressure liquid chromatography (HPLC) (Shuker *et al.*, 1987a, 1987b; Stillwell *et al.*, 1989). This method was used in studies to show that (a) a precursor–product relationship existed between administered methylating agents or precursors and urinary 3-MeAde (Shuker *et al.*, 1987a) and (b) both experimental animals and humans excreted 3-MeAde when not, apparently, exposed to methylating carcinogens (Shuker *et al.*, 1987b; Stillwell *et al.*, 1989).

In order to develop more rapid analytical methodology, it was decided to develop an antiserum to 3-MeAde, and this was successful using a novel haptenic derivative of 3-MeAde (Shuker and Farmer, 1988). Because of the complex mixture of purines with closely related structures present in urine, direct analysis using an enzyme-linked immunosorbent assay (ELISA) was not possible. However, the antiserum was used to prepare immunoaffinity chromatography columns that permit the extraction of 3-MeAde from urine in a sufficiently clean fraction that can be derivatized directly and used for quantitation by GC-MS (Shuker *et al.*, 1990, in press).

A recent improvement in the methodology has been the development of a monoclonal antibody against 3-MeAde, which is not susceptible to interference from other urinary purines (V. Prévost *et al.*, 1990). Thus a wholly immunochemical, i.e., affinity chromatography followed by ELISA, method for urinary 3-MeAde is now available that is rapid and reliable.

Recent studies in various groups of human subjects have indicated that a large part of urinary 3-MeAde may be derived from exogenous sources, and this was confirmed by a series of dietary intervention experiments (V. Prévost *et al.*, 1990). Model studies in volunteers have shown that a stable low background of 3-MeAde excretion can be achieved that permits the detection of ''normal'' methylating agent exposures, such as cigarette smoking (D. E. G. Shuker and V. Prévost, unpublished observations).

3. APPLICATIONS IN CLINICAL AND EPIDEMIOLOGICAL STUDIES

3.1. Environmental Exposures, Possibly Related to Specific Cancer Sites

3.1.1. Oral cavity

Tobacco smokers, chewers of tobacco and of betel quid (often containing tobacco), and snuff dippers are all exposed to increased levels of nitrosamine precursors (such as nicotine, nitrate, nitrite, and NO_x) and nitrosation modifiers (thiocyanate) (Hoffmann and Hecht, 1985). In these subjects, endogenous NOC synthesis occurs at a higher rate than in persons without such habits (Hoffmann and Brunnemann, 1983; Bartsch *et al.*, 1984; Ladd *et al.*, 1984; Tsuda *et al.*, 1986; Nair *et al.*, 1986), thus adding to the body burden of preformed carcinogenic tobacco-specific nitrosamines (TSNA) (ingested or inhaled). Endogenous NPRO formation in smokers was partially inhibited by daily addition of 1 g ascorbic acid to their diet (Hoffmann and Brunnemann, 1983). There is now good evidence that TSNA are major carcinogenic agents responsible for tobacco-associated cancers in smokers and, to an even greater degree, users of smokeless tobacco products (Hoffmann and Hecht, 1985). From these results, one can also infer that a substantial fraction (as yet undetermined) of tobacco-related and possibly other nitrosamines is synthesized within the body. Vitamin C, an effective inhibitor of nitrosation in humans, also inhibits endogenous NOC synthesis in smokers, possibly explaining why regular consumption of vegetables and fresh fruits (a source of vitamin C) has some protective effect against tobacco-associated malignancies (Tuyns, 1983).

The causal correlation between oral cancer and chewing of betel quid (containing areca nut pieces, slaked lime, and often tobacco) in India and other Southeast Asian countries is well established (IARC, 1985). Nitrosation *in vitro* of areca nut–specific alkaloids (such as arecoline) leads to the formation of areca nut–specific nitrosamines (ASNA) of which at least one, 3-(methylnitrosamino)propionitrile, is carcinogenic in experimental animals (Wenke and Hoffmann, 1983; Propkopczyk *et al.*, 1987). To examine the extent to which such nitrosation of betel-nut constituents occurs *in vivo* (e.g., in the oral cavity of chewing subjects), saliva and urine samples were collected from chewers of betel quid with or without tobacco, from tobacco chewers, from cigarette smokers, and from people without such habits (Nair *et al.*, 1985). Saliva samples were analyzed for TSNA such as N'-nitrosonornicotine and N-nitrosoanatabine, and for ASNA such as N-nitrosoguvacoline and N-nitrosoguvacine. TSNA and ASNA were detected (ng/ml) in saliva of chewers of betel quid with tobacco; ASNA were detected in chewers of betel quid without tobacco. Precursor alkaloids from tobacco and areca nut (nicotine and arecoline), as well as nitrite and thiocyanate, were detected (at μg/ml levels) in the saliva. A good positive correlation was found between TSNA and nicotine content in saliva. Urinary NPRO levels tended to be higher in chewers of betel quid, tobacco, or both, and in smokers than in the no-habit group, suggesting increased endogenous nitrosation. Urinary levels of tobacco-related compounds were similar in chewers of tobacco with or without betel quid to those in smokers, implying similar intake. That certain TSNA and ASNA can be formed in subjects chewing betel quid, with or without tobacco, was indicated by an increase in nitrosamine levels after *in vitro* nitrosation of betel-quid extracts at pH 7.4. We further demonstrated that nitrosation also occurs in the oral cavity of betel-quid chewers, as shown by increased NPRO formation during a 20-min chewing period. In this study, quids either unmodified or supplemented with proline were chewed by volunteers, and saliva samples were analyzed for NPRO, TSNA, and ASNA (Nair *et al.*, 1986, 1987). All chewers of betel quid with or without tobacco had increased saliva levels of NPRO after proline supplementation.

These data suggest that in betel-quid and tobacco chewers a substantial fraction (as yet undetermined) of TSNA and ASNA appears to be synthesized *in vivo*, i.e., in the oral cavity or in the acidic gastric environment when saliva is swallowed. TSNA and/or ASNA may, therefore, be causative agents in cancers associated with betel quid and tobacco, either alone or in combination with other betel-quid ingredients (Hoffmann and Hecht, 1985).

A number of TSNA, e.g., *N*-nitrosonornicotine, are known to be metabolized to methylating agents, and O^6-methyldeoxyguanosine (O^6-MedG) has been detected in human tissues obtained from smokers (Foiles *et al.*, 1988). Urine samples are being collected from users of betel quid with different chewing habits, including those who chew betel quid with tobacco, and 3-MeAde analyses will be carried out to determine if there is a correlation between urinary 3-MeAde and use of tobacco products. However, due to the confounding effect of dietary 3-MeAde, more specific indicators of betel quid and/or tobacco specific nitrosamines are being sought. Recent developments include the preparation of immunoaffinity columns capable of retaining a range of 3-alkyladenines that are then quantified by GC-MS (D. E. G. Shuker, V. Prévost, M. D. Friesen, G. Eberle and M. F. Rajewsky, unpublished observations).

3.2.1. Stomach

Intragastric formation of NOC, particularly *N*-nitrosamides, has been implicated in the etiology of human stomach cancer (Weisburger *et al.*, 1980; Mirvish, 1983). However, no convincing epidemiological evidence has previously been presented, mainly due to a lack of reliable data for assessing endogenous nitrosation in humans. In a recently completed study, exposure to NOC ingested in foods or formed endogenously was compared in inhabitants in high- and low-risk areas for stomach cancer in northern Japan by determining urinary levels of NAA as exposure indices. Three different samples of 24-h urine were collected from each of 104 inhabitants of high-risk (Akita) and low-risk (Iwate) areas for stomach cancer according to protocol B (see Sec. 2.1.2) (i) when they were undosed, (ii) after ingestion of proline three times a day (Kamiyama *et al.*, 1987), and (iii) after ingestion of proline together with vitamin C three times a day. These samples were analyzed for NAA, nitrate, and chloride ion as indices of the exposure. The median values of NPRO and *N*-nitroso-2-methylthiazolidine 4-carboxylic acid were significantly higher in subjects of the high-risk area; the latter NAA is formed by nitrosating agents much more rapidly than NPRO (Ohshima *et al.,* 1984b). Salt intake estimated from the level of chloride ion in urine did not differ between the two areas. After intake of proline, the NPRO level increased significantly only in subjects of the high-risk area, but not in those of the low-risk area; intake of vitamin C inhibited this increase in NPRO excretion and lowered the levels of other NAA only in the high-risk subjects. Conversely, the urinary level of nitrate was higher in subjects of the low-risk area than in those of the high-risk area; nitrate levels correlated well with the amounts of vegetables consumed. These results indicate that, although subjects in the high-risk area have lower nitrate intake, their potential for intragastric nitrosation is higher, suggesting the occurrence of some inhibitory factor(s) for nitrosation, probably in the diet of the low-risk area subjects. Independently, previous studies on comparison of food habits and analysis of mutagens in foods in these two areas have suggested that some vegetables are protective against development of stomach cancer (Shimada, 1980).

We conclude from this study that determination of nitrate and nitrite alone in saliva, urine, or gastric juice is insufficient to assess the complex *in vivo* nitrosation process in man. This may explain some other conflicting results showing no or even inverse correlations between nitrate exposure and the incidence of stomach cancer.

The role of nitrite and subsequent nitrosation in the etiology of gastric cancer remains elusive. Other clinical studies (see Table 1) have examined the model of gastric carcinogenesis based on endogenous nitrosation as proposed by Correa *et al.* (1975). Progress has been made in explaining some steps in the model, but several controversies remain. Although bacterial strains possessing enzymes that catalyze nitrosamine formation at neutrality have been isolated from gastric juice of subjects with an achlorohydric stomach, their precise role in gastric carcinogenesis remains to be clarified.

Measurement of excreted DNA adducts, such as 3-MeAde, in urine in future studies may permit a more ready interpretation of the influence of various factors, since they are markers of the end result of the sequence of N-nitrosation, subsequent metabolism, and reactions with DNA. For example, a recent study in Japan (Kamiyama *et al.,* 1987) showed a positive correlation between vegetable consumption and urinary nitrate excretion in areas of high and low incidence of gastric cancer, with the lowest-risk group excreting more nitrate as a consequence of eating more vegetables. The complex interaction between increased nitrate consumption from vegetables along with a corresponding increased level of ascorbic acid, an inhibitor of nitrosation, may be better resolved using a marker of DNA alkylation such as urinary alkylpurines.

3.1.3. Esophagus
Esophageal cancer is a prevalent disease in northern China; Lin-xian

county (Hunan province) has the highest age-adjusted mortality rates of 151 per 100,000 men and 115 per 100,00 women. An extensive search for the causative factors of this cancer was begun in 1972, and subsequent studies suggested that environmental factors such as NOC and their precursors, together with micronutrient deficiencies, may be important (Yang, 1980; Coordinating Group for Research on the Etiology of Esophageal Cancer of North China, 1975).

A collaborative study was initiated with Chinese scientists in 1982. For this purpose, a total of 238 samples of 24-h urine was collected from inhabitants of high-risk (Lin-xian) and low-risk (Fan-xian) areas for esophageal cancer in northern China, according to protocol B (see Sec. 2.1.2.) (Lu *et al.*, 1986): (i) from undosed subjects, (ii) from subjects who had ingested 100 mg proline three times a day 1 h after each meal, and (iii) from subjects who had ingested proline together with vitamin C three times a day. As an index of individual exposure to NOC or their precursors, ingested in food and/or formed endogenously, the levels of four urinary NAA and nitrate were determined. The amounts of NPRO and the sum of nitrosamino acids and nitrate excreted in 24-h urine of undosed subjects in Lin-xian were significantly higher than those in Fan-xian, indicating a higher NOC exposure. Ingestion of L-proline resulted in a marked increase in urinary NPRO levels in inhabitants of both areas, suggesting endogenous nitrosation may occur to a larger extent when appropriate amine precursors are ingested in foods. Intake of moderate doses of ascorbic acid by high-risk subjects effectively reduced the urinary levels of NAA to those found in undosed subjects in the low-risk area, an observation similar to that found in the Japanese high-risk population for gastric cancer.

Lu *et al.* (1987) compared the nitrosation potential in eight populations at different risks for esophageal cancer in China and showed that the amount of NPRO and other NAA excreted in the 24-h urine of eight different populations positively correlated with the mortality rates for esophageal cancer. In a recently completed study, urine samples were collected in 1984 from 1035 subjects living in 26 counties in the People's Republic of China, selected on the basis of contrasting mortality rates for cancers of the esophagus, stomach, and liver. Samples of 12-h overnight urine were collected from approximately 20 male adults in two communes in each of the 26 counties. Two 12-h urine specimens were collected from each subject, one after a dose of proline and ascorbic acid, and another after a dose of proline only. The pooled aliquots of urine samples were then analyzed for nitrate and nitrite, NAA, and thioethers (Chen *et al.*, 1987). There was an excellent within-county correlation for two communes for NPRO. Results from urine analyses were correlated with cancer mortality in male subjects in the truncated age range 35–64. No clear correlations were seen for stomach or liver cancer and endogenous NOC exposure. For esophageal cancer, however, the cancer mortality rates were moderately associated positively with the endogenous nitrosation and negatively with background ascorbate levels in plasma. Thus NOC may be implicated, but the specific carcinogenic compound(s) responsible still need to be identified. This project was done within a wider ecological study of diet and cancer in China, covering 65 counties (Chen *et al.*, 1987), which is under evaluation. Finally, our demonstration on the inhibitory effect of moderate doses of ascorbic acid on endogenous nitrosation offers a rational basis for long-term intervention studies through dietary modifications.

Further support for the involvement of NOC in cancer etiology and also for the validity of the NPRO test in identifying highly exposed subjects comes from recent findings by Umbenhauer *et al.* (1985): levels of O^6-methyldeoxyguanosine were elevated in the DNA of esophageal and stomach mucosa in specimens removed surgically from cancer patients in Lin-xian. These results suggested that exposure to methylating NOC derived from diet may be a major risk factor in the development of esophageal cancer in this area. These analyses were carried out on DNA samples extracted from tissue obtained at surgery, and this obviously limits the number of samples available for analysis. The use of urinary alkylpurine determination may provide an alternative means to assess exposure to alkylating *N*-nitroso compounds in the various risk groups. Methods are currently being developed to detect benzyl adducts that may be derived from methyl benzyl nitrosamine, a potent esophageal carcinogen, which has recently been shown to be present in gastric contents of individuals from high-risk groups (Lu *et al.*, 1987).

3.1.4. Urinary tract

Bacterial or parasitical infection of the urinary bladder can result in an elevated risk of developing bladder cancer, and a number of studies have suggested a role for *N*-nitrosamines synthesized *in situ* as etiological agents (Radomski *et al.*, 1978; Hicks *et al.*, 1982). Since the first observation of bacterially mediated nitrosation (Sander, 1968), there have been many studies aimed at determining if this pathway is relevant in human cancer etiology (Hill, 1986). However, it was only recently shown that various bacterial strains isolated from urinary-tract infections are capable of catalyzing nitrosation of secondary amines at neutrality *in vitro* (Calmels *et al.*, 1985, 1987a). This nitrosation capacity in *E. coli*, for example, is closely related to nitrate–reductase activity and was shown to be linked to the expression of the nitrate–reductase gene *narGHI* (Calmels *et al.*, 1987b, 1988; Ralt *et al.*, 1988). In contrast, in *Pseudomonas aeruginosa* (a denitrifying bacterial strain that shows extremely high nitrosation activity compared with *E. coli*), the nitrosating activity was induced by nitrite as well as by nitrate, suggesting that an additional pathway can operate in this strain (Leach *et al.*, 1987; Calmels *et al.*, 1988).

A recent epidemiological study has shown a positive association between urinary-tract infection and bladder cancer, particularly squamous-cell carcinoma (Kantor *et al.*, 1984). To investigate the role of bacterially mediated NOC formation in humans, urine samples from 31 patients with urinary-tract infections and from 31 controls were analyzed. The levels of volatile nitrosamines, NAA, total NOC, and nitrite were elevated in patients with bacteriuria (Ohshima *et al.*, 1987). Twelve of 14 bacterial species that were isolated from the infected samples catalyzed nitrosation of morpholine *in vitro* (Calmels *et al.*, 1987a). These results strongly support the hypothesis that NOC can be formed *in vivo* in the bladder of patients with urinary-tract infections, and can increase the body burden of carcinogens through subsequent absorption of the compounds from the urinary bladder into the bloodstream.

The excretion of urinary alkylpurines may be a good marker for damage to bladder epithelium, and current studies incorporate recently developed analytical methodology, in particular, for the determination of 3-MeAde, as well as other 3-alkyladenines.

3.1.5. Other sites:

Subjects living in some provinces of Northeast Thailand that are infested with liver fluke *(Opisthorchis viverrini)* are at high risk to develop cholangiocarcinoma. Srianujata *et al.* (1987) found that nitrite and/or nitrate levels in saliva and total nitrate NPRO excreted in urine were higher in subjects with liver fluke than in those without. A second study has confirmed a severalfold increase in NPRO excretion in infested subjects (Srivanatakul *et al.*, 1990). The results suggest that chronic inflammatory processes lead via activated macrophages to increased nitrite or nitrate and nitrosamine synthesis; the latter could play a role in the etiology of this cancer, as in Syrian golden hamsters, where administration of *N*-nitrosodimethylamine, at the same time as liver fluke infestation, resulted in the production of cholangiocarcinomas in a synergistic fashion (Thamavit *et al.*, 1978).

In a prospective case control study in Thailand, a cohort of hepatitis B surface antigen positive carriers is being followed with respect to incidence of primary hepatocellular carcinoma (PHC). Samples of urine and blood are being collected at yearly intervals. When cases appear they will be matched with controls and analyses will be performed in the appropriate stored samples. Among a number of parameters, including mycotoxin exposure, 3-MeAde will be measured in order to see if the cases had elevated levels prior to onset of the disease (Bosch and Muñoz, 1988). It is of interest to note that patients with liver cirrhosis excreted a higher level of background NPRO and other NAA, as compared with healthy controls (Table 1).

3.2. Medicinal Exposures

A number of anticancer drugs are carcinogenic to humans (Schmähl and Kaldor, 1986). Their use in treatment is a unique case of human exposure to known quantities of alkylating agents at precisely measured times. Among the alkylating drugs, there are several methylating agents, including procarbazine and *N*-nitrosomethylurea.

Preliminary studies are currently underway in which urinary 3-methyladenine levels are being correlated with methyl adducts in lymphocyte DNA. This permits validation of the idea of using urinary excretion of alkyl adducts as a marker of DNA alkylation. The cytotoxicity of many alkylating drugs is thought to be mediated via the formation of DNA adducts. However, it is not clear if responsiveness to therapy is related to level or type of alkylation. Monitoring of adduct levels may offer a short-term clinical advantage as a tool to monitor efficacy of treatment as well as individual response to a given quantity of drug.

4. CONCLUSIONS

A sensitive procedure to estimate exposure of humans to exogenous and endogenous NOC (the NPRO test) has been developed. Results of the studies carried out in human subjects allowed the following conclusions to be drawn.

Endogenous nitrosation of amino proline precursors occurs in the human body after ingestion of amounts of precursors (nitrate, amines), which are considered as normal dietary intake. Increasing intake of nitrate leads to an exponential increase in the amount of nitrosation products formed *in vivo*.

Nitrosation inhibitors, including vitamin C, significantly reduced the yield of NPRO and other NAA formed in healthy human subjects.

Increased excretion of NPRO and other NAA was observed in cigarette-smoking subjects, which may be attributable to higher levels of thiocyanate (a catalyst of nitrosation) in saliva of smokers and to higher exposure to aldehydes and nitrosating agents (for example, NO_x) present in cigarette smoke.

The process of endogenous nitrosation in humans was found to be highly complex and influenced by many factors. Therefore determination of nitrate and nitrite alone in body fluids is insufficient to assess the *in vivo* nitrosation process in man; individual monitoring is required, rather than measurement of precursor intake, if endogenous formation of NOC is to be associated with cancer at specific sites.

Formation of endogenous NOC was assessed by the NPRO test in (i) subjects living in high- and low-incidence areas for stomach cancer or esophageal cancer in northern Japan and in the People's Republic of China, (ii) subjects with different habits of betel-quid chewing and tobacco use,

(iii) European patients with urinary bladder infection, and (iv) subjects infested with liver fluke in Thailand. In all instances (Table 1), higher exposures to endogenous NOC were found in high-risk subjects, but individual exposure was greatly affected by dietary modifiers or disease state. Vitamin C efficiently lowered the body burden of intragastrically formed NOC.

Although the evidence that endogenously formed NOC are involved in human cancers is far from being proven, it is at least cohesive and supported by our results (Table 1); it also justifies the reduction of exposure to NOC as a preventive measure. The demonstrated efficacy of nitrosation inhibitors, e.g., vitamin C, also provides a plausible interpretation of epidemiological studies that have shown protective effects of fruits and vegetables (sources of vitamin C or other vitamins and polyphenols) against various malignancies and particularly stomach cancer (Mirvish, 1983, 1986).

The NPRO test, however, has certain limitations, and therefore the adequacy of proline and other amino compounds as a substrate for nitrosation mediated by bacteria or macrophages or for nitrosation by NO_x in the lung needs to be investigated. Additional markers for exposure to nitrosating agents are therefore developed and validated. One such approach involves the determination of urinary 3-alkyladenines as a marker of human exposure to alkylating carcinogens, as part of which analytical methods have been developed for 3-MeAde. 3-MeAde itself may be of limited use as a marker in subjects eating uncontrolled diets due to its apparent natural occurrence, but is extremely useful in model studies. Other 3-alkyladenines, for which there should be a much lower dietary component, are currently being studied.

ACKNOWLEDGMENTS

We wish to thank J.-C. Béréziat, I. Brouet, M.-C. Bourgade, and J. Michelon for technical assistance and Z. Schneider for secretarial work. One of the thermal energy analyzer detectors was provided on loan by the National Cancer Institute of the United States under Contract No. NO1 CP-55715. V. Prévost is thanked for work related to urinary alkylpurines. Part of the work reported on the 3-methyladenine assay was undertaken during the tenure of a Visiting Scientist's Award given to D. Shuker by the International Agency for Research on Cancer. Support from the U.S. National Cancer Institute is also gratefully acknowledged (Grant No. CA 48473).

The authors gratefully acknowledge the scientific contributions and collaborative efforts of N. Muñoz and J. Kaldor (IARC, Lyon, France), S. V. Bhide and J. Nair (Cancer Research Institute, Tata Memorial Centre, Bombay, India), T. C. Campbell (Cornell University, Ithaca, NY), J. Chen (Chinese Academy of Medical Sciences, Beijing, People's Republic of China), P. B. Farmer (MRC Toxicology Unit, Carshalton, U.K.), S. Kamiyama

(Akita University, Akita, Japan), S. H. Lu (Cancer Institute, Academy of Medical Sciences, Beijing, People's Republic of China), and R. Peto (ICRF Cancer Studies Unit, University of Oxford, U.K.).

REFERENCES

H. Bartsch and R. Montesano, Commentary: Relevance of nitrosamines to human cancer, *Carcinog.* **5,** 1381–1393 (1984).

H. Bartsch, H. Ohshima, N. Muñoz, M. Crespi, V. Cassale, V. Ramazotti, R. Lambert, Y. Minaire, J. Forichon, and C. L. Walters, *In-vivo* nitrosation, precancerous lesions and cancers of the gastrointestinal tract. Ongoing studies and preliminary results. In: *N-Nitroso Compounds: Occurrence, Biological Effects and Relevance to Human Cancer,* IARC Scientific Publication No. 57, eds. I. K. O'Neill, R. C. von Borstel, C. T. Miller, J. E. Long, and H. Bartsch, IARC, Lyon, France, pp. 957–964 (1984).

X. F. Bosch and N. Muñoz, Prospects for epidemiological studies on hepatocellular cancer as a model for assessing viral and chemical interactions, In: *Methods for Detecting DNA Damaging Agents in Humans: Applications to Cancer Epidemiology and Prevention,* IARC Scientific Publication No. 89, eds. H. Bartsch, K. Hemminki, I. K. O'Neill, IARC, Lyon, France, pp. 92–96 (1988).

S. Calmels, H. Ohshima, and H. Bartsch, Nitrosamine formation by denitrifying and non-denitrifying bacteria: implication of nitrite reductase and nitrate reductase in nitrosation catalysis, *J. General Microbiol.* **134,** 221–226 (1988).

S. Calmels, H. Ohshima, M. Crespi, H. Leclerc, C. Cattoen, and H. Bartsch, Nitrosamine formation by microorganisms isolated from human gastric juice and urine; biochemical studies on the bacterial catalysed nitrosation. In: *The Relevance of N-Nitroso Compounds to Human Cancer: Exposure and Mechanisms,* IARC Scientific Publication No. 84, eds. H. Bartsch, I. K. O'Neill, and R. Schulte-Hermann, IARC, Lyon, France, pp. 391–395 (1987a).

S. Calmels, H. Ohshima, H. Rosenkranz, E. McCoy, and H. Bartsch, Biochemical studies on the catalysis of nitrosation by bacteria, *Carcinog.* **8,** 1985–1987 (1987b).

S. Calmels, H. Ohshima, P. Vincent, A.-M. Gounot, and H. Bartsch, Screening of microorganisms for nitrosation catalysis at pH 7 and kinetic studies on nitrosamine formation from secondary amines by *E. Coli* strains, *Carcinog.* **6,** 911–915 (1985).

J. Chen, H. Ohshima, H. Yang, J. Li, C. T. Campbell, and H. Bartsch, A correlation study on urinary excretion of N-nitroso compounds and cancer in the People's Republic of China: Interim results. In: *The Relevance of N-Nitroso Compounds to Human Cancer: Exposure and Mechanisms,* IARC Scientific Publication No. 84, eds. H. Bartsch, I. K. O'Neill, and R. Schulte-Hermann, IARC, Lyon, France, pp. 503–506 (1987).

C. Chu and P. N. Magee, Metabolic fate of nitrosoproline in the rat, *Cancer Res.* **41,** 3653–3657 (1981).

Coordinating Group for Research on the Etiology of Esophageal Cancer of North China, *Sci. Sin.* **18,** 131–148 (1975).

P. Correa, W. Haenszel, C. Cuello, S. Tannenbaum, and M. Archer, A model for gastric cancer epidemiology, *Lancet* **ii,** 58–60 (1975).

V. M. Craddock and P. N. Magee, Effects of the administration of dimethylnitrosamine on urinary 7-methylguanine, *Biochem. J.* **104,** 435–440 (1967).

M. Crespi, H. Ohshima, V. Ramazzotti, N. Muñoz,, A. Grassi, V. Casale, H. Leclerc, S. Calmels, C. Cattoen, C. J. Kaldor, and H. Bartsch, Intragastric nitrosation and precancerous lesions of the gastrointestinal tract: testing of an etiological hypothesis. In: *The Relevance of N-Nitroso Compounds to Human Cancer: Exposure and Mechanisms,* IARC Scientific Publication No. 84, eds. H. Bartsch, I. K. O'Neill, and R. Schulte-Hermann, IARC, Lyon, France, pp. 511–517 (1987).

R. E. Dailey, R. C. Braunberg, and A. M. Blaschka, The absorption, distribution and excretion of [^{14}C]-nitrosoproline by rats, *Toxicol.* **3,** 23–28 (1975).

H. Druckrey, R. Preussmann, S. Ivankovic, and D. Schmähl, Organotrope carcinogene Wirkungen bei 65 verschiedenen N-Nitroso-Verbindungen in BD-Ratten, *Z. Krebsforsch.* **69,** 103–201 (1967).

P. B. Farmer, D. E. G. Shuker, and I. Bird, DNA and protein adducts as indicators of *in vivo* methylation by nitrosatable drugs, *Carcinog.* **7,** 49–52 (1986).

P. G. Foiles, L. M. Miglietta, S. A. Akerkar, R. B. Everson, and S. S. Hecht, Detection of O^6-methyldeoxyguanosine in human placental DNA, *Cancer Res.* **48,** 4184–4188 (1988).

W. A. Garland, W. Kuenzig, F. Rubio, H. Kornychuk, E. P. Norkus, and A. H. Conney, Studies on the urinary excretion of nitrosodimethylamine and nitrosoproline in humans: interindividual and intraindividual differences and the effect of administered ascorbic acid and alpha-tocopherol, *Cancer Res.* **46,** 5392–5400 (1986).

C. N. Hall, J. S. Kirkham, and T. C. Northfield, Urinary N-nitrosoproline excretion: a further revaluation of the nitrosamine hypothesis of gastric carcinogenesis in precancerous conditions, *Gut* **28,** 216–220 (1987a).

C. N. Hall, D. Darkin, N. Viney, A. Cook, J. S. Kirkham, and T. C. Northfield, Evaluation of the nitrosamine hypothesis of gastric carcinogenesis in man. In: *The Relevance of N-Nitroso Compounds to Human Cancer: Exposure and Mechanisms,* IARC Scientific Publication No. 84, eds. H. Bartsch, K. I. O'Neill, and R. Schulte-Hermann, IARC, Lyon, France, pp. 527–530 (1987b).

C. Hanski and P. D. Lawley, Urinary excretion of 3-methyladenine and 1-methylnicotinamide by rats following administration of [methyl-^{14}C]-methylmethanesulphonate and comparison with (^{14}C)methionine or formate, *Chem. Biol. Interact.* **55,** 225–234 (1985).

R. M. Hicks, M. M. Ismall, C. L. Walters, P. T. Beecham, M. T. Rabie, and M. A. El-Alamy, Association of bacteruria and urinary nitrosamine formation with schistosma haematobium infection in the Qalyub area of Egypt, *Trans. R. Soc. Trop. Med. Hyg.* **76,** 519–527 (1982).

M. J. Hill, *Microbes and Human Carcinogenesis,* Edward Arnold Publishers, London, U.K. (1986).

D. Hoffman and K. Brunnemann, Endogenous formation of N-nitrosoproline in cigarette smokers, *Cancer Res.* **43,** 5570–5574 (1983).

D. Hoffmann and S. S. Hecht, Nicotine-derived N-nitrosamines and tobacco related cancer: current status and future directions, *Cancer Res.* **45,** 935–942 (1985).

P. W. J. Houghton, S. Leach, R. W. Owen, N. J. McC. Mortensen, M. J. Hill, and R. C. N. Williamson, Use of a modified N-nitrosoproline test to show intragastric nitrosation in patients at risk of gastric cancer, *Br. J. Cancer* **60,** 231–234 (1989).

IARC (1978) IARC Monographs on the Evaluation of the Carcinogenic Risk of Chemicals to Humans, Vol. 17, Some N-Nitroso Compounds, Lyon, France, pp. 303–308.

IARC, IARC Monographs on the Evaluation of the Carcinogenic Risk of Chemicals to Humans, Vol. 33, *Polynuclear Aromatic Compounds, Part 2, Carbon Blacks, Mineral Oils and some Nitroarenes,* IARC, Lyon, France, pp. 167–222 (1984).

S. Kamiyama, H. Ohshima, A. Shimada, N. Saito, M.-C. Bourgade, P. Ziegler, and H. Bartsch, (1987) Urinary excretion of N-nitrosamino acids and nitrate by inhabitants in high- and low-risk areas for stomach cancer in northern Japan. In: H. Bartsch, I. K. O'Neill and R. Schulte-Hermann (eds), *The Relevance of N-Nitroso Compounds in Human Cancer: Exposures and Mechanisms,* pp 497–502. IARC Scientific Publications No 84. International Agency for Research on Cancer, Lyon, France.

A. F. Kantor, P. Hartge, R. N. Hoover, A. S. Narayana, J. W. Sullivan, and J. F. Fraumeni, (1984) Urinary tract infection and risk of bladder cancer. *American Journal of Epidemiology,* **119,** 510–515.

P. Karran and T. Lindahl, Cellular defence mechanisms against alkylating agents, *Cancer Surveys* **4,** 585–599 (1985).

K. F. Ladd, H. L. Newmark, and M. C. Archer, N-nitrosation of proline in smokers and nonsmokers, *J. Natl. Cancer Inst.* **73,** 83–87 (1984).

P. D. Lawley, Methylation of DNA by carcinogens; some applications of chemical analytical methods. In: *Screening Tests in Chemical Carcinogenesis,* IARC Scientific Publication No. 12, eds. R. Montesano, H. Bartsch, L. Tomatis, and W. Davis, IARC, Lyon, France, pp. 181–208 (1976).

S. A. Leach, A. R. Cook, B. C. Challis, M. J. Hill, and M. H. Thompson, Bacterially mediated N-nitrosation reactions and endogenous formation of N-nitroso compounds. In: *The Relevance of N-Nitroso Compounds to Human Cancer: Exposure and Mechanisms,* IARC Scientific Publication No. 84, eds. H. Bartsch, I. K. O'Neill, and R. Schulte-Hermann, IARC, Lyon, France, pp. 396–399 (1987).

C. D. Leaf, A. J. Vecchio, D. A. Roe, and J. H. Hotchkiss, Influence of ascorbic acid dose on N-nitrosoproline formation in humans, *Carcinog.* **8,** 791–795 (1987).

S. H. Lu, H. Ohshima, H.-M. Fu, Y. Tian, F.-M. Li, M. Blettner, J. Wahrendorf, and H. Bartsch, Urinary excretion of N-nitrosamino acids and nitrate by inhabitants of high- and low-risk areas for esophageal cancer in northern China: endogenous formation of nitrosoproline and its inhibition by vitamin C, *Cancer Res.* **46,** 1485–1491 (1986).

S. H. Lu, W. X. Yang, L. P. Guo, F. M. Li, G. J. Wang, J. S. Zhang, and P. Z. Li, Determination of N-nitrosamines in gastric juice and urine and a comparison of endogenous formation of N-nitrosoproline and its inhibition in subjects from high- and low-risk areas for esophageal cancer. In: *The Relevance of N-Nitroso Compounds to Human Cancer; Exposure and Mechanisms,* IARC Scientific Publication No. 84, eds. H. Bartsch, I. K. O'Neill, and R. Schulte-Hermann, IARC, Lyon, France, pp. 538–543 (1987).

M. A. Marletta, Mammalian synthesis of nitrite, nitrate, nitric oxide and N-nitrosating agents, *Chem. Res. Toxocol.* **1**, 249–257 (1988).

S. S. Mirvish, The etiology of gastric cancer. Intragastric nitrosamide formation and other theories, *J. Natl. Cancer* **71**, 629–647 (1983).

S. S. Mirvish, Effects of vitamins C and E on N-nitroso compound formation, carcinogenesis and cancer, *Cancer* **58**, 1842–1850 (1986).

S. S. Mirvish, O. Bulay, R. G. Runge, and K. Patil, Study of the carcinogenicity of large doses of dimethylnitrosamine, N-nitroso-L-proline and sodium nitrite administered in drinking water to rats, *J. Natl. Cancer Inst.* **64**, 1435–1442 (1980).

M. Miwa, D. J. Stuehr, M. A. Marletta, J. S. Wishnok, and S. R. Tannenbaum, Nitrosation of amines by stimulated macrophages, *Carcinog.* **8**, 955–958 (1987).

J. Nair, U. J. Nair, H. Ohshima, S. V. Bhide, and G. Bartsch, Endogenous nitrosation in the oral cavity of chewers while chewing betel quid with or without tobacco. In: *The Relevance of N-Nitroso Compounds to Human Cancer: Exposure and mechanisms,* IARC Scientific Publication No. 84, eds. H. Bartsch, I. K. O'Neill, and R. Schulte-Hermann, IARC, Lyon, France, pp. 465–469 (1987).

J. Nair, H. Ohshima, M. Friesen, A. Croisy, S. V. Bhide, and H. Bartsch, Tobacco-specific and betel nut-specific N-nitroso compounds: Occurrence in saliva and urine of betel quid chewers and formation *in vitro* by nitrosation of betel quid, *Carcinog.* **6**, 295–303 (1985).

J. Nair, H. Ohshima, B. Pignatelli, M. Friesen, C. Malaveille, S. Calmels, and H. Bartsch, Modifiers of endogenous carcinogen formation: studies on *in vivo* nitrosation in tobacco users. In: *Mechanisms in Tobacco Carcinogenesis,* eds. D. Hoffmann and C. C. Harris, Banbury Report No. 23, CSH Press, New York, pp. 45–61 (1986).

National Research Council, *The Health Effects of Nitrate, Nitrite and N-Nitroso Compounds,* Part 1 of a 2-part study, National Academy Press, Washington, D.C. (1981).

H. Ohshima and H. Bartsch, Quantitative estimation of endogenous nitrosation in humans by monitoring N-nitrosoproline excreted in the urine, *Cancer Res.* **41**, 3658–3662 (1981).

H. Ohshima and H. Bartsch, Urinary N-nitrosamino acids as an exposure index to N-nitroso compounds. In: *Methods for Detecting DNA Damaging Agents in Humans: Applications in Cancer Epidemiology and Prevention,* IARC Scientific Publication No. 89, eds. H. Bartsch, K. Hemminki, and I. K. O'Neill, IARC, Lyon, France, pp. 83–91 (1988).

H. Ohshima, J.-C. Béréziat, and H. Bartsch, Monitoring N-nitrosamino acids excreted in the urine and faeces of rats as an index for endogenous nitrosation, *Carcinog.* **3**, 115–120 (1982a).

H. Ohshima, J.-C. Béréziat, and H. Bartsch, (1982b) Measurement of endogenous N-nitrosation in rats and humans by monitoring urinary and faecal excretion of N-nitrosamino acids. In: H. Bartsch, I. K. O'Neill, M. Castegnaro, and M. Okada, eds. *N-Nitroso Compounds: Occurrence and Biological Effects* (IARC Scientific Publications No. 41), Lyon, France, International Agency for Research on Cancer, pp. 397–411.

H. Ohshima, S. Calmels, B. Pignatelli, P. Vincent, and H. Bartsch, *N*-Nitrosamine formation in urinary-tract infections. In: *The Relevance of N-Nitroso Compounds to Human Cancer: Exposure and mechanisms,* IARC Scientific Publication No. 84, eds. H. Bartsch, I. K. O'Neill, and R. Schulte-Hermann, IARC, Lyon, France, pp. 384–390 (1987).

H. Ohshima, M. Friesen, C. Malaveille, I. Brouet, A. Hautefeuille, and H. Bartsch, Formation of direct-acting genotoxic substances in nitrosated smoked fish and meat products: Identification of simple phenolic precursors and phenyldiazonium ions as reactive products, *Food Chem. Toxicol.* **27**, 193–203 (1989).

H. Ohshima, G. A. T. Mahon, J. Wahrendorf, and H. Bartsch, Dose-response study of *N*-nitrosoproline formation in rats and a deduced kinetic model for predicting carcinogenic effects caused by endogenous nitrosation, *Cancer Res.* **43**, 5072–5076 (1983).

H. Ohshima, I. K. O'Neill, M. Friesen, J. C. Béréziat, and H. Bartsch, Occurrence in human urine of new sulphur-containing *N*-nitroso acids *N*-nitrosothiazolidine 4-carboxylic acid and its 2-methyl derivative, and their formation, *J. Cancer Res. Clin. Oncol.* **108**, 121–128 (1984b).

H. Ohshima, I. K. O'Neill, M. Friesen, B. Pignatelli, and H. Bartsch, Presence in human urine of new sulfur-containing *N*-nitrosamino acids: *N*-nitrosothiazolidine 4-carboxylic acid and *N*-nitroso 2-methylthiazolidine 4-carboxylic acid. In: *N-Nitroso Compounds Occurrence, Biological Effects and Relevance to Human Cancer,* IARC Scientific Publication No. 57, eds. I. K. O'Neill, P. C. Von Borstel, J. E. Long, C. T. Miller, and H. Bartsch, IARC, Lyon, France, pp. 77–85 (1984a).

R. Preussmann and B. W. Stewart, *N*-nitroso carcinogens. In: *Chemical Carcinogenesis,* ACS Symposium Series 182, ed. C. E. Searle, American Chemical Society, Washington D.C., pp. 643–828 (1984).

V. Prévost, D. E. G. Shuker, H. Bartsch, R. Pastorelli, W. R. Stillwell, L. J. Trudel, and S. R. Tannenbaum, (1990) The determination of urinary 3-methyladenine by immunoaffinity chromatography - monoclonal antibody based ELISA: use in human dosimetry studies. *Carcinogenesis* (in press).

B. Propkopczyk, A. Rivenson, P. Bertinato, K. D. Brunnemann, and D. Hoffmann, A study of betel quid carcinogenesis. V. 3-(Methylnitrosamino)propionitrile: occurrence in saliva, carcinogenicity and DNA methylation in F344 rats, *Cancer Res.* **47**, 467–471 (1987).

J. L. Radomski, D. Greenwald, W. L. Hearn, N. L. Block, and F. M. Woods, Nitrosamine formation in bladder infections and its role in the etiology of bladder cancer, *J. Urol.* **120**, 48–50 (1978).

D. Ralt, J. S. Wishnok, R. Fitts, and S. R. Tannenbaum, Bacterial catalysis of nitrosation: involvement of the *nar* operon of *Escherichia coli, J. Bacteriol.* **170**, 359–364 (1988).

J. Sander, Nitrosaminosynthese durch Bakterien, *Hoppe Seyler's Z. Physiol. Chem.* **349**, 429–432 (1968).

D. Schmähl and J. M. Kaldor (eds), Carcinogenicity of alkylating cytostatic drugs (IARC Scientific Publication No. 78) International Agency for Research on Cancer, Lyon, France (1986).

S. E. Shephard, C. Schlatter, and W. K. Lutz, Assessment of the risk of formation of carcinogenic *N*-nitroso compounds from dietary precursors in the stomach, *Food Chem. Toxicol.* **25**, 91–108 (1987).

H. Shimada, A study on dietary habits of populations in Akita and Iwate prefectures with different gastric cancer risks, *Akita J. Med.* **7**, 153–178 (1980).

D. E. G. Shuker and P. B. Farmer, Urinary excretion of *N*-3-methyladenine in humans as a marker of nucleic acid methylation. In: *Methods for Detecting DNA Damaging Agents in Humans: Applications to Cancer Epidemiology and Prevention,* IARC Scientific Publication No. 80, eds. H. Bartsch, K. Hemminki, and I. K. O'Neill, IARC, Lyon, France, pp. 92–96 (1988).

D. E. G. Shuker, E. Bailey, and P. B. Farmer, Excretion of methylated nucleic acid bases as an indicator of exposure to nitrosatable drugs. In: *The Relevance of N-Nitroso Compounds to Human Cancer: Exposure and Mechanisms,* IARC Scientific Publication No. 84, eds. H. Bartsch, R. Schulte-Hermann, and I. K. O'Neill, IARC, Lyon, France, pp. 407–410 (1987a).

D. E. G. Shuker, E. Bailey, A. Parry, J. Lamb, and P. B. Farmer, The determination of urinary 3-methyladenine in humans as a potential monitor of exposure to methylating agents, *Carcinog.* **8**, 959–962 (1987b).

D. E. G. Shuker, M. D. Friesen, L. Garren, and V. Prévost, A rapid GC/MS method for the determination of urinary 3-methyladenine: applications to human studies. In: *Relevance to Human Cancer of N-Nitroso Compounds, Tobacco Smoke and Mycotoxins,* IARC Scientific Publication No. 105, eds. I. K. O'Neill, J. S. Chen, and H. Bartsch, IARC, Lyon, France (1990, in press).

S. Srianujata, S. Tonbuth, S. Bunyaratvej, A. Valyasevi, N. Promvanit, and W. Chaivatsagul, High urinary excretion of nitrate and *N*-nitrosoproline in opisthorchiasis subjects. In: *The Relevance of N-Nitroso Compounds to Human Cancer: Exposures and Mechanisms,* IARC Scientific Publication No. 84, eds. H. Bartsch, I. K. O'Neill, and R. Schulte-Hermann, IARC, Lyon, France, pp. 544–546 (1987).

P. Srivanatakul, H. Ohshima, M. Khlat, M. Parkin, S. Sukaryodhin, I. Brouet, and H. Bartsch, Endogenous nitrosamines and liver fluke as risk factors for cholangiocarcinoma in Thailand. In: *Relevance to Human Cancer of N-Nitroso Compounds, Tobacco Smoke and Mycotoxins,* IARC Scientific Publication No. 105, eds. I. K. O'Neill, J. S. Chen, and H. Bartsch, IARC, Lyon, France (1990, in press).

S. Stillwell, H. X. Xu, J. A. Adkins, J. S. Wishnok, and S. R. Tannenbaum, Analysis of methylated and oxidized purines in urine by capillary gas chromatography-mass spectrometry, *Chem. Res. Toxicol.* **2**, 94–99 (1989).

D. J. Stuehr and M. A. Marletta, Mammalian nitrate biosynthesis: mouse macrophages produce nitrite and nitrate in response to *Escherichia coli* lipopolysaccharide, *Proc. Natl. Acad. Sci. USA* **82**, 7738–7742 (1985).

W. Thamavit, N. Bhamarapraviti, S. Sahapong, S. Vajarasthira, and S. Angsubhakorn, Effect of dimethylnitrosamine on induction of cholangio carcinoma in opisthorchis viverrini-infected syrian golden hamsters, *Cancer Res.* **38**, 4634–4639 (1978).

M. Tsuda, J. Niitsuma, S. Sato, T. Hirayama, T. Kakizoe, and T. Sugimura, Increase in the levels of *N*-nitrosoproline, *N*-nitrosothioproline and *N*-nitroso-2-methylthioproline in human urine by cigarette smoking, *Cancer Lett.* **30**, 117–124 (1986).

A. J. Tuyns, Protective effect of citrus fruit on esophageal cancer, *Nutr. Cancer* **5,** 195–200 (1983).

D. Umbenhauer, C. P. Wild, R. Montesano, R. Saffhill, J. M. Boyle, N. Huh, U. Kirstein, J. Thomale, M. F. Rajewsky, and S. H. Lu, O^6-Methyldeoxyguanosine in esophageal DNA among individuals at high risk of esophageal cancer, *Int. J. Cancer* **36,** 661–665 (1985).

J. H. Weisburger, H. Marquardt, N. Hirota, H. Mori, and G. M. Williams, Induction of cancer of the glandular stomach in rats by an extract of nitrite-treated fish, *J. Natl. Cancer Res.* **64,** 163–167 (1980).

G. Wenke and J. Hoffmann, A study of betel quid carcinogenesis. On the in-vitro *N*-nitrosation of arecoline, *Carcinog.* **4,** 169–172 (1983).

C. S. Yang, Research on esophageal cancer: a review, *Cancer Res.* **40,** 2633–2644 (1980).

Molecular Dosimetry of Human Aflatoxin Exposures

John D. Groopman
The Johns Hopkins University
School of Hygiene and Public Health
Department of Environmental Health Sciences
Baltimore, Maryland

Gabriele Sabbioni
Institute of Pharmacology and Toxicology
University of Würzburg
Würzburg, Germany

Christopher P. Wild
Unit of Mechanisms of Carcinogenesis
International Agency for Research on Cancer
Lyon, France

1. INTRODUCTION

In the last twenty-five years there have been extensive efforts to investigate the association between aflatoxin exposure and human liver cancer. Studies using standard epidemiological methods have been hindered by the lack of adequate dosimetry data on aflatoxin intake, excretion, and metabolism, as well as by the general poor quality of worldwide cancer morbidity and mortality statistics. Despite these difficulties, the aflatoxins are among the environmental carcinogens for which quantitative risk assessments have been attempted. These efforts have spurred a number of investigators, in the last few years, to develop reliable, fast, and accurate techniques to assess individual human exposure to this carcinogen.

This chapter will discuss many of the classical and molecular epidemiological studies used to associate dietary aflatoxin exposure with human liver cancer and cited by the International Agency for Research on Cancer (IARC) in its reclassification of aflatoxin B_1 to a category I carcinogen (IARC, 1987). The literature on the toxicology of aflatoxins has been extensively covered by Busby and Wogan (1984), and reviews focused on biological monitoring and epidemiological considerations have been published by Groopman *et al.* (1988) and Bosch and Muñoz (1988). It is well documented in this literature that biological risk of exposure to aflatoxins is much lower in technologically developed countries than in developing ones. Parenthetically, prevention of dietary exposure to aflatoxins will improve the general health status of a developing nation's population. In fact, limiting the intake of aflatoxins is a public health goal requiring the allocation of appropriate economic resources to assure minimal mold contamination of foods and grains.

The development of a human tumor is modulated by many factors, both biological and chemical in nature. Since initiation, promotion, and progression-like events are required prior to clinical diagnosis of a tumor, no one agent can be responsible or present at all critical stages during the growth of a tumor. However, by developing methods to permit the monitoring of an individual's aflatoxin-induced genotoxic burden, the identification of people at high risk for developing disease long before clinical manifestation could be accomplished. In this article, particular attention will be paid to the current state of monitoring individual exposure to aflatoxins. Dosimetry data obtained using the techniques of monoclonal antibody affinity chromatography, polyclonal sera–based enzyme-linked immunoassay, synchronous fluorescence spectroscopy, and other immunological methods will be addressed. The technological advances of the last few years indicate that great strides will be made in the near future to answer not only whether aflatoxin is a cause of human disease but also how can we apportion resources to prevent exposure to this human toxin.

2. AFLATOXIN AND HUMAN LIVER CANCER: RECENT EPIDEMIOLOGICAL STUDIES

Primary liver cancer is one of the leading causes of cancer mortality in Asia and Africa. In the People's Republic of China, this disease accounts for 120,000 deaths per year. This malignancy is the third leading cause of cancer mortality in males behind cancer of the esophagus and stomach as reported by the National Cancer Office of the Ministry of Public Health, People's Republic of China (1980). In contrast, liver cancer incidence in the United States is much lower, and liver cancer rates vary worldwide by at least 100-fold. Several epidemiological studies were conducted 15 to 20 years ago to obtain information on the relationship of estimated dietary intake of aflatoxin with the incidence of primary human liver cancer in different parts of the world (reviewed by Groopman *et al.*, 1988). These investigations showed that increased aflatoxin ingestion corresponded with increased liver cancer incidence. While these early studies could not account for confounding factors such as hepatitis B virus (HBV) infection, this information provided a strong motivation to investigate further the circumstantial relationship between aflatoxin ingestion and liver cancer incidence.

Within this decade, epidemiologic studies have been published on the association of aflatoxin exposure and liver cancer. In one of these reports, Bulatao-Jayme *et al.* (1982) compared the dietary intakes of confirmed primary liver cancer cases in the Philippines against age–sex matched controls. By using dietary recall, the frequency and amounts of food items consumed were calculated into units of aflatoxin load per day. These calculations revealed that the mean aflatoxin load per day of the liver cancer cases was 4.5 times higher than the controls. Alcohol intake as a risk factor was also analyzed by subjectively allocating the subjects into heavy and light aflatoxin exposure groups. These researchers combined aflatoxin load and alcohol intake and determined that a synergistic and statistically significant effect on relative risk with aflatoxin exposure and alcohol intake occurs. These findings suggest that alcohol may have a direct effect as a liver-damaging agent, together with aflatoxin exposure, in the development of liver cancer.

Van Rensburg and his collaborators (1985) studied the occurrence and potential etiologies of hepatocellular carcinoma for the period 1968–1974 in the province of Inhambane, Mozambique. These incidence rates were compared with those observed in South Africa among mine workers migrating from Inhambane. Food samples were randomly collected and aflatoxin content determined in six districts of Inhambane as well as from Manhica-Magude, a region of lower liver cancer incidence. A third set of food samples was taken in Transkei, where an even lower incidence of liver cancer had been recorded. Analysis of the data showed the mean aflatoxin dietary intake values were significantly correlated to the varied liver cancer rates. These studies provide evidence for a dose-dependent increase of aflatoxin intake associated with increased liver disease.

A criticism of the above studies was the failure to consider HBV carrier status in the individuals or populations studied. Three more recent reports have addressed this question. Peers *et al.* (1987) extended the database for Africa and published an epidemiological study conducted in Swaziland. The data collected were analyzed for the relationship between aflatoxin exposure,

HBV status, and the incidence of liver-cell carcinoma. The levels of aflatoxin intake were evaluated in dietary samples from households across the country and crop samples taken from farms. The prevalence of HBV markers was estimated from the serum of blood donors. Liver cancer incidence was recorded for the years 1979–1983 through a national system of cancer registration. Across four broad geographic regions, there was a more than fivefold variation in the estimated daily intake of aflatoxin, ranging from 3.1 to 17.5 μg. The proportion of HBV-exposed males was very high, but varied relatively little by geographic region. However, liver cancer incidence varied over a fivefold range and was strongly associated with estimated levels of aflatoxin. In an analysis involving ten smaller subregions, aflatoxin exposure emerged as a more important determinant of the variation in liver cancer incidence than the prevalence of hepatitis infection.

The epidemiology of liver cancer in China was recently reviewed by Yeh and Shen (1986). This article describes twenty years of research on the epidemiology of liver cancer in China. Their discussion covers the putative role of hepatitis B infection, dietary aflatoxin exposure, and other potential etiologies such as polluted drinking water, pesticide exposure, and nitrosamine contamination. One of the most extensive investigations described was done in Guangxi Province. The staple food of people living in this region is corn, and much of the corn crop was determined to have high levels of aflatoxin. In the heavily contaminated areas, aflatoxin content in corn ranged from 53.8 to 303 ppb, while the lightly contaminated regions showed aflatoxin levels in grains of less than 5 ppb. It is important that in the lightly contaminated areas rice was the predominant dietary grain. After five to eight years of follow-up studies, liver cancer incidence was determined for these two aflatoxin contamination regions. Several thousand person-years were observed for this study. Those individuals who were HBV surface antigen positive and found to have heavy aflatoxin exposure had a liver cancer incidence of 649.35 cases per 100,000 compared with 65.92 cases per 100,000 in lightly contaminated aflatoxin areas. Those people who were HBV surface antigen negative and eating heavily contaminated aflatoxin diets had a liver cancer rate of 98.57 per 100,000 compared with 0 cases detected in the lightly contaminated area. While these data indicate a strong interaction, in terms of relative risk, for those people with HBV surface antigen exposure and aflatoxins in the diet, these data also indicate that aflatoxin plays a significant risk role in developing the disease in the absence of HBV infection.

In a follow-up analysis, Yeh *et al.* (1988) examined the roles of the hepatitis B virus and aflatoxin B_1 (AFB_1) in the development of liver cancer in a cohort of 7917 men aged 25 to 64 years in southern Guangxi Province. After accumulating 30,188 man-years of observation, 149 deaths were observed, 76 (51%) of which were due to primary hepatocellular carcinoma (PHC). Ninety-one percent (69 of 76) of PHC deaths were hepatitis B surface antigen (HBsAg) positive at enrollment into the study in contrast to 23% of all members of the cohort (relative risk = 38.6). Three of the four patients who died of liver cirrhosis also were HBsAg positive at enrollment. There was no association between HBsAg positiveness state and other causes of death. Within the cohort, there was a 3.5-fold difference in PHC mortality by site of residence. When estimated aflatoxin B_1 levels in the subpopulations were plotted against the corresponding mortality rates of PHC, a positive and almost perfectly linear relationship was observed. On the other hand, no significant association was observed when the prevalence of HBsAg positiveness in the subpopulations was compared with their corresponding rates of PHC mortality. However, several questions remain, specifically as to whether an interaction or synergy between aflatoxin and hepatitis B virus occurs and as to the public health impact of aflatoxin exposure in the absence of chronic hepatitis B virus infection. The latter question is of particular relevance as countrywide HBV vaccination programs are being established to reduce chronic HBV infection.

The above questions can be addressed in animal models, such as ducks infected with duck HBV, and by the methods available to measure individual human exposure to aflatoxin as described below.

3. AFLATOXIN METABOLISM, DNA AND PROTEIN ADDUCT FORMATION: THE MOLECULAR BASIS FOR MOLECULAR EPIDEMIOLOGICAL STUDIES

The aflatoxins are primarily metabolized by the microsomal mixed-function oxygenase system. These enzymes catalyze the oxidative metabolism of AFB_1, resulting in the formation of various hydroxylated derivatives as well as an unstable, highly reactive epoxide metabolite, the ultimate carcinogen. Detoxification of AFB_1 is accomplished by enzymatic conjugation of the hydroxylated metabolites with sulfate or glucuronic acid to form water-soluble sulfate or glucuronide esters that are excreted in urine or bile together with the unconjugated compounds such as AFB_1 and AFM_1. In addition, during lactation, AFM_1 can be excreted in the milk. An alternative route for removal of AFB_1 from the organism involves the enzyme-catalyzed reaction of the epoxide metabolite with glutathione and its subsequent excretion in the bile. Some of the known detoxification pathways of AFB_1 metabolism have been summarized in Figure 1.

Recently Shimada and Guengerich (1989) reported an *in vitro* study with human liver indicating that the major cytochrome P-450 involved in the bioactivation of aflatoxin B_1 to its genotoxic epoxide derivative is cytochrome $P-450_{NF}$ ($P-450^{NF}$). This is a previously characterized protein that also catalyzes the oxidation of nifedipine and other dihydropyridines, quini-

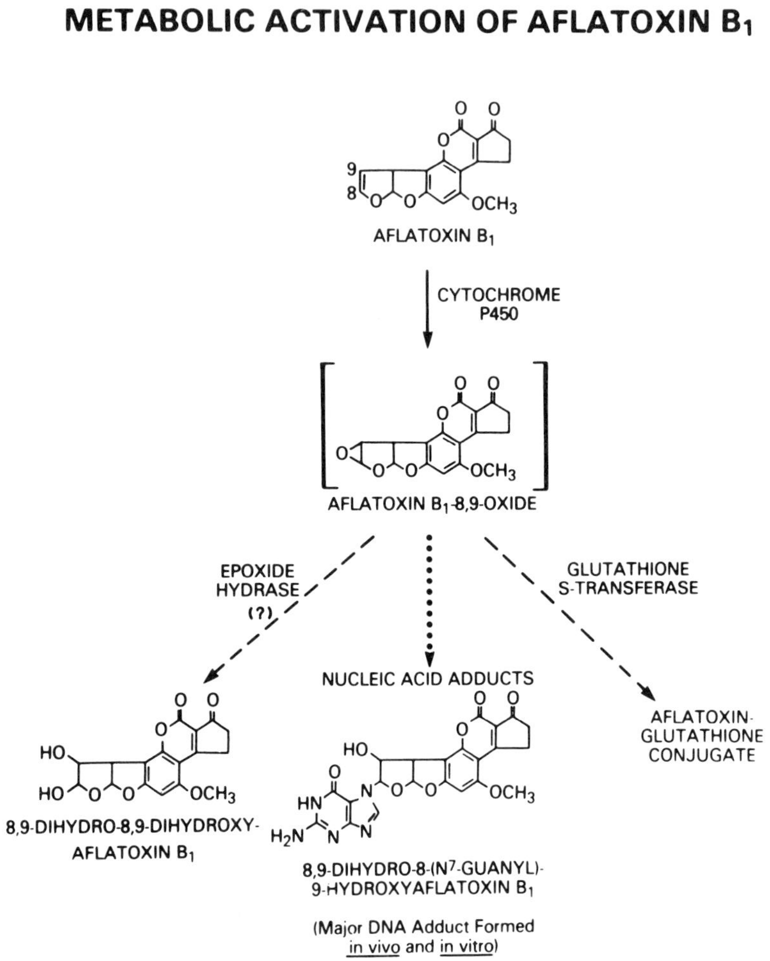

METABOLIC ACTIVATION OF AFLATOXIN B₁

Figure 1. Summary of the major detoxication pathways of aflatoxin B1 metabolism.

dine, macrolide antibiotics, various steroids, and other compounds. Evidence was obtained using activation of AFB_1 as monitored by *umu* C gene expression response in *Salmonella typhimurium* TA1535/pSK1002, enzyme reconstitution, immunochemical inhibition, correlation of response with levels of $P-450_{NF}$ and nifedipine oxidase activity in different liver samples. Liver samples with increasing levels of $P-450_{NF}$ also produced higher amounts of 2,3-dihydro-2-$(N^7$-guanyl)-3-hydroxyaflatoxin B_1 formed in DNA *in vitro*. Several drugs and conditions are known to influence the levels and activity of $P-450_{NF}$ in human liver, and the activity of the enzyme can be estimated by noninvasive assays. These findings provide another avenue for understanding the molecular basis of aflatoxin induced carcinogenesis.

The primary AFB_1–DNA adduct was identified by Essigmann *et al.* (1977) as 2,3-dihydro-2-(N^7-guanyl)-3-hydroxy-AFB_1 (AFB_1-N^7-Gua, Fig. 1) the major product liberated from DNA modified *in vitro* with AFB_1, and its presence was subsequently confirmed *in vivo* (Croy *et al.*, 1978). The binding of AFB_1 residues to DNA *in vivo* was essentially a linear function of dose (Appleton *et al.*, 1982). A number of other DNA components including AFB-dihydrodiol were isolated from nucleic acid hydrolysates activated *in vivo* and *in vitro* with AFB_1 (Lin *et al.*, 1977; Hertzog *et al.*, 1982). These adducts were identified as 2,3-dihydro-2-(N^5-formyl-2,5,6-triamino-4-oxopyrimidin-N^5-yl)-3-hydroxy AFB_1 (AFB_1-FAPyr), a formamidopyrimidine derivative of AFB_1-N^7-Gua, which contained an opened imidazole ring. At the present time, it appears that between 95% and 98% of the aflatoxin residues bound to DNA have been accounted for by these chemical structures.

The investigation of the interactions and biological consequences of aflatoxin B_1 with DNA has been an intensive area of study. Since DNA-adduct formation is probably a requisite event in the initiation of carcinogenesis by aflatoxin B_1, the efforts of many investigators to study aflatoxin B_1–DNA adduct formation is well justified on mechanistic grounds. Despite the suggestions of many years ago, protein–carcinogen interactions, sometimes called an epigenetic mechanism of initiation, has not been a research area favored by many laboratories. The interest to develop serum-screening methods to assess human exposure to dietary aflatoxins has rejuvenated investigations into the mechanisms of aflatoxin binding to proteins. AFB_1 binds to hemoglobin with only a very low yield. Albumin is the only protein in serum that binds AFB_1 to any significant extent in monkeys and rats (Dalezios, 1971; Wild *et al.*, 1986a). In a single dose, 1% to 3% is bound to serum albumin after 24 h in rats. A constant relationship between AFB_1 bound to plasma albumin and liver DNA has been observed in Wistar rats following single (3.5–200 µg/kg) and multiple doses (3.5 µg/kg) of AFB_1 (Wild *et al.*, 1986a). This suggested that albumin-bound aflatoxin might be a particularly valuable indicator of DNA damage in the hepatocyte, with albumin being modified in the hepatocyte itself. At present it is not known whether this is the case or whether the albumin may be modified in the circulation by, for example, the 8,9-dihydro-8,9-dihydroxy AFB_1.

Sabbioni *et al.* (1987) have elucidated the structure of the major aflatoxin–albumin adduct found *in vivo*. According to the major adducts that AFB_1 forms with guanine (Croy *et al.*, 1978) and glutathione (Moss *et al.*, 1983) *in vivo*, the analog substitution product 8,9-dihydro-8-(N-α-lysyl)-9-hydroxy-AFB_1 would be expected. However, the collected spectroscopic data from nuclear magnetic resonance (NMR) and mass spectrometry indicated a different final product. The adduct formed with serum albumin by the binding of the epoxide with subsequent formation of the dihydrodiol, sequential oxidation to the dialdehyde, and condensation with the ε-amino group of lysine. This adduct is a Schiff base, which undergoes Amadori rearrangement to an α-amino ketone (Fig. 2). This protein adduct is a completely modified aflatoxin structure retaining only the coumarin and cyclopentenone rings of the parent compound. A human monitoring study will be described in the following section, exploiting the structural knowledge of this adduct.

This extensive knowledge of the metabolism and adduct formation of AFB$_1$ has enabled a rationale to be developed for the monitoring of human exposure. The approaches are summarized and examples of their development and application are given below.

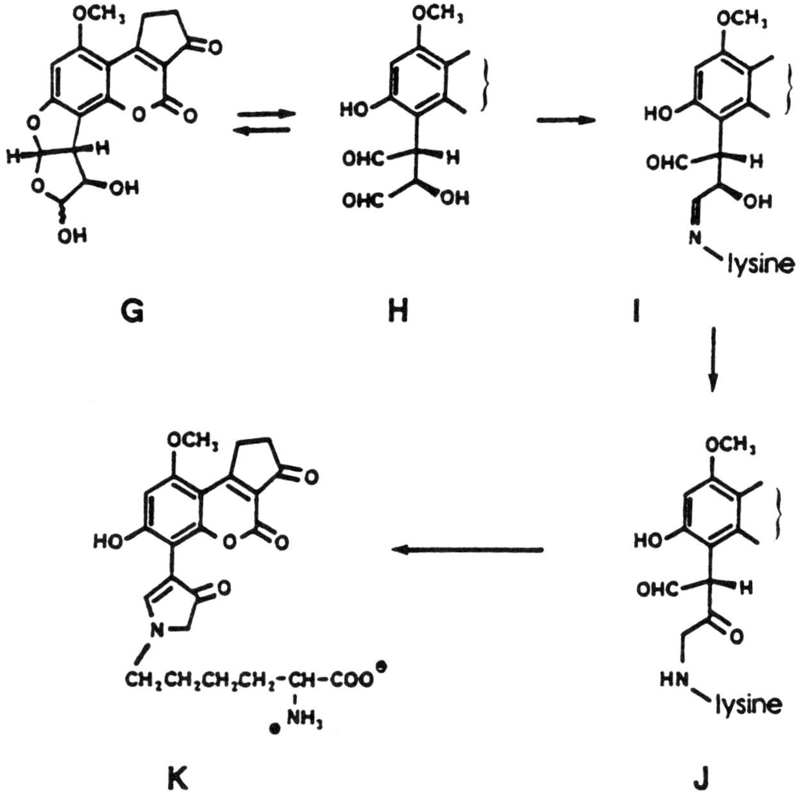

Figure 2. Summary of the pathway leading to the formation of the major aflatoxin B1 serum albumin adduct.

4. MOLECULAR DOSIMETRY: AFLATOXIN AND LIVER CANCER

4.1. Urine

Initial studies used thin-layer chromatography to detect the presence of AFM_1 in human urine (Campbell *et al.*, 1970), and various other groups reported the presence of metabolites in human urine (see Garner *et al.*, 1985). Most of these studies were of a preliminary nature and relied on co-chromatography of the aflatoxin metabolites with marker compounds. In contrast, Autrup *et al.*, (1983; 1987) used synchronous fluorescence spectroscopy (SFS) for the analysis of aflatoxin–DNA adducts in urine. Synchronous fluorescence spectroscopy relies upon the sensitive and specific measurement of physical chemical properties of chemical compounds. By monitoring differences in excitation and emission energies for a specific agent, very sensitive quantitative analyses can be performed. Data on 2,3-dihydro-2-(N^7-guanyl)-3-hydroxyaflatoxin in B_1 in urine samples from the Murang'a district, Kenya, were determined by this method. By using high-pressure liquid chromatographic (HPLC) methods and photon-counting fluorescence spectrophotometry, over 1000 urine samples were analyzed, and 12.6% of the urines were positive for excretion of aflatoxin B_1–guanine adducts. These researchers found a regional variation in the excretion levels of aflatoxin–DNA adducts.

The availability of monoclonal antibodies has permitted the development of immunologically based purification and quantitation methods for specific chemicals, and several research groups have applied these methods to purify and quantitate aflatoxin adducts and metabolites in urine and serum (see below).

An example of the application of enzyme-linked immunosorbent assay (ELISA) methods to human urine analysis is found in the study of ZHU *et al.* (1987). This research group analyzed a total of 252 urine samples from 32 households in Fushui county of the Guangxi autonomous region of the People's Republic of China. A good correlation between total dietary AFB_1 intake and total AFM_1 excretion in human urine was observed during a three-day study, and a linear regression equation of AFB_1 consumed compared with excretion of AFM_1 could be generated. Between 1.2% and 2.2% of dietary AFB_1 was found to be present as AFM_1 in human urine. A good correlation was also observed between the AFB_1 concentration in corn and the AFM_1 concentration in human urine. The percentage of AFM_1 excreted into the urine of these people living in Guangxi Province was similar to data collected by Campbell *et al.* (1970).

A complementary study of aflatoxin M_1 excretion into urine was reported by Nyathi *et al.* (1988). Over 1200 urine samples were collected from different areas of Zimbabwe and were analyzed for aflatoxin content. The urine samples were extracted with chloroform, and the resultant aflatoxins quantified by thin-layer chromatography and high-performance liquid chromatography. The most commonly observed contaminant was aflatoxin M_1, at an average concentration of 4.2 ng/ml of urine. Although the national average of urine samples contaminated was 4.3%, there were areas in which up to 10% of the urine samples were contaminated. These data suggest that aflatoxin M_1 is a predominant metabolite in human urine.

Sun *et al.* (1986) employed monoclonal antibody affinity column and HPLC techniques to the analysis of aflatoxin M_1 in human urine samples. These researchers were among the first to use this new technique, and they reported the measurement and quantitation of AFM_1 in people dietarily exposed to this carcinogen. Their data also provide support for the use of aflatoxin M_1 as a dosimeter for recent exposure to the carcinogen.

Groopman and co-workers, in collaboration with Dr. Chen Jun-shi (Chinese Academy of Preventive Medicine), have been studying people living in Guangxi Province, People's Republic of China, for aflatoxin exposure. These studies were started five years ago and now are beginning to yield the requisite data for exploring both dietary intake of the parent compound, aflatoxin B_1, and the urinary output of aflatoxin metabolites in the same person. These pharmacokinetic data are essential to conduct the assessments that will address the question of the relationship between aflatoxin exposure and liver cancer.

People exposed to AFB_1 from dietary sources were identified for pilot studies (Groopman *et al.*, 1985, 1987). These urine samples were used to gain preliminary evidence of the applicability of the monoclonal antibody affinity column technique and HPLC analysis procedures for monitoring individuals for exposure to aflatoxins. For the initial study, 20 individuals were selected and two 25-ml aliquots of urine obtained from a morning voiding for each individual. The intake of AFB_1 from the diet, primarily corn contaminated with aflatoxin B_1 from 20 to 200 ppb (μg/kg), from the previous day (24 h) was calculated. The exposures ranged from 13.4 to 87.5 μg AFB_1. Urine samples from four individuals who had been exposed to the highest level (87.5 μg) the previous day were prepared with the antibody affinity column and then measured by analytical HPLC (Fig. 3). HPLC analysis demonstrated the presence of the major AFB_1–DNA adduct, AFB_1–N^7-Gua, at levels representing between 7 and 10 ng of the adduct. These data indicate that the monoclonal antibody columns, coupled with HPLC, can quantify aflatoxin–DNA adducts in human urine samples obtained from environmentally exposed people. The urinary findings indicated that knowledge of the dietary intake of aflatoxins is important in the assessment of the quality of the urine data.

The experience from the first China samples stimulated a more extensive

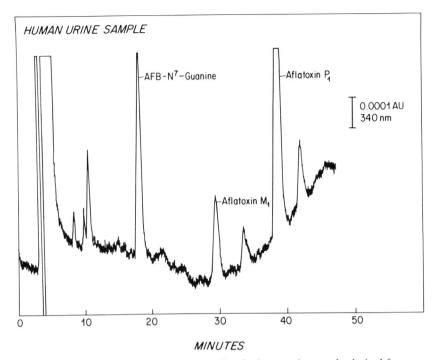

Figure 3. Analytical reverse phase HPLC profile of a human urine sample obtained from an aflatoxin B1 exposed individual.

study in Guangxi Province. To facilitate learning about the relationship between dose and excretion of AFB_1 and its adducts in chronically exposed people, the following protocol was developed. The diets of 30 males and 12 females, ages ranging from 25 to 64 years, were monitored for one week and total aflatoxin intake determined for each day. Urine was obtained in two 12-h fractions for three consecutive days during the one-week period. These urine samples were obtained only after dietary aflatoxin levels had been measured for at least three consecutive days. Therefore the urine collections were initiated on the fourth day of the protocol. The average male intake of AFB_1 was 48.4 µg per day for a total exposure over the 7-day period of 276.8 µg. The average female daily intake was 92.4 µg per day. Immunoassays were performed on aliquots of the 12-h urine following clean-up of the samples by C18 Sep-Pak and monoclonal antibody affinity chromatography.

Total AFB excretion for each 12-h sample period was calculated by multiplying the urine volume by the concentration of AFB determined in the aliquot of urine. Figure 4 depicts a scatter-plot comparison of aflatoxin intake with aflatoxin metabolite excretion. All the male and female data were combined for this analysis. The aflatoxin intake data represent the total integrated ingestion by an individual for the day prior to urine collection and during the three days of the urine collection. The excretion data are the composite of all aflatoxin metabolites excreted into the urine during the three days of urine sampling. These data reveal that despite a 20-fold range of aflatoxin B_1 intake, the amount of aflatoxin excreted generally varies only over a 3-fold range.

We also performed HPLC analysis of the urine samples for AFM_1, AFP_1,

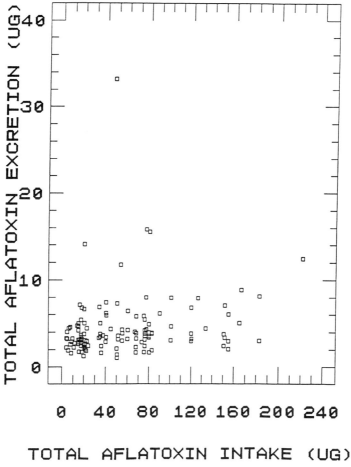

Figure 4. A scatterplot comparison of the relationship between total aflatoxin metabolites in urine and aflatoxin B1 intake data.

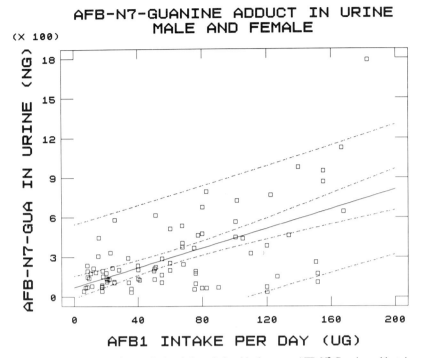

Figure 5. Linear regression analysis of the relationship between AFB-N[7]-Guanine adduct in urine and daily dietary exposure top aflatoxin B1.

and the major AFB–DNA adducts. The data shown in Figure 5 represent the aflatoxin–DNA adduct excretion levels in urine plotted against the aflatoxin B_1 intake amount in μg per day. Unlike the scatter diagram for total aflatoxin excretion, when an individual metabolite is measured, a dose-dependent excretion is seen. Because the aflatoxin–DNA adduct has a relatively short half-life in DNA, this dose-dependent excretion pattern will reflect exposures during the previous few days. The correlation coefficient for this association of adduct excretion and intake is 0.6 with a P significance (P) value of less than 0.00001. Taken together, it appears that aflatoxin–DNA adducts in urine is a valid compartment to sample for aflatoxin exposure, but more data must be collected for developing a risk model for people.

It is important to note that the affinity column technique is also applicable to the analysis of food samples. Naturally contaminated corn and peanut samples were obtained, and the monoclonal antibody affinity method previously described was modified to analyze these samples. Methanol–water (60%:40%, vol/vol) extracts of rehydrated food samples were made by blending the food sample in a Waring blender for one minute. An aliquot of extract was passed through the monoclonal antibody affinity column and then washed and eluted. The eluate was then dried and analyzed by reversed-phase HPLC. In Figure 6 is depicted the HPLC profiles from one of the peanut samples. The peanut sample was analyzed to contain 943 ppb (μg/kg) total aflatoxins. All four of the naturally occurring aflatoxins were detected. The corn samples were found to contain only aflatoxin B_1. These data indicate that different strains of *Aspergillus flavus* and *parasiticus* grow on peanuts and corn. The range of aflatoxin contamination in the corn samples (10 tested) was 0 to 35 ppb with a mean of 9 ppb. The range of contamination in the peanut samples was 0 to 943 ppb with a mean of 146 ppb. Human consumption of this peanut crop would result in heavily contaminated diet.

Wild *et al.* (1986b, 1988) have also used immunopurification to extract aflatoxins from urine but have used a polyclonal antibody to quantitate in a highly sensitive ELISA. In a large collaborative study in The Gambia (ancillary study of the Gambia Hepatitis Intervention Program), 20 individuals were assessed for dietary intake of aflatoxin over an eight-day period, and excretion in 24-h urine samples from days 4 to 8 was determined. In contrast to the above radioimmunoassay approach used in the People's Republic of China, there was evidence that such an analysis of total urinary aflatoxin excretion reflected aflatoxin intake in a dose-dependent manner (Wild and Montesano unpublished data). This may result from a different range of metabolites being extracted and quantitated in this laboratory compared with studies in the People's Republic of China, and thus the validity of measuring 'total' aflatoxin in the urine by immunoassay is probably dependent upon the antibody specificity and should be established for each antibody individually.

This ELISA approach has been used for analysis of urine from several populations of high and low aflatoxin exposure, and the data are presented in

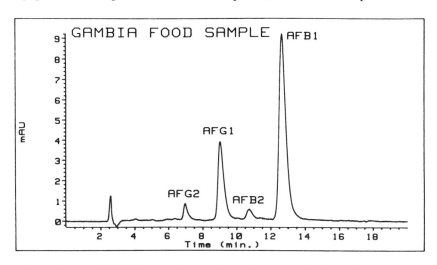

Figure 6. Analytical HPLC reverse phase profile of aflatoxins in a food sample.

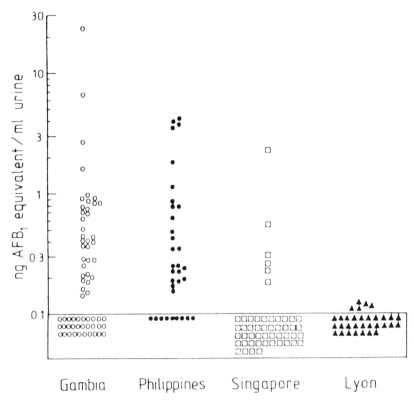

Figure 7. Competitive ELISA results comparing total aflatoxin metabolite levels in different populations having high and low aflatoxin exposures.

Figure 7. At a population level it was clear that (i) the assays are sensitive enough to detect environmental levels of exposure, and (ii) clear differences in populations can be observed. The ease of analysis suggests that this may be a useful screening method to establish the level of exposure in a population.

The second application and value of urine analysis, as evidenced by the work in the People's Republic of China and The Gambia, is that the pattern of metabolites in the urine can be examined (see Fig. 5) and compared at an individual level. Thus in the study in The Gambia, analyses are being performed on chronic HBV carriers compared to noncarriers. This may give information about the interaction of AFB_1 and HBV in terms of AFB_1 metabolism and may be complementary to the approach discussed above regarding the identification of the cytochrome P-450 responsible for aflatoxin metabolism in man.

4.2. Milk

Aflatoxin metabolites are excreted not only in urine but also in lactating animals and can be found in breast milk. Similar ELISA methodologies have thus been applied to the problem of monitoring human milk. The vertical transmission of carcinogenic aflatoxin M_1 from mother's milk to newborn is as yet an unassessed route of exposure. Given the increased sensitivity of newborns to the toxic effects of aflatoxins, an ELISA and complementary fluorescence-based HPLC method has been developed to quantify aflatoxin M_1 in human milk (Wild *et al.*, 1987). A correspondence between these methods was found at levels up to 40 pg AFM_1 per ml milk. 54 milk samples were obtained from women living in Zimbabwe, and 11% of those samples were found to contain up to 50 pg AFM_1 per ml milk. No positive samples were found in 42 milk samples obtained from women living in France. Therefore the ELISA technique can be applied to human samples as part of an overall integrated effort to measure aflatoxin metabolites in human samples.

4.3. Serum Albumin

Some measurements have been made on the presence of free AFB_1 in serum. For example, Tsuboi *et al.* (1984) in Japan and Denning *et al.* (1988) in Nigeria have used ELISA to detect AFB_1 at levels of up to 3 ng per ml serum. Assuming a total serum volume of 2800 ml in a 70-kg human, this would represent 8.4 μg of circulating AFB_1, suggesting an initial exposure orders of magnitude higher, as a small proportion of intake ($<1\%$) is found as free AFB_1 in blood (see Busby and Wogan, 1984). Similar levels (up to 1.5 ng/ml) have been reported by HPLC-fluorescence techniques in sera from Sudanese children (Hendrickse *et al.*, 1982). These approaches, however, measure only free aflatoxins, and a more promising approach would appear to be the measurement of aflatoxin–protein adducts. The significance of this latter work is twofold. First, these adducts will represent an integrated exposure over a number of weeks to months, based on the 20-day half-life of albumin in humans, and second, as discussed in Sec. 3, the adduct level may reflect genotoxic damage in the liver.

To date, both HPLC-fluorescence and immunological approaches have been taken to measure aflatoxin bound to albumin. In an initial study, Gan *et al.* (1988) examined serum albumins from individuals for whom AFB_1 intake and urine excretion were already available (see above) from Guangxi. Serum albumin was isolated from blood by affinity chromatography on Reactive Blue 2–Sepharose and subjected to enzymatic proteolysis using Pronase. Immunoreactive products were purified by immunoaffinity chromatography and quantified by competitive radioimmunoassay. A highly significant correlation of adduct level with intake ($R = 0.69$, $P < 0.000001$)

was also observed (Fig. 8). From the slope of the regression line for the adduct level as a function of intake, it was calculated that 1.4%–2.3% of ingested AFB_1 becomes covalently bound to serum albumin, a value very similar to that observed when rats are administered AFB_1. These studies indicate serum–albumin adduct formation and DNA-adduct excretion can be used as a screen to identify populations at risk and that measurements of these adducts become useful dosimeters in exposed people.

Wild *et al.* (1990a) have recently developed an ELISA for an AFB–lysine adduct that can reliably detect 1 fmol of this adduct. Albumin is isolated from serum or plasma by precipitation and hydrolyzed with proteinase K. The hydrolysate is then passed through via a Sep-Pak cartridge and the bound AF–amino acid residues are quantitated by competitive ELISA. This method requires as little as 50μl serum or plasma and can be applied to large numbers of samples. In addition, the approach has been combined with HPLC-fluorescence techniques for AFB–lysine adducts where, after hydrolysis, an immunoaffinity chromatography step is used to purify the adduct prior to fluorescence detection. A good correlation has been found between the two techniques using human sera from Kenya (Fig. 9). The ELISA method was proved valuable in comparing exposure among various popula-

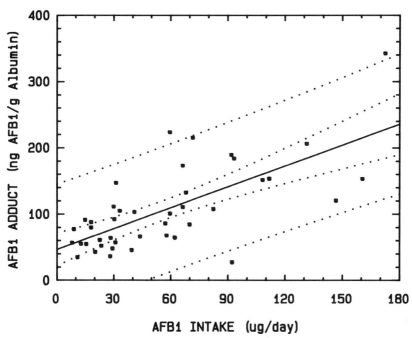

Figure 8. Linear regression analysis of the relationship between aflatoxin serum albumin adduct levels and aflatoxin B1 exposures.

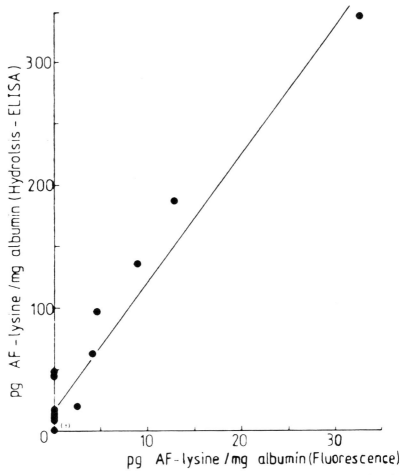

Figure 9. Linear regression analysis comparing two different methods to determine aflatoxin serum albumin adducts in exposed individuals.

tions in Africa, Southeast Asia, and Europe, demonstrating the sensitivity and specificity of the method, and this technique should find widespread application in epidemiological studies.

4.4. Tissue

The use of immunological methods is not only limited to quantitative immunoassays, but can also be applied to immunolocalization techniques. Hseih *et al.* (1988) reported on the use of a monoclonal antibody recognizing aflatoxin–DNA adducts and applied to the analysis of aflatoxin–DNA adducts in human liver tissue. The liver tissue was obtained from surgical samples in Taiwan. The samples contained both tumor tissue and adjacent normal liver. In 9 patient samples examined, AFB–DNA adducts were detectable in 2 of 8 adjacent-normal and 7 of 7 tumor tissues. These intriguing data indicate that persistent adducts can be detected in human liver samples through the use of immunocytochemistry. It will remain for future studies to determine how this technique can be added to the other molecular methods described, to help in our understanding of the role of aflatoxin in human liver cancer.

Garner *et al.* (1988) have reported the use of quantitative immunological methods to examine human liver for the presence of aflatoxin–DNA adducts. Eight liver samples obtained from Czechoslovakian surgical patients with primary hepatocellular carcinoma were studied, seven of which had detectable antiaflatoxin inhibitory material. Values ranged between 0.63 and 3.51 pmol aflatoxin per mg DNA. These preliminary results support the view that immunological methods can be used to examine human tissue DNA for aflatoxin–DNA adducts.

In addition to analyses on extracted DNA, methods have also been established to measure AF–DNA adducts *in situ,* using immunohistochemical techniques (Wild *et al.,* 1990b). In this case, a detection of 2 adducts per 10^7 nucleotides was established in rat liver and this is 10-fold lower than the levels reported in human liver by Hseih *et al.* (1988) and Garner *et al.* (1988). Thus it may be possible in the future to adapt this approach to human exposure assessment. The stability of the AFB-FaPyr in DNA possibly offers the longest-term marker of aflatoxin exposure and may offer a valuable approach within a case-control context. The major limitation is the fact that surgical intervention in liver cancer is relatively rare.

5. SUMMARY

Primary hepatocellular carcinoma is one of the most lethal and common cancers in the world. It is particularly prevalent on the continents of Africa and Asia. A number of epidemiological studies have associated the exposure status of people to aflatoxin B_1 as being important in the etiology of liver cancer. However, to date these studies have relied upon the criteria of presumptive intake data, rather than relying upon quantitative analyses of aflatoxin–DNA adduct and metabolite content obtained by monitoring biological fluids from exposed people. Information obtained by monitoring exposed individuals for specific DNA adducts and metabolites will define the pharmacokinetics of aflatoxin B_1 in people, thereby facilitating risk assessments. However, findings do support the concept that measurement of the major, rapidly excised AFB-N^7-Gua adduct in tissues and fluids and the more persistent aflatoxin–albumin adduct are appropriate dosimeters for estimating exposure status and risk in individuals consuming this mycotoxin. Finally, while there is a large body of scientific data about the presumptive role of aflatoxins in the etiology of human liver cancer, one must not lose sight of the importance of providing the resources necessary to protect the food supply from aflatoxin contamination as a readily obtainable and important goal for protecting the public's health in high-exposure regions of the world.

ACKNOWLEDGMENT

This work was supported in part by a grant from the U.S. Public Health Service, No. 1 U01 CA48409.

REFERENCES

S. Appleton, M. P. Coetschius, and T. C. Campbell, Linear dose-response curve for hepatic macromolecular binding of alfatoxin B1 in rats at very low exposures, *Cancer Res.* **42**, 3659–3662 (1982).

H. Autrup, K. A. Bradley, A. K. M. Shamsuddin, J. Wakhisi, and A. Wasunna, Detection of putative adduct with fluorescence characteristics identical to 2,3-digydro-2-(7'-guanyl)-3-hydroxyaflatoxin B1 in human urine collected in Murang'a district, Kenya, *Carcinog.* **4**, 1193–1195 (1983).

H. Autrup, T. Seremet, J. Wakhisi, and A. Wasunna, Aflatoxin exposure measured by urinary excretion of aflatoxin B_1-Guanine adduct and hepatitis B virus infection in areas with different liver cancer incidence in Kenya, *Cancer Res.* **47**, 3430–3433 (1987).

F. X. Bosch and N. Muñoz, Prospects for epidemiological studies on hepatocellular cancer as a model for assessing viral and chemical interactions. In: *Methods for Detecting DNA Damaging Agents in Humans: Applications to Cancer Epidemiology and Prevention,* IARC Scientific Publication No. 89, eds. H. Bartsch, K. Hemminki, and I. K. O'Neill, IARC, Lyon, France, pp. 427–438 (1988).

J. Bulatao-Jayme, E. M. Almero, C. A. Castro, T. R. Jardeleza, and L. Salamat, A case-control dietary study of primary liver cancer risk from aflatoxin exposure, *Int. J. Epidemiol.* **11**, 112–119 (1982).

W. F. Busby and G. N. Wogan, Aflatoxins. In: *Chemical Carcinogens, 2nd ed.,* ed. C. E. Searle, American Chemical Society, Washington, D.C., pp. 945–1136 (1984).

T. C. Campbell, J. P. Caedo, J. P. Bulatao-Jayme, L. Salamat, and R. W. Engel, Aflatoxin M_1 in human urine, *Nature* **227**, 403–404 (1970).

R. G. Croy, J. M. Essigmann, V. N. Reinhold, and G. N. Wogan, Identification of the principal aflatoxin B1-DNA adduct formed *in vivo* in rat liver, *Proc. Natl. Acad. Sci. USA* **75**, 1745–1749 (1978).

J. I. Dalezios, Aflatoxin P1: A new metabolite of aflatoxin B1. Its isolation and identification, Ph.D. thesis, Massachusetts Institute of Technology, Cambridge, MA (1971).

D. W. Denning, J. K. Onwubalili, A. P. Wilkinson, and M. R. Morgan, Measurement of aflatoxin in Nigerian sera by enzyme-linked immunosorbent assay, *Trans. R. Soc. Trop. Med. Hyg.* **82,** 169–171 (1988).

J. M. Essigmann, R. G. Croy, A. M. Nadzan, W. F. Busby, Jr., V. N. Reinhold, G. Buchi, and G. N. Wogan, Structural identification of the major DNA adduct formed by aflatoxin B1 *in vitro, Proc. Natl. Acad. Sci. USA* **74,** 1870–1874 (1977).

L. S. Gan, P. L. Skipper, X.-C. Peng, J. D. Groopman, J. S. Chen, G. N. Wogan, and S. R. Tannenbaum, Serum albumin adducts in the molecular epidemiology of aflatoxin carcinogenesis: correlation with aflatoxin B1 intake and urinary excretion of aflatoxin M1, *Carcinog.* **9,** 1323–1325 (1988).

R. C. Garner, R. Ryder, and R. Montesano, Monitoring of aflatoxins in human body fluids and application to filed studies, *Cancer Res.* **45,** 922–928 (1985).

R. C. Garner, I. Dvorackova, and F. Tursi, Immunoassay procedures to detect exposure to aflatoxin B1 and benzo(a)pyrene in animals and man at the DNA level, *Int. Arch. Occup. Environ. Health.* **60,** 145–150 (1988).

J. D. Groopman, P. R. Donahue, J. Zhu, J. Chen, and G. N. Wogan, Aflatoxin metabolism in humans: Detection of metabolites and nucleic acid adducts in urine by affinity chromatography, *Proc. Natl. Acad. Sci. USA* **82,** 6492–6497 (1985).

J. D. Groopman, P. R. Donahue, J. Zhu, J. Chen, and G. N. Wogan, Temporal patterns of aflatoxin metabolites in urine of people living in Guangxi Province, P.R.C., *Proc. Am. Assoc. Cancer Res.* **28,** 36 (1987).

J. D. Groopman, L. G. Cain, and T. W. Kensler, Aflatoxin exposure in human populations: Measurements and relationships to cancer, *CRC Crit. Rev. Toxicol.* **19,** 113–145 (1988).

R. G. Hendrickse, J. Coulter, S. Lamplugh, S. MacFarlane, T. Williams, M. Omer, and G. Suliman, Aflatoxins and kwashiorkor: a study in Sudanese children, *Br. Med. J.* **285,** 843–846 (1982).

P. J. Hertzog, J. R. Lindsay Smith, and R. C. Garner, Production of monoclonal antibodies to guanine imidazole ring-opened aflatoxin B1-DNA, the persistent DNA adduct *in vivo, Carcinog.* **3,** 723–725 (1982).

L. L. Hsieh, S.-W. Hsu, D.-S. Chen, and R. M. Santella, Immunological detection of aflatoxin B1-DNA adducts formed in vivo, *Cancer Res.* **48,** 6328–6331 (1988).

IARC Monographs on the evaluation of carcinogenic risks to humans, Supplement 7, pp. 83–87, 1987.

J. K. Lin, J. A. Miller, and E. C. Miller, 2,3-Dihydro-2-(guan-7-yl)-3-hydroxy-aflatoxin B1, a major acid hydrolysis product of aflatoxin B1-DNA or-ribosomal RNA adducts formed in hepatic microsome mediated reactions in rat liver *in vivo, Cancer Res.* **37,** 4430–4438 (1977).

E. J. Moss, D. J. Judah, M. Przyblyski, and G. E. Neal, Some mass-spectral and nmr analytical studies of a glutathione conjugate of aflatoxin B1, *Biochem. J.* **210,** 227–233 (1983).

C. B. Nyathi, C. F. Mutiro, J. A. Hasler, and C. J. Chetsanga, A survey of urinary aflatoxin in Zimbabwe, *Int. J. Epidemiol.* **16,** 516–519 (1987).

National Cancer Office of the Ministry of Public Health, People's Republic of China, *Studies on Mortality Rates of Cancer in China,* People's Publishing House, Beijing (1980).

F. Peers, X. Bosch, J. Kaldor, A. Linsell, and M. Pluijmen, Aflatoxin exposure, hepatitis B virus infection and liver cancer in Swaziland, *Int. J. Cancer* **39**, 545–553 (1987).

G. Sabbioni, P. Skipper, G. Buchi, and S. R. Tannenbaum, Isolation and characterization of the major serum albumin adduct formed by aflatoxin B1 *in vivo* in rats, *Carcinog.* **8**, 819–824 (1987).

T. Shimada and F. P. Guengerich, Evidence for cytochrome P-450NF, the nifedipine oxidase, being the principal enzyme involved in the bioactivation of aflatoxins in human liver, *Proc. Natl. Acad. Sci. USA* **86**, 462–465 (1989).

T. T. Sun, Y.-R. Chu, C.-C. Hsia, Y.-P. Wei, and S.-M. Wu, Strategies and current trends of etiologic prevention of liver cancer. In: *Biochemical and Molecular Epidemiology,* ed. C. C. Harris, Alan R. Liss, New York, pp. 283–292 (1986).

S. Tsuboi, T. Nakagawa, M. Tomita, T. Soo, H. Ono, K. Kawamura, and N. Iwamura, Detection of aflatoxin B_1 in serum samples of male Japanese subjects by radioimmunoassay and high-performance liquid chromatography, *Cancer Res.* **44**, 1231–1234 (1984).

S. J. Van Rensburg, P. Cook-Mozaffari, D. J. Van Schalkwyk, J. J. Van der Watt, T. J. Vincent, and I. F. Purchase, Hepatocellular carcinoma and dietary aflatoxin in Mozambique and Transkei, *Br. J. Cancer* **51**, 713–726 (1985).

C. P. Wild, R. G. Garner, R. Montesano, and F. Tursi, Aflatoxin B1 binding to plasma albumin and liver DNA upon chronic administration to rats, *Carcinog.* **7**, 853–858 (1986a).

C. P. Wild, D. Umbenhauer, B. Chapot, and R. Montesano, Monitoring of individual human exposure to aflatoxins (AF) and *N*-nitrosamines (NNO) by immunoassays, *J. Cell. Biochem.* **30**, 171–179 (1986b).

C. P. Wild, F. Pionneau, R. Montesano, C. F. Mutiro, and C. J. Chetsanga, Aflatoxin detected in human breast milk by immunoassay, *Int. J. Cancer* **40**, 328–333 (1987).

C. Wild, B. Chapot, E. Scherer, L. Den Engelse, and R. Montesano, The application of antibody methodologies to the detection of aflatoxin in human body fluids. In: *Methods for Detection of DNA Damaging Agents in Humans: Applications in Cancer Epidemiology and Prevention,* IARC Scientific Publication No. 89, eds. H. Bartsch, K. Hemminki, and I. K. O'Neill, IARC, Lyon, France, pp. 67–74 (1988).

C. Wild, Y. Jiang, G. Sabbioni, B. Chapot, and R. Montesano, Evaluation of methods for quantitation of aflatoxin-albumin adducts and their application to human exposure assessment, *Cancer Res.* **50**, 245–251 (1990a).

C. P. Wild, R. Montesano, J. Vanbenthem, E. Scherer, L. Denengelse, Intercellular variation in levels of adducts of aflatoxin-B1 and aflatoxin-G1 in DNA from rat tissues—A quantitative immunocytochemical study, *J. Cancer Res. Clin. Oncol.* **116(2)**, 134–140 (1990b).

F.-S. Yeh and K.-N. Shen, Epidemiology and early diagnosis of primary liver cancer in China, *Adv. Cancer Res.* **47**, 297–329 (1986).

F.-S. Yeh, M.-C. Yu, C.-C. Mo, S. Luo, M.-J. Tong, and B. E. Henderson, Hepatitis B virus, aflatoxins, and hepatocellular carcinoma in southern Guangxi, China. *Cancer Res.* **49**, 2506–2509 (1989).

J.-Q. Zhu, L.-S. Zhang, X. Hu, Y. Xiao, J.-S. Chen, Y.-C. Xu, J. Fremy, and F. S. Chu, Correlation of dietary aflatoxin B_1 levels with excretion of aflatoxin M_1 in human urine, *Cancer Res.* **47**, 1848–1852 (1987).

Monitoring Exposure to Tobacco Products by Measurement of Nicotine Metabolites and Derived Carcinogens

Stephen S. Hecht, Nancy J. Haley,
and Dietrich Hoffmann
American Health Foundation
Valhalla, New York

1. INTRODUCTION

Nicotine and its metabolites are specific indicators of exposure to tobacco and tobacco smoke. Nicotine is also a precursor to tobacco-specific
nitrosamines, an important group of carcinogens in tobacco and tobacco
smoke. These nitrosamines bind to cellular macromolecules, and the resulting adducts are potentially important indicators of an individual's ability to
metabolically activate them to carcinogenic intermediates. In this chapter,
we discuss methods for measuring exposure to tobacco products by assessing
levels of nicotine metabolites and tobacco-specific nitrosamines–macromolecule adducts. These approaches have public health significance because of
the major role of tobacco products as causes of cancer as well as other
diseases of the respiratory and circulatory systems.

Tobacco has been used for over 16 centuries. *Nicotiana* seeds have been
uncovered in archeological excavations in Mexico and Peru while Mayan

carvings from 600 to 900 A.D. document the chewing of tobacco leaves (Davis, 1988). Independently, Australian natives cultivated a variety of the solanaceous shrub *Duboisia hopwoodii,* which contains up to 5.3% nicotine as its major alkaloid (Watson, 1988). With the beginnings of international exploration and trade, tobacco was one of the first products taken from the native populations of America to Europe. By the 1600s, smoking had become widespread throughout the world [International Agency for Research on Cancer (IARC), 1986].

The search for the chief active ingredient of tobacco began with the discovery of an "essential oil" in 1807 by Cerioli and the characterization of this material by Vauquelin in 1809. Within the next 25 years, the presence of this active substance was confirmed in 16 species of *Nicotiana* and the empirical formula determined (Holmstedt, 1988).

Research into the effects of this powerful tobacco alkaloid generally consisted of self-experimentation by chemists and ended with what is known as severe nicotine poisoning or "green tobacco sickness." These early experiments did help determine, however, that nicotine, through its pharmacologic actions, is a major reason why people chew or smoke tobacco. There are nine other known pyridine alkaloids in cigarette smoke, but nicotine constitutes at least 85% of the total alkaloids in U.S. cigarettes (Fig. 1) (IARC, 1986). Compared with the amount of research devoted to nicotine, the biological information on the related alkaloids is cursory and incomplete. However, Mattila and Saarnivaara determined the relative activities of nicotine and eight of its analogs, several of which are illustrated in Fig. 1, on the following parameters in different species of animals: the median lethal dose (LD_{50}), sedation, analgesia, presser effect, tremor, and ear twitch as well as antidiuresius in hydrated rats. Nicotine showed much greater potency than any of the analogs (Holmstedt, 1988).

Throughout the 20th century, and especially during the 1970s and 1980s, research has investigated the role of nicotine in the reinforcement of smoking behavior. While the habitual use of tobacco is, in part, maintained by psychological and social needs, nicotine is the critical pharmacologic agent for maintenance of tobacco dependence (U.S. Surgeon General, 1988). Nicotine induces a wide variety of stimulant and depressant effects involving the central and peripheral nervous systems as well as the cardiovascular, gastrointestinal, and endocrine systems. The heterogeneous effects of this drug are partly responsible for the self-administration of tobacco, for the development of tobacco tolerance, and for withdrawal phenomena associated with cessation of tobacco use.

In addition to its pharmacologic properties, nicotine provides researchers with a marker for evaluation of tobacco use and smoking behavior. Nicotine and its metabolites provide specific biomarkers for the use of tobacco products. Current methodology allows sensitive measurement of these compounds

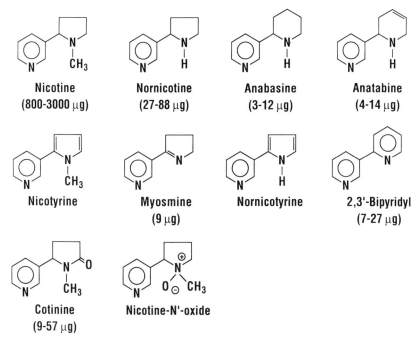

Figure 1. Some pyridine alkaloids present in tobacco smoke. Numbers in parentheses are the amounts per cigarette in mainstream smoke (Schmeltzl and Hoffmann, 1977).

in body fluids. Nicotine is not a carcinogen *per se*, but reactions of this compound with sodium nitrite under mild conditions have produced carcinogenic nitrosamines as discussed in Sec. 5 (IARC, 1986).

2. NICOTINE TOXICITY AND HABITUATION

Understanding the targets and toxic actions of nicotine is important in the elucidation of the overall health effects of tobacco use. Nicotine is soluble in water as well as in hydrophobic solvents. It can be readily absorbed via the skin, gastrointestinal tract, and respiratory system in the form of a free base and when present in alkaline solutions may cross mucus membranes. In evaluating toxicity data, it is important to remember that the route of administration and the rate of uptake are important determinants of effect. Intravenous injections of nicotine produce the highest toxicity at the lowest doses while oral administration requires high doses to produce its effects. The amount of nicotine absorbed per day in chronic dosing over several hours as often seen in smokers could be lethal if administered as a single injection.

The pH of smoke from air-cured tobacco such as pipe tobacco, cigars,

**TABLE 18-1. Human Pharmacokinetics of
Nicotine and Cotinine in Cigarette Smokers***

	Nicotine	Cotinine
Half-life	120 min	18 hours
Volume of Distribution	180 liters	88 liters
Total Clearance	1300 ml/min	72 ml/min
Renal Clearance	200 ml/min (acid urine)	12 ml/min
Nonrenal Clearance	1100 ml/min	60 ml/min

*From Benowitz, 1988.

**TABLE 18-2. Steady-State
Distribution of Nicotine***

Tissue	Ratio: Tissue to Blood
Blood	1.0
Brain	3.0
Heart	3.7
Muscle	2.0
Adipose	0.5
Kidney	21.6
Liver	3.7
Lung	2.0
Gastrointestinal	3.5

*From U.S. Surgeon General, 1988.

and a few brands of cigarettes as well as the extracts of some chewing tobaccos and snuff is alkaline. Unprotonated nicotine from these products is absorbed through the oral mucosa (Armitage and Turner, 1970). The portion that is swallowed is absorbed through the gastrointestinal tract. The pH of smoke of blended cigarettes and cigarettes made from flue-cured tobacco is below 6.2, and nicotine is delivered in mostly the monoprotonated form (Brunnemann and Hoffmann, 1974). Therefore there is little buccal absorption of nicotine from these cigarettes with most of the uptake occurring in the small airways of the lung. These differences in uptake have significant impacts on the bioavailability and metabolism of this pharmacoactive alkaloid.

2.1. Target Organs

After absorption into the blood, which is at pH 7.4, about 69% of the nicotine is ionized and 31% nonionized. The drug is distributed extensively to body tissues with a steady-state volume of distribution of about 180 liters or 2.6 times body weight (Table 1) (Benowitz, 1988). Tissue distribution has been studied in animals (Table 2). The results indicate that the spleen, liver, lung, and brain have high affinity for nicotine while that of adipose tissue is relatively low (U.S. Surgeon General, 1988).

Nicotine is taken up rapidly by the brain and elicits a number of cardiovascular (U.S. Surgeon General, 1988) and neuroendocrine (Fuxe *et al.,* 1988) responses several times each day in active cigarette smokers. Nicotine exerts a direct effect on ganglion cells, producing transient excitation followed by depression or transmission blockade. At higher doses, nicotine also acts on the peripheral nervous system and induces the secretion of adrenal catecholamines (Hill and Synder, 1974).

The cardiovascular responses to nicotine, in general, parallel those that follow stimulation of the sympathicoadrenal system. Because nicotine has both stimulant and depressant effects, the cardiovascular responses represent the total of several different modes of action of this compound.

2.2. Physiologic Effects, Tolerance, and Habituation

Nicotine exerts its actions on the cardiovascular, respiratory, skeletal motor, and gastrointestinal systems through stimulation of peripheral cholinergic neurons via afferent chemoreceptors and ganglia of the autonomic nervous system (Ginzel, 1967). Increased heart rate and blood pressure are accompanied by cutaneous vasoconstriction of the hands and feet as well as relaxation of skeletal muscles (Ginzel, 1973; 1988). Nicotine can enhance respiration but at high doses can also cause respiratory blockade and death (Domino, 1979). Additional stimulation comes from the release of acetylcholine (ACh) and other neurotransmitters from cholinergic nerve terminals. There is evidence that this occurs via a presynaptic mechanism (Domino, 1979) and is accompanied by release of serotonin, endogenous opiate peptides, pituitary hormones, catecholamines, and vasopressin (Fuxe, *et al.,* 1988). These chemicals may be involved in the reinforcing effects of nicotine on the tobacco habit.

Other hormonal systems are affected by nicotine, but the functional significance of these perturbations is poorly understood. A single injection of 200 μg/kg of nicotine was found to have no effect on thyroid function by Cam and Bassett (1983) while Sepkovic *et al.* (1984c) found a dose–response relationship between blood levels of cotinine, a major metabolite, and circulating levels of thyroid hormones. The latter authors extended their studies to animals, demonstrating the thyroid effects of chronic nicotine metabolite dosing and cigarette smoking in rats (1984b) and baboons (1988).

Testosterone levels have been found to be variable among smokers. Shaarawy and Mahmoud (1982) observed reduced serum testosterone in smokers compared to nonsmokers. Chronic use of high tar and nicotine products produced decreased serum testosterone levels in male beagles (U.S. Surgeon General, 1988). Epidemiological studies in male smokers suggest that testosterone levels might be "high normal."

Cigarette smoking is associated with antiestrogenic effects in women, including earlier menopause, and possibly lower incidence rates of breast and endometrial cancer as well as increased osteoporosis. MacMahon *et al.* (1982) observed reduced luteal-phase urinary estrogen excretion in premenopausal women who were current smokers. Michnovicz *et al.* (1986) noted a significant increase in estradiol 2-hydroxylation in premenopausal heavy smokers. This could help explain the antiestrogenic effects of smoking by decreasing the bioavailability of estrogens. However, this concept has not been evaluated with regard to nicotine alone, in animal studies, or by nicotine replacement in humans.

Acute tolerance to the effects of cigarettes has been demonstrated by Russell (1988) and correlated with concomitant blood levels of nicotine. After an overnight fast from smoking, fewer puffs of a cigarette are required to elicit a pleasurable response than at the end of the day. Tolerance can develop rapidly to the effects of nicotine. Studies by Lee *et al.* (1987) have demonstrated significantly greater cardiovascular responses to infused nicotine after 18-hour or 7-day abstinence than during hourly infusions. The longer the abstinence, the greater were the changes in heart rate and blood pressure. However, one hour of intermittant infusions of nicotine reestablished responses seen before tobacco deprivation in these subjects.

Chronic tolerance is difficult to quantify since acute tolerance occurs every day for most smokers and increasing nicotine delivery beyond the customary dose can result in the establishment of a new baseline. Studies have demonstrated that subjects can continue to show physiological responses to their customary cigarette if nicotine delivery is increased. Compensation studies in which smokers are shifted to cigarettes with lower deliveries of tar and nicotine have demonstrated that smokers acquire the desire for a maintained nicotine delivery per day and adjust their smoking behavior to attain their level of nicotine intake (Russell *et al.*, 1975; Sepkovic *et al.*, 1983). Smokers may change their nicotine input by adjusting the number of cigarettes smoked per day or by altering the manner in which each cigarette is smoked such as by modifying their puff volume or depth of inhalation (Herning *et al.*, 1981; Haley *et al.*, 1985; Puustinen *et al.*, 1987).

Tolerance to a drug might be due to an increase in the rate of drug metabolism or to a decrease in sensitivity of the target tissue. Considerable differences exist in metabolism of nicotine by smokers and nonsmokers. Nicotine is metabolized more quickly in smokers (Kyerematen *et al.*, 1982) and cotinine elimination is slower in nonsmokers exposed to nicotine via environmental tobacco smoke than in active smokers inhaling mainstream cigarette smoke (Sepkovic *et al.*, 1986b; Haley *et al.*, 1989b).

Habituation to tobacco use may arise from psychosocial or behavioral needs; however, it arises primarily from nicotine-derived effects that are perceived as desirable by the tobacco user. The view that nicotine is a substance of dependence has been the subject of years of research culminating in the 1988 U.S. Surgeon General's Report on the Health Consequences of Smoking.

3. OVERVIEW OF NICOTINE METABOLISM

Nicotine is metabolized primarily in the liver and to some extent in the lung and other tissues. A large proportion of nicotine can also be excreted unchanged, and that percentage is dependent on urinary pH. Typically, however, about 5% to 10% of nicotine is cleared by the kidneys in habituated cigarette smokers and subjects receiving infused nicotine (Rosenberg *et al.*, 1980).

Nicotine metabolism is summarized in Fig. 2 (Gorrod and Jenner, 1975; Pool *et al.*, 1984; Mattammal *et al.*, 1987; Kyerematen *et al.*, 1988; Shigenaga *et al.*, 1988, and references therein). The major primary metabolites of nicotine are cotinine *(14)* and the nicotine N'-oxides *1* and *2* (two of the four diastereomers that can be formed from racemic nicotine are shown). Cotinine is formed in the liver in a two-step process, the first of which involves oxidation of position 5 of the pyrrolidine ring in a cytochrome P-450– mediated process to nicotine–Δ-$1'$,$5'$-iminium ion *(9)*. In the second step, the iminium ion is metabolized by a cytosolic aldehyde oxidase to cotinine. Approximately 10% of cotinine formed by liver metabolism of nicotine in the rat is excreted unchanged in urine. Further metabolism of cotinine to 4- (3-pyridyl)-4-(N-methylamino)butyric acid *(15)*, cotinine-1-N-oxide *(16)*, *trans*-$3'$-hydroxycotinine *(17)*, $5'$-hydroxycotinine *(18)*, and norcotinine *(20)* has been documented. These metabolites, in turn, lead to the acids *22–24*. Urinary *trans*-$3'$-hydroxycotinine accounts for 4%–6% of the dose of nicotine in the rat, and recently Neurath and colleagues (1987) proposed that it is the major metabolite of nicotine to be found in smokers' urine. Its plasma concentration in smokers, however, is lower than that of the other nicotine metabolites (Aklkofer *et al.*, 1988).

Substantial quantities of nicotine are metabolized to nicotine-N'-oxides by N-oxidation (Booth and Boyland, 1970). The *cis:trans* ratio is governed by the enantiomeric composition of the starting nicotine (Jenner *et al.*, 1973a). The formation of the N'-oxides is important for two reasons. First, nicotine-N'-oxides are rapidly back-converted to nicotine both *in vitro* and *in vivo* by gut flora as well as intestinal and hepatic microsomal enzymes (Jenner *et al.*, 1973b; Dajani *et al.*, 1972). These metabolites may represent a nicotine reservoir that could play a role in reinforcing the tobacco habit. Second, nicotine N'-oxides are relatively easily nitrosated to form the tobacco-specific carcinogen, N'-nitrosonornicotine (NNN) (Klimisch and Stadler, 1979).

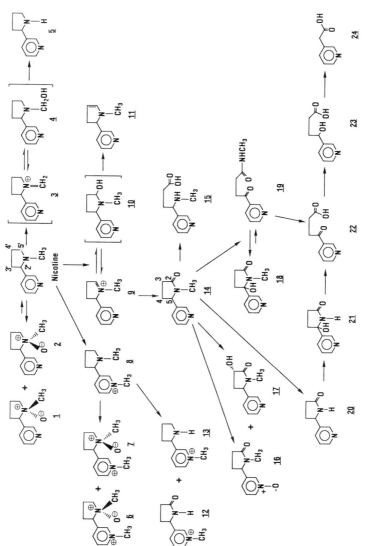

Figure 2. Mammalian methabolism of racemic nicotine, compiled from Gorrod nand Jennifer, 1975; Pool *et al.*, 1986; Mattammal *et al.*, 1987; Kyerematen *et al.*, 1988; Shigenaga *et al.*, 1988, and references therein. Some pathways are under stereochemical control, e.g., metabolites 6–8, 12, and 13 are formed in the guinea pig exclusively from R(+) nicotine (Pool *et al.*, 1986), and the *cis:trans* ratio of the N′-oxides 1 and 2 depends on whether R(+) nicotine or S(−) nicotine

The rate of nicotine metabolism can be determined by measuring blood levels after administration of a known dose. In one study by Benowitz *et al.* (1982), total and renal clearances were computed for cigarette smokers following intravenous infusions of nicotine. Total clearance averaged 1300 ml/min while nonrenal clearance averaged 1100 ml/min, which represents about 70% of liver blood flow (Table 1). Since nicotine is metabolized mainly by the liver, about 70% of the drug is extracted from the blood in each pass through the liver.

Nicotine is excreted by the kidney. The excretion rate of metabolites reflect the rate of metabolite formation and is less dependent on pH than is the excretion of nicotine itself. While most of the metabolites can be recovered in urine, the measurements of nicotine and cotinine are the most generally employed to evaluate uptake and absorption of nicotine from tobacco products or nicotine-containing smoking cessation aids.

4. MONITORING NICOTINE AND METABOLITES IN RELATION TO TOBACCO-PRODUCT EXPOSURE

Trans-3′-hydroxycotinine is measured by gas chromatography following formation of the heptafluoropropionate derivative (Neurath *et al.*, 1987). If its quantitation becomes easier, it should have potential as a urinary biomarker for nicotine metabolism in smokers and passively exposed nonsmokers. Currently, however, nicotine and especially its metabolite, cotinine, are the most widely analyzed compounds for measuring nicotine exposure (Sepkovic and Haley; 1985, 1987).

The determination of nicotine and cotinine in body fluids of active smokers and passively exposed nonsmokers is done primarily by gas chromatography (GC) with a nitrogen-sensitive detector (D Feyerabend, 1987; Jacobs III *et al.*, 1981) or by radioimmunoassay (RIA) (Langone *et al.*, 1973; Haley *et al.*, 1983; Van Vunakis *et al.*, 1987). High-pressure liquid chromatography (HPLC) methods have been developed for quantitation of cotinine in saliva and serum, but have not yet been successfully applied to urine samples (Machacek and Jaing, 1986). An emerging technique, HPLC with dual electrochemical detection, has not yet been applied to nicotine analyses of samples from environmental tobacco smoke (ETS)–exposed individuals but has been used to quantitate nicotine in body fluids of active smokers (Chen *et al.*, 1988). Another analytic technique is enzyme-linked immunosorbent assay (ELISA); Bjercke *et al.*, 1986), which has potential for nicotine and cotinine analyses from ETS-exposed individuals but has not been demonstrated to date.

Polyclonal and monoclonal antibodies are specific for cotinine and have been tested against approximately 100 compounds of tobacco origin or those sharing structural similarity with the analyte. So far, however, these RIA techniques have been applied in relatively few laboratories due to the complex synthesis of antigenic conjugates that are necessary for the raising of antibodies (Langone *et al.*, 1973) and the care that must be exerted in checking for cross-reactions with each new antibody production.

Currently, the GC and RIA methods are the most widely used for assaying nicotine and cotinine in active as well as passive smokers primarily because of their specificity and sensitivities. An interlaboratory comparison study by 11 laboratories from 6 countries demonstrated that GC and RIA techniques can quantitate nicotine and cotinine in urine and plasma samples (Biber *et al.*, 1987). An excellent correlation between these methods was observed in plasma samples from smokers and nonsmokers and in urine samples to which cotinine had been added. In urine samples with only endogenous cotinine, however, the RIA values were higher than the GC values in varying degrees depending upon the laboratory source of antibody. Higher RIA values may reflect cross-reaction of anti-cotinine with *trans*-5'hydroxycotinine or another nicotine metabolite. Alternatively, lower GC values might reflect incomplete extraction from urine. The former possibility is more likely, however. In the interlaboratory comparison study, all methods made a clear distinction between smokers and nonsmokers (Biber *et al.*, 1987). Nicotine and cotinine can, therefore, be determined in saliva, blood, or urine by a variety of analytic procedures with limits of detection in the picogram per milliliter range.

Nicotine has a half-life in blood of about 2 h while cotinine has a much longer residence time, which extends its utility as a biomarker for active smoking and for ETS uptake (Benowitz *et al.*, 1982). The half-life of cotinine elimination in active smokers has been shown to be about 18 h while the half-elimination-time in nonsmokers is longer (Sepkovic *et al.*, 1986b; U.S. Surgeon General, 1988). In a recent comparison study of cotinine elimination in smokers and nonsmokers, five cigarette smokers were asked to abstain from tobacco use for 5 days and were then given nicotine gum for 3 days. After another 8 days of abstinence from nicotine, the volunteers were exposed to sidestream smoke along with a group of people who never smoked. The mean $t_{1/2}$ for urinary cotinine in smokers was 16.5 h; nicotine gum users, 18.2 h; and those who never smoked but were exposed to ETS, 27.3 h (Haley *et al.*, 1989b). These results suggest that the residence time of nicotine and cotinine as well as other tobacco alkaloids could be related to the route of nicotine uptake as well as to possible differences in nicotine metabolism between smokers and nonsmokers. The longer residence time of nicotine metabolites in nonsmokers could conceivably increase the possibility of endogenous formation of carcinogenic tobacco-specific nitrosamines (Hoffmann and Hecht, 1985; Hecht and Hoffmann, 1988).

4.1. Active Smoking

Until the last decade, smoking control research had relied upon self-report for information regarding smoking behavior. As smoking became less acceptable socially, denying and minimizing the extent of smoking became common practices (Gillies *et al.*, 1982). Biochemical validation presents a more objective alternative to questionnaire data. Because of its relatively long half-life and its specificity for tobacco products, cotinine in body fluids has emerged as a good biomarker for determining smoking status. The choice of body fluids includes plasma, saliva, and urine. In active smokers, each of the above contain sufficient quantities of cotinine for validation of nicotine intake. Additionally, cotinine has been quantitated in cervical mucus (Schiffman *et al.*, 1987) and human milk (Schwartz-Bickenbach *et al.*, 1987; Piazza *et al.*, 1987), as well as breast fluid from nonlactating women (Hill and Wynder, 1979).

A recent study from Japan (Muronaka *et al.*, 1988) suggests that plasma cotinine or urinary cotinine:creatinine ratios provide the most suitable uptake variables for field studies in active smoking or passive exposure. Comparing their data with those collected in an American population (Hill *et al.*, 1983) shows that the levels of nicotine and cotinine in Japanese smokers are lower than those in Americans despite the similarity in numbers of cigarettes smoked per day and nicotine strength of the products. The more moderate inhalation of Japanese smokers might help explain the attenuation of lung cancer rates among smokers in that country. Within the United States, black smokers have been found to have higher plasma cotinine levels than white smokers when they are matched for cigarettes used per day and nicotine content (Wagenknecht *et al.*, 1988). While it is possible that differences in inhalation are responsible for the higher cotinine levels in blacks, a cotinine:thiocyanate ratio (thiocyanate providing a marker for gas-phase inhalation) remains higher for blacks compared to whites at all levels of cigarette use. This suggests that there may be differences in nicotine metabolism between racial groups.

Nicotine tolerance and compensation for changes in nicotine delivery has been investigated in several studies (U.S. Surgeon General, 1988). In our laboratory, plasma nicotine has been found to be a good marker for acute changes in nicotine intake from a single cigarette and cotinine a marker for longer changes in smoking behavior (Sepkovic *et al.*, 1984a). The monitoring of nicotine and cotinine levels in smokers who change to low-yield products has increased understanding of smoker compensation mechanisms and its relationship to tobacco-related disease.

The consumption of low-yield (i.e., low tar and low nicotine) cigarettes has increased in the United States in the 1970s and 1980s, due primarily to changes in product marketing and increased public perception that low-yield products mean less hazardous products (Folsom *et al.*, 1984). Smokers use a variety of methods to adjust to reduced nicotine delivery when they switch. These include increasing the cigarettes smoked per day or changing the topographical parameters of smoking behavior such as depth of inhalation or puff frequency (Russell *et al.*, 1988; Sepkovic *et al.*, 1983, 1984a; Haley *et al.*, 1985; Puustinen *et al.*, 1987). Compensation mechanisms may explain the failure of low-yield products to result in the reduction of coronary heart disease in men (Aronow, 1981). However, the introduction of filter cigarettes has assisted in reducing the delivery of some carcinogenic compounds, leading to somewhat lower lung cancer rates for long-term filter cigarette smokers compared to smokers of nonfilter cigarettes (Wynder and Stellman, 1979; IARC, 1986).

Compensation for nicotine can be complete or incomplete depending on the magnitude of reduction in the effective yield of nicotine. Because complete compensation is not possible when nicotine availability is greatly reduced, attempts to compensate may be more forceful, resulting in a greater body burden of gas-phase constituents since the latter are not received to the same extent as nicotine (Hill *et al.*, 1983).

An example of smoker compensation is illustrated in Fig. 3. Smokers received increased or decreased nicotine delivery cigarettes over an 8-week period. Individuals receiving increases in nicotine delivery reached higher levels of plasma cotinine and did not alter smoking behavior as evidenced by cigarettes used per day or by inhalation patterns (measured by carboxyhemoglobin and thiocyanate). In those smokers receiving reduced nicotine delivery, plasma cotinine levels dropped despite increases in cigarettes smoked per day or increased inhalation. When these subjects were returned to their usual level of nicotine delivery, blood levels of smoking-related biomarkers increased as smokers inhaled using their new smoking patterns acquired as they attempted to compensate for nicotine delivery (Sepkovic *et al.*, 1984a).

4.2. Passive Smoking

Passive smoking has been associated with pulmonary dysfunction in healthy adults and chronically ill adults and children, and has been implicated in the development of lung cancer in several epidemiological studies, especially in nonsmokers married to smokers (National Research Council, 1986). Some investigators have not noted an increased lung cancer risk in nonsmoking wives of smokers (Kabat and Wynder, 1984), and a recent report by Koo *et al.* (1988) has provided data on demographics, health status, and dietary differences between women married to smokers and to nonsmokers. Such studies point out the potential problems of confounders in establishing relative-risk ratios by using questionnaire data alone, and illustrate the need for biomarkers to evaluate the actual uptake of ETS by nonsmokers.

The degree of exposure to ETS depends on several factors including

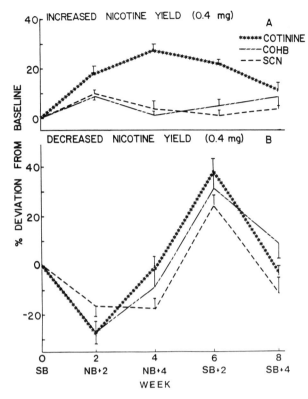

Figure 3. Plasma cotinine, thiocyanate (⁻SCN), and blood carboxyhemoglobin (COHb) in smokers who have shifted from their standard brands of either high-nicotine-yield cigarettes (A) or lower-nicotine-yield cigarettes (B) and then returned to their standard brand. Each point represents the mean ± standard error of the mean. SB = Standard brand; NB = New brand.

length of time spent in a smoke-polluted room, the number of smokers in the enclosed area, the degree of ventilation, and the respiratory rate of the exposed individual. Thus optimal assessment of ETS exposure should be achieved by analysis of physiological fluids for tobacco-smoke–associated compounds. Measurement of absorbed nicotine and its metabolites provides a means for measuring uptake of ETS by nonsmokers.

While not as important when determining smoking or nonsmoking status, the choice of body fluid for measurement of ETS-absorbed nicotine must provide the greatest sensitivity of measurement. Recent data by Wall *et al.* (1988) suggest that urine might be the fluid of choice for quantitating passive smoking. These researchers found that subjects who reported no significant tobacco-smoke exposure had detectable cotinine in their urine, but that the levels in reported passive smokers tended to be higher. In contrast, saliva and plasma measurements from these subjects were generally lower than the detection limit of the assay.

In epidemiological studies, it is impractical to collect a 24-h urine sample for validation of ETS uptake. A ratio of cotinine to creatinine in a given sample is often used to allow for differences in urine dilution. Urinary creatinine is fairly constant from day to day within individuals, but does vary among individuals and is significantly different between sexes. Caution should be used in applying a creatinine correction to a cohort of subjects including males and females of various ages. In studies with subjects of similar age, the ratio of cotinine to creatinine has afforded the best separation of nonsmokers into categories according to reported ETS exposure (Greenberg *et al.*, 1984; Henderson *et al.*, 1987).

Exposure to ETS has also been estimated with the use of salivary cotinine concentrations in free-living subjects. Early studies by Jarvis *et al.* (1984) demonstrated the utility of cotinine measurements for passive smoking surveys. They reported that the exposure of the passive smoker to nicotine is a few percent or less of the amount inhaled by the active smoker, and positive trends were found between salivary cotinine and reported exposure. A detailed study by Coultas *et al.* (1987) documented a high prevalence of passive smoking among adults and children; salivary cotinine was detected in 80% of children living with smoking parents. The average level of salivary cotinine increased with the numbers of smokers in the home, but significant dose–response relationships were not documented.

How well does an individual judge his or her exposure to ETS? To address this question, survey data on exposure at home, in the workplace, and on social occasions were collected from 319 employed subjects and correlated with levels of cotinine in a random urine sample (Haley *et al.*, 1989a). Mean urine cotinine:creatinine levels were found to be higher for women than for men and might be due to differences in creatinine excretion between the sexes as well as the fact that 94% of the women were employed indoors. Higher urine cotinine levels (ng/mg creatinine) were noted in both men and women who lived with a smoker than in those subjects who did not report living with a smoker (13.3 ± 2.4 vs 5.1 ± 0.4 in men and 13.9 ± 1.9 vs 5.6 ± 0.6 in women). Differences in the prevalence of exposure at home existed between sexes (males 13.5% vs females 29.2%). Levels of cotinine found from different exposures indicate (Table 3) that home exposure has a more pronounced effect on urine cotinine than does workplace exposure. These data on home versus workplace exposure confirm those obtained in Bremen, Federal Republic of Germany, as a part of an international comparison of passive smoking (Becker *et al.*, 1987).

These and several other studies have demonstrated the presence of a nicotine metabolite in body fluids of nonsmokers and relative correlations between these levels and reported exposures. Such measurements should be only utilized as estimates for the exposure to ETS nicotine and not as direct indicators of cancer risk. Several factors including differences in cotinine

TABLE 18–3. Self-Reported ETS Exposure in Various Environments

Home	Work	Social	n	Percent of Population	Urinary Cotinine (ng/mg creatinine)
+*	+	+	19	6.3	$13.3 \pm 2.6^{\ddagger}$
+	+	$-^{\dagger}$	1	0.3	12.2 ± 0
+	−	+	29	9.5	14.7 ± 2.1
+	−	−	5	1.6	6.8 ± 2.8
−	+	+	30	9.9	5.8 ± 1.0
−	−	+	160	52.6	5.5 ± 1.0
−	+	−	2	0.7	2.4 ± 2.4
−	−	−	16	5.3	4.2 ± 0.8

*+ = Yes.
† − = No.
‡Each value represents the mean ± the standard error of the mean (SEM).

elimination, physicochemical differences between mainstream and side-stream smoke, and variance in the mode of uptake of carcinogenic compounds suggest caution before extrapolation to cigarette equivalents of uptake can be applied to passive smokers based upon a single measurement of a single compound from inhaled ETS. Additional investigations especially with markers of particulate-phase components of ETS are needed to determine relative uptake by smokers and nonsmokers of carcinogens and cocarcinogens from ETS.

5. CARCINOGENIC NICOTINE-DERIVED TOBACCO-SPECIFIC NITROSAMINES: FORMATION AND OCCURRENCE

As discussed above, nicotine has toxic effects and is the major dependence-producing substance that leads to continued tobacco use. However, nicotine is also a precursor to powerful carcinogenic nitrosamines that are present in tobacco and tobacco smoke.

Reactions of nicotine and sodium nitrite under mild conditions produced N'-nitrosonornicotine (NNN), 4-(methylnitrosamino)-1-(3-pyridyl)-1-butanone (NNK) and 4-(methylnitrossamino)-4-(3-pyridyl)butanal (NNA) in yields ranging from 0.1% to 7% (Hoffmann *et al.*, 1974; Hecht *et al.*, 1978a). Since nitrate and nitrite are normal constituents of tobacco, and nitrogen oxides are present in tobacco smoke, it seemed likely that these nitrosamines might be found in both unburned tobacco and tobacco smoke. Extensive analyses of tobacco products have confirmed and extended this hypothesis (Hoffmann and Hecht, 1985; Hecht and Hoffmann, 1988). The most prevalent tobacco-specific nicotine-derived nitrosamines in tobacco and tobacco smoke are NNN and NNK. The aldehyde NNA has not been detected. However, products of its reduction and oxidation, iso-NNAL and iso-NN-acid, are present in tobacco (Brunnemann *et al.*, 1987b; Djordjevic *et al.*, 1988). The structures of the tobacco-specific nitrosamines are illustrated in Fig. 4. Minor alkaloids of tobacco—nornicotine, anabasine, and anatabine—also are readily nitrosated. Thus the NNN in tobacco and tobacco smoke is formed from both nicotine and nornicotine, while NAB and NAT arise by nitrosation of anabasine and anatabine. Table 4 summarizes data from some recent analyses of these nitrosamines in tobacco products (Hecht and Hoffmann, 1988). Furthermore, tobacco-specific nitrosamines have been detected in the saliva of snuff dippers. In the saliva of U.S. snuff dippers, NNN ranged from 26 to 420 ng/ml (mean 126 ng/ml) and NNK between <10 and 96 ng/ml (mean 17 ng/ml; Hoffmann and Adams, 1981; Palladino *et al.*, 1986). Surprisingly, the saliva of Inuits (Eskimos in the Northwest Territories of Canada) had significantly higher levels of NNN (272–2600 ng/ml; mean 380 ng/ml and of NNK (0–201 ng/ml; mean 56 ng/ml; Brunnemann *et al.*, 1987a).

As discussed below, bioassays in laboratory animals have demonstrated that NNN and NNK are potent carcinogens. The data in Table 4 are significant because they demonstrate that the levels of these carcinogenic nitrosamines in tobacco products are remarkably higher than the levels of carcinogenic nitrosamines in all other consumer products (Hecht and Hoffmann , 1988). Nitrosamine levels in most consumer products are regulated. For example, nitrosamines in beer or bacon cannot exceed 5 ppb. The levels of NNN and NNK in smokeless tobacco products are up to 10,000 times higher than these regulated amounts. The amounts of NNN and NNK in cigarette mainstream smoke are typically 100 times greater than the nitrosamine levels in 1 g of bacon or 1 ml of beer. The lack of regulation of nitrosamine levels in tobacco products is a legislative failure that presents an unacceptable risk to the consumer.

Since nitrite, nicotine, and nornicotine are present in the saliva of tobacco consumers, and nitrogen oxides are present in tobacco smoke, the endogenous formation of NNN and NNK is likely. Although this has not been unambiguously demonstrated for NNN and NNK, analyses of smokers' urine have shown that levels of *N*-nitrosoproline are higher than those in nonsmokers (Hoffmann and Brunnemann, 1983; Hoffmann *et al.*, 1984a; Ladd *et al.*, 1984; Ohshima *et al.*, 1985). Thus the amounts of nitrosamines listed in Table 4 are likely to represent minimum values in terms of human exposure. In addition, analyses of tobacco and tobacco smoke are carried out under certain standard laboratory conditions that do not duplicate individual tobacco smoking and chewing practices. For these reasons and others discussed below, it is desirable to develop markers of individual uptake of

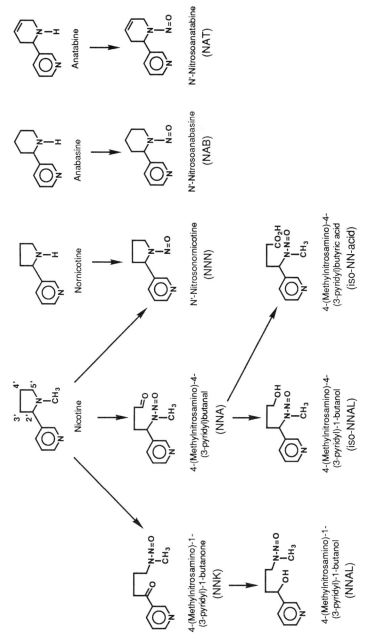

Figure 4. Structures of tobacco-specific nitrosamines. With the exception of NNA, all have been detected in tobacco and/or tobacco smoke (see Table 4).

TABLE 18–4. Some Recent Analyses of Tobacco-Specific Nitrosamines in Smokeless Tobacco and Mainstream and Sidestream Cigarette Smoke*

	NNN	NNK	NAB	NAT	iso-NNAL
Smokeless Tobacco (ppm)					
Chewing tobacco	0.67– 4.1	0.03– 0.38	0.03–0.14	0.33– 2.3	NDe[†]
Moist snuff	5.8 –64	0.1 – 3.1	0.2 –6.5	3.5 –214	0.08–2.5
Dry snuff	9.4 –55	1.9 –14	7.2 –1.2	1.9 – 40	0.07–0.14
Tobacco used in betel quid	0.47– 0.85	0.13– 0.23	0.03–0.07	0.30– 0.45	NDe
Cigarette Smoke (µg/cigarette)					
Mainstream	0.066–1.01	0.017–0.43	0.10–0.74[‡]		ND[§]
Sidestream	0.19 –0.86	0.39 –1.44	0.13–0.78		ND

*Hecht and Hoffmann, 1988.
[†]NDe, Not determined.
[‡]Combined NAB and NAT.
[§]ND, not detected.

nicotine-derived nitrosamines. Such markers could potentially give an accurate measure of an individual's total exposure to these carcinogens.

6. CARCINOGENICITY OF NICOTINE-DERIVED NITROSAMINES

Bioassays in laboratory animals have demonstrated that NNK and NNN are the most carcinogenic of the nitrosamines illustrated in Fig. 4. Representative bioassay data are summarized in Table 5. The lung is the main target organ for tumor induction by NNK in bioassays carried out in mice, rats, or hamsters. Tumors of the pancreas, liver, and nasal cavity are also observed. The potency of NNK is demonstrated by its ability to induce significant incidences of respiratory tract tumors in hamsters after a single dose of only 1 mg, or to produce significant numbers of lung tumors in rats over a dose range of 0.07 to 0.7 mmol/kg, corresponding to chronic administration in the drinking water of 0.5–5 ppm. Whereas NNK induces mainly lung tumors in rats, NNN causes tumors of the esophagus and nasal cavity when administered in the drinking water, but primarily nasal cavity tumors when given by injection.

The versatile carcinogenic activities of NNK and NNN in laboratory animals support their role as human carcinogens. NNK produces both squamous-cell carcinoma and adenocarcinoma of the lung, as seen in smokers. It is the only constituent of tobacco smoke known to induce tumors of the exocrine pancreas in laboratory animals (Rivenson *et al.*, 1988). This is significant because tobacco smoking is an important cause of pancreatic cancer in man. A mixture of NNK and NNN applied to the rat oral cavity induced significant incidences of local tumors. This is important because NNK and NNN are the only known constituents of smokeless tobacco that have been shown to cause oral cavity tumors in experimental animals.

The amounts of NNK and NNN that produce tumors in experimental animals are similar to the estimated doses to which tobacco consumers are exposed. Based on a total amount of NNK and NNN in mainstream cigarette smoke of 440 ng/cigarette, a smoker of one pack per day would be exposed to approximately 8 μg. The total dose in 40 years of approximately 117 mg or 0.01 mmol/kg is similar to the 0.05 mmol/kg of NNK that produces tumors in hamsters or the 0.07 mmol/kg that gives tumors in rats. Snuff dippers using the most popular products marketed in the U.S. would be exposed to approximately 67 μg of NNK and NNN per gram of snuff. In 40 years of dipping, a user of 10 g tobacco per day would be exposed to approximately 9.8 g or 0.7 mmol/kg of these carcinogens. This can be compared to the 1.6 mmol/kg amount that produced oral cavity tumors in rats. These estimates certainly represent unacceptable risks, which are far

TABLE 18–5. Representative Bioassays of Nicotine-Derived Nitrosamines

Nitrosamine	Species and Strain	Route of Administration	Approximate total dose (mmol/kg)	Target Tissue and Incidence Data	Reference
NNK	A/J mouse	i.p.	0.8	Lung (100%) 7.2 tumors per animal	Hecht *et al.*, 1988a
	SENCAR mouse	Topical	0.2–1.1	Skin (24%–59%) Lung (10%–63%)	LaVoie *et al.*, 1987
	F-344 rat	s.c.	0.3–9	Lung (48%–77%) Nasal Cavity (22%–93%) Liver (13%–37%)	Hoffmann *et al.*, 1984b; Hecht *et al.*, 1986a
		p.o.	0.07–0.7	Lung (11%–90%) Liver (0%–40%) Exocrine Pancreas (6%–11%)	Rivenson *et al.*, 1988
	Syrian golden hamster	s.c.	0.04–6	Lung (20%–80%) Nasal Cavity (10%–55%) Trachea (5%–35%)	Hoffmann *et al.*, 1981; Hecht *et al.*, 1983a
NNN	A/J mouse	i.p.	4	Lung(83%) 1.8 tumors per animal	Hecht *et al.*, 1988a
	SENCAR mouse	Topical	1.1	Inactive	LaVoie *et al.*, 1987

Compound	Species	Route	Dose	Target Organ	Reference
	F-344 rat	s.c.	1–9	Nasal Cavity (50%–93%) Esophagus (4%–24%)	Hoffmann et al., 1984b
		p.o.	3	Esophagus (83%) Nasal Cavity (60%)	Castonguay et al., 1984
	Syrian golden hamster	s.c.	6–13	Trachea (5%–78%)	Hilfrich et al., 1977 Hoffmann et al., 1981
NNK + NNN	F-344 rat	Oral swabbing	0.2	Oral Cavity (27%) Lung (17%)	Hecht et al., 1986b
NNAL	A/J mouse	i.p.	4	Lung (100%) 26.3 tumors per animal	Castonguay et al., 1983a
	F-344 rat	p.o.	0.7	Lung (87%) exocrine Pancreas (27%)	Rivenson et al., 1988
NNA	A/J mouse	i.p.	4	Inactive	Hecht et al., 1978b
iso-NNAL	SENCAR mouse	Topical	1.1	Inactive	LaVoie et al., 1987

outside the standards for any consumer product other than tobacco. The development of internal dosimeters for NNK and NNN would allow a more reliable quantification of cancer risk associated with exposure to these compounds. These dosimeters would provide necessary information on risk to the individual tobacco consumer, and would also be likely to provide new insights on the quantitative aspects of the relationship of carcinogen exposure to tumor development.

7. METABOLISM OF NNK AND NNN

Figure 5 summarizes known metabolic pathways of NNK and NNN. These have been determined by using subcellular fractions and organ cultures of rodent and human tissues and by studies of *in vivo* metabolism in rats (Hecht *et al.*, 1983b; Hoffmann and Hecht, 1985; Hecht and Hoffmann, 1988).

In hepatic microsomes, the initial pathways of NNK metabolism are pyridine-*N*-oxidation to NNK-*N*-oxide (1), carbonyl reduction to NNAL, and α-hydroxylation to the unstable metabolites *6* and *7*, which spontaneously decompose to the keto aldelyde *11* and formaldehyde *13*. The other products of this spontaneous reaction are the diazohydrozides *12* and *14*. The former is a methylating agent and reacts with DNA to give methylated bases, which are involved in the initiation of tumorigenesis. The latter reacts with both DNA and globin to give adducts that can be hydrolyzed to the keto alcohol *20*. These adducts form the basis of some of the dosimetry methods described below. The initial products of microsomal metabolism such as keto aldehyde *11* and keto alcohol *20* are further metabolized *in vivo* to the keto acid *17*, which is a major urinary metabolite of NNK.

NNN is metabolized by rat hepatic microsomes to NNN-1-*N*-oxide *(2)* and the four possible products of pyrrolidine ring oxidation *4, 5, 9,* and *10*. The latter two compounds, formed by α-hydroxylation, are unstable and spontaneously yield the diazohydroxides *14* and *16* as well as myosmine *(15)*. Diazohydroxide *14* is common to both NNN and NNK and is involved in macromolecular binding by both compounds. Hydrolysis of *14* and *16* gives the metabolites *20* and *19*, respectively. These undergo further oxidation *in vivo*, producing the major urinary metabolites of NNN, keto acid *17* and hydroxy acid *22*.

The key pathways involved in NNK and NNN carcinogenesis are the α-hydroxylation pathways because these result in modification of DNA leading to miscoding, oncogene activation, and cancer. The extents to which these pathways occur in particular individuals would be one factor influencing the probability of tumor development. Research with cultured human tissues has demonstrated that there are wide interindividual variations in the extents to

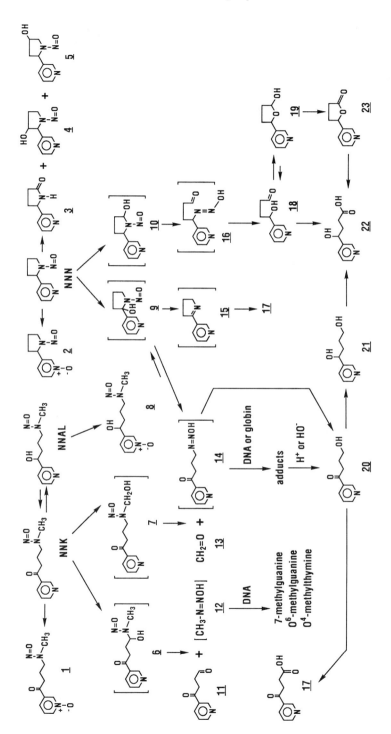

Figure 5. Mammalian metabolism of NNK and NNN. For further details, see Hecht *et al.*, 1983[b].

which α-hydroxy metabolites of NNK and NNN are formed (Castonguay *et al.*, 1983b). Dosimetric methods that focus on these pathways would perhaps be the most useful in establishing a quantitative indicator of risk. A potential problem inherent in this approach is the similarity of NNK and NNN metabolites to those of nicotine. For example, the keto acid *17* and hydroxy acid *22* are also formed from nicotine (see structures *22* and *23* of Fig. 2). The concentration of nicotine in tobacco and tobacco smoke is generally at least 1000 times higher than that of NNK and NNN. Thus compounds *17* and *22* would clearly not be suitable markers for metabolism of NNK and NNN in tobacco consumers. A possible way to circumvent this problem is the use of macromolecular adducts as dosimeters, since these are not likely to be formed from nicotine.

8. MONITORING INDIVIDUAL UPTAKE AND METABOLIC ACTIVATION OF NNK AND NNN

8.1 Why are Dosimeters for Nicotine-Derived Nitrosamines Necessary?

Section 4 of this chapter describes the use of nicotine and cotinine as objective markers of exposure to tobacco products. Other markers such as thiocyanate, carboxyhemoglobin, *N*-nitrosoproline, thioethers, and urinary mutagenicity are also available, but none is specific for exposure to tobacco products (IARC, 1986). Heavy elements such as strontium may be useful as markers for validation of smokeless tobacco use (Robertson and Bray, 1988). Other chapters in this book describe dosimeters for tobacco-smoke carcinogens such as 4-aminobiphenyl and benzo[a]pyrene. What advantages would be presented by dosimeters for nicotine-derived nitrosamines? Although nicotine is a tobacco-specific dosimeter, it is not a carcinogen. None of the other carcinogens for which methods have been developed are tobacco-specific. Thus dosimeters for the nicotine-derived nitrosamines NNK and NNN would provide objective markers of uptake and metabolic activation of carcinogens that are associated *only* with tobacco products. Such information would be useful in epidemiologic studies on topics such as passive smoking, and for assessing risk for cancer development specifically associated with exposure to tobacco products. Within this context, several specific points need to be reemphasized:

1. Measurements of amounts of NNK and NNN in tobacco products are based on standard machine smoking conditions or, for tobacco, solvent extractions. These measurements do not take into account individual differences in inhalation patterns or chewing practices. They

also give no indication of exposure to ETS or to one's own sidestream smoke.

2. The formation of nicotine-derived nitrosamines in the body is likely but the amounts formed are not known and at present are difficult to estimate.

3. The extent of metabolic activation of NNK and NNN will vary in individuals. An internal dosimeter is necessary to determine the extent to which this key process occurs.

4. Macromolecular adducts of nicotine-derived nitrosamines could give an indication of the degree of chronic exposure to these carcinogens.

Thus dosimetric methods for nicotine-derived nitrosamines would bypass many of the uncertainties inherent in estimates of carcinogen exposure associated specifically with tobacco products.

8.2. Protein Adducts as Dosimeters

The binding to hemoglobin of NNK and NNN was investigated in rats treated with these compounds labeled with tritium at the 5 position of the pyridine ring ([5-^{3}H]NNK and [5-^{3}H]NNN) (Carmella and Hecht, 1987). Fig. 6 shows the relationship between total binding to globin and dose of [5-^{3}H]NNK over a 100-fold dose range for rats given a single injection and sacrificed 24 h later. A linear relationship was observed. The amount of binding corresponded to approximately 0.1% of the dose.

Fig. 7 shows that during chronic treatment of rats with [5-^{3}H]NNK, a steady-state concentration of tritium in globin is reached after approximately 35 days. After cessation of treatment, tritium disappears from globin with apparent biphasic kinetics.

Mild basic hydrolysis of the globin released the keto alcohol *20*, Fig. 5. The amount released corresponded to 15%–25% of the bound tritium, depending on dose. This is believed to occur by hydrolysis of an adduct formed by reaction of diazohydroxide *14* with globin. The structure of the adduct is not known. Similar results have been obtained in experiments with [5-^{3}H]NNN, consistent with the generation of *14* in its metabolism, as shown in Fig. 5.

The keto alcohol *20* is a potentially useful dosimeter for NNK and NNN binding to globin. Analysis of keto alcohol *20* is facilitated by its release from globin with mild base, thus separating it from the constituents of the polymer. A method has been developed to enrich keto alcohol *20*, derivatize it with pentafluorobenzoyl chloride, and analyze it by GC-MS with negative-ion chemical ionization (NICI) detection and specific-ion monitoring. The detection limit for the pentafluorobenzoate of keto alcohol *20* is approximately 1 fmol per injection, or 20 fmol per gram of hemoglobin. With use

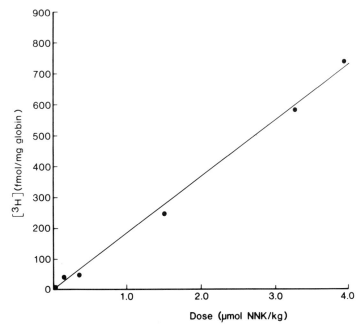

Figure 6. Presence of tritium in globin 24 h after i.p. treatment of F-344 rats with various doses of [5-³H]NNK.

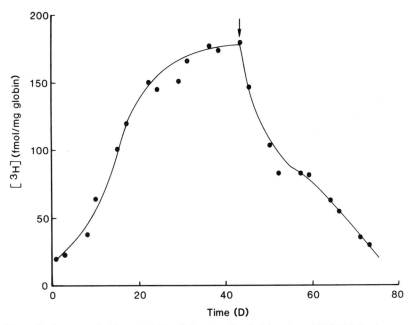

Figure 7. Presence of tritium in globin during chronic administration of [5-³H]NNK. Arrow, termination of [5-³H]NNK treatment.

of this method, the keto alcohol *20* has been detected in base hydrolyzates of globin isolated from the blood of smokers and snuff dippers. Current research is focused on the quantitation of keto alcohol *20*.

The globin adducts of NNK and NNN, as measured by the release of keto alcohol *20,* would appear to provide a potential dosimeter for exposure to tobacco-specific carcinogens. Further research is necessary to validate this approach. It would be important to identify the structure of the adduct that gives rise to keto alcohol *20* on base hydrolysis in order to establish whether the mechanism for its formation is indeed proceeding *via* the diazohydroxide *14*. This is important because the latter is considered to be an ultimate carcinogen of NNK and NNN. We also need to investigate whether keto alcohol *20* can be released from globin of animals treated with nicotine. Although this would not be expected based on the known pathways of nicotine metabolism illustrated in Fig. 2, it may occur by minor pathways, possibly even endogenous nitrosation. If keto alcohol *20* is formed from nicotine, the interpretation of data from people exposed to tobacco products would perhaps be more complex than if NNK and NNN were the only precursors. In addition, it is important to determine the relationship between globin adducts and DNA adducts in target tissues. This would facilitate interpretation of the globin-adduct data with respect to risk for cancer, since DNA adducts but not globin adducts are involved in the initiation of the carcinogenic process.

As indicated in Fig. 5, NNK is also a methylating agent. Treatment of rats with [C^{3}H$_3$]NNK resulted in incorporation of tritium in globin, in amounts that generally exceeded those observed with [5-^{3}H]NNK. Other methylating agents such as N-nitrosodimethylamine have been shown to yield S-methylcysteine in globin, and its formation from NNK would be expected. The utility of S-methylcysteine as a dosimeter is limited by its natural occurrence in hemoglobin (Bailey *et al.,* 1981). In addition, sources other than tobacco contribute to methylation of globin. Nevertheless, data on methylation of globin in individuals exposed to tobacco may be useful when combined with corresponding data obtained by measuring release of keto alcohol *20*.

8.3. DNA Adducts as Dosimeters

Acid hydrolysis or neutral thermal hydrolysis of DNA isolated from tissues of rats treated with [5-^{3}H]NNK or [5-^{3}H]NNN gave the keto alcohol *20,* as illustrated in Fig. 5 (Hecht *et al.,* 1988b). As in the case of globin, the structure of the adduct precursor to *20* is not known, but is believed to be formed by reaction with DNA of diazohydroxide *14*. These results are consistent with pyridyloxobutylation of DNA by NNK or NNN. Methods developed for monitoring levels of *20* released by hydrolysis of globin should also be applicable for DNA. The GC-MS-NICI method described above would appear to have the requisite sensitivity for analysis of DNA isolated from white blood cells or exfoliated oral cavity cells. At present it is not known whether data obtained in this way would be more suitable for dosimetry than measurements of keto alcohol *20* released from globin.

The development of monoclonal antibodies specific for O^6-methyldeoxyguanosine has allowed analysis of DNA isolated from human tissues. It has been detected in esophageal DNA samples obtained from patients with cancer of the esophagus, who lived in Linxian county, China, a high-risk area for esophageal cancer (Umbenhauer *et al.*, 1985). It has also been detected in DNA from human placenta, in a study that compared smokers and nonsmokers (Foiles *et al.*, 1988). Among 20 subjects, there was no apparent relationship with smoking. This was perhaps not surprising because there are several environmental agents other than NNK that can methylate DNA. Nevertheless, more extensive studies of DNA methylation in tobacco users might provide further insights, especially when combined with data on pyridyloxobutylation of DNA.

8.4. Metabolites as Dosimeters

The development of standard methods for analysis of urinary NNK and NNN or their metabolites would be another approach for assessing individual uptake and rates of metabolic activation of nicotine-derived nitrosamines. NNK and NNN are extensively metabolized in laboratory animals, and only a small percentage is excreted unchanged in urine. Preliminary studies using gas chromatography and thermal energy analysis indicated that this method did not have the requisite sensitivity to determine NNN in human urine. GC-MS approaches to analysis of urinary NNK and NNN have not been extensively studied and might be feasible.

Compounds *3, 17,* and *22* of Fig. 5, which are urinary metabolites of NNK and NNN in laboratory animals, are also metabolites of nicotine and are therefore unsuitable as dosimeters of NNK and NNN (see Fig. 2). Compounds *11, 15, 19, 20, 21,* and *23* are converted to *17* and/or *22 in vivo,* and consequently are also unsuitable. NNAL, a major metabolite of NNK, would be a suitable marker for its uptake. The presence of a hydroxyl group suggests methods for its derivatization and analysis by GC-MS. The *N*-oxides *1, 2,* and *8* might also be appropriate dosimeters for detoxification of NNK and NNN in humans. Further research is needed on the development of analytical methods for detection of these metabolites in human urine.

9. SUMMARY AND CONCLUSIONS

Tobacco is in worldwide use by hundreds of millions of people. In 1982, 6.7 million tons of tobacco were produced and the U.S. usage share of this product yielded a per person annual consumption of 3500 cigarettes. In many developed countries, sizable decreases in cigarette consumption have occurred, although increasing use has been noted in developing countries (IARC, 1986).

The uptake of tobacco-smoke constituents by persons is practically universal due to exposure of nonsmokers to tobacco-polluted environments as well as actual user consumption. Such uptake in tobacco chewers and in both active and passive smokers can be monitored by quantitation of a tobacco-specific substance in body fluids of tobacco users and tobacco smoke–exposed persons. Nicotine and its major metabolites have provided sensitive and specific markers of tobacco use. Monitoring nicotine uptake has allowed the toxicological effects of this pharmacoactive alkaloid to be evaluated and has provided important information on pharmacologic dependency to nicotine. Nicotine is not only the dependence-producing alkaloid in tobacco but is also responsible for several of the toxicological effects of tobacco use such as cardiovascular responses and metabolic and endocrine perturbations.

Nicotine itself is not carcinogenic. However, nicotine-derived compounds, namely the tobacco-specific nitrosamines NNK and NNN, are powerful animal carcinogens. In sharp contrast to tobacco, the nitrosamine contents of other consumer products are monitored by regulatory agencies. The nitrosamine content of a serving of bacon or beer is about 100-fold less than the NNN or NNK content of mainstream cigarette smoke yet is carefully controlled.

Monitoring nicotine and its metabolites as well as derived carcinogens through their macromolecular adducts provides information on the uptake, absorption, and metabolic activation of these tobacco constituents. It also allows estimates of exposure to other tobacco constituents, but does not address the carcinogenic potential of other chemicals contained in tobacco smoke. Continued research on the formation and metabolism of other carcinogens such as polycyclic aromatic hydrocarbons, aromatic amines, and gas-phase constituents will lead to increasing knowledge of the health risks associated with cigarette smoking. Reduction of tar and nicotine content of cigarettes has occurred over the last 20 years, but the product remains our number one health hazard.

Why is the monitoring of exposure to tobacco-smoke products necessary since several health hazards of their use are well established? Part of the answer is the continued production and intensive marketing of these highly profitable but dangerous products. The main reason for the continued consumer demand for tobacco products is the pharmacologically active alkaloid, nicotine. Monitoring will provide insights on susceptibility to disease of the millions of tobacco users and on approaches to prevention of tobacco-related diseases.

REFERENCES

F. Adlkofer, G. Scherer, L. Jarczyk, W. D. Heller, and G. B. Neurath, Pharmaco-kinetics of 3′-hydroxycotinine. In: *The Pharmacology of Nicotine,* eds. M. J. Rand and K. Thurau, IRL Press, Washington D.C., pp. 25–28 (1988).

A. K. Armitage and D. H. Turner, Absorption of nicotine in cigarette and cigar smoke through the oral mucosa, *Nature* **226**, 1231–1232 (1970).

W. S. A. Aronow, Effect of cigarette smoking and of carbon monoxide on coronary heart disease. In: *Smoking and Arterial Disease,* ed. R. M. Greenhalgh, Pitman Biomedical, Bath, England, pp. 226–238 (1981).

E. Bailey, T. A. Conners, P. B. Farmer, S. M. Gorf, and J. Rickard, Methylation of cysteine in hemoglobin following exposure to methylating agents, *Cancer Res.* **41**, 2514–2517 (1981).

H. Becker, I. Jahn, and K.-H. Jöckel, *Passivrauchen als Gusundheitsrisiko,* Minister für Arbeit, Gesundheit und für Soziales des Landes Nordrhein-Westfalen, Herford, FRG, pp. 8–10 (1987).

N. L. Benowitz, Nicotine pharmacokinetics and dynamics. In: *The Pharmacology of Nicotine,* eds. E. J. Rand and K. Thurau, IRL Press, Washington, D.C., pp. 3–18 (1988).

N. L. Benowitz, P. Jacob, R. T. Jones, and J. Rosenberg, Interindividual variability in the metabolism and cardiovascular effects of nicotine in man, *J. Pharmacol Exp. Ther.* **221**, 368–371 (1982).

A. Biber, G. Scherer, I. Hoepfner, F. Adlkofer, W.-D. Heller, J. E. Haddow, and G. Knight, Determination of nicotine and cotinine in human serum and urine: an interlaboratory study, *Toxicol. Lett.* **35**, 45–52 (1987).

R. J. Bjercke, G. Cook, N. Roychlik, H. B. Gjika, H. Van Vunakis, and J. J. Langone, Stereospecific monoclonal antibodies to nicotine and cotinine and their use in enzyme-linked immunosorbent assays, *J. Immunol. Methods* **90**, 203–213 (1986).

J. Booth and E. Boyland, The metabolism of nicotine into two optically active stereoisomers of nicotine-1′-oxide by animal tissues *in vitro* and by cigarette smokers, *Biochem. Pharmacol.* **19**, 733–742 (1970).

K. D. Brunnemann and D. Hoffmann, The pH of tobacco smoke, *Food Cosmetic Toxicol.* **12**, 115–124 (1974).

K. D. Brunnemann, L. Genoble, and D. Hoffmann, Identification and analysis of a new tobacco-specific N-nitrosamine, 4-(methylnitrosamino)-4-(3-pyridyl)-1-bu-tanol, *Carcinog.* **8**, 465–459 (1987b).

K. D. Brunnemann, A. P. Hornby, and H. F. Stich, Tobacco-specific nitrosamines in the saliva of Inuit snuff dippers in the Northwest Territories of Canada, *Cancer Lett.* **37**, 7–16 (1987a).

G. R. Cam and J. R. Bassett, The effect of acute nicotine administration on plasma levels of thyroid hormones and corticosterone in the rat, *Pharmacol. Biochem. Behav.* **19**, 559–561 (1983).

S. G. Carmella and S. S. Hecht, Formation of hemoglobin adducts upon treatment of F344 rats with the tobacco-specific nitrosamines 4-(methylnitrosamino)-1-(3-pyridyl)-1-butanone and N′-nitrosonornicotine, *Cancer Res.* **47**, 2626–2630 (1987).

A. Castonguay, D. Lin, G. D. Stoner, P. Radok, K., Furuya, S. S. Hecht, H. A. J. Schut, and J. E. Klaunig, Comparative carcinogenicity in A/J mice and metabolism by cultured mouse peripheral lung of N'-nitrosonornicotine, 4-(methyl nitrosamino)-1-(3-pyridyl)-1-butanone and their analogues, *Cancer Res.* **43**, 1223–1229 (1983a).

A. Castonguay, A. Rivenson, N. Trushin, J. Reinhardt, S. Stathopoulos, C. J. Weiss, B. Reiss, and S. S. Hecht, Effects of chronic ethanol consumption on the metabolism and carcinogenicity of N'-nitrosonornicotine in F344 rats, *Cancer Res.* **44**, 2285–2290 (1984).

A. Castonguay, G. D. Stoner, H. A. J. Schut, and S. S. Hecht, Metabolism of tobacco-specific N-nitrosamines by cultured human tissues, *Proc. Natl. Acad. Sci. USA* **80**, 6694–6697 (1983b).

C. Y. Chen, J. N. Deona, and P. A. Crooks, Determination of nicotine in plasma by HPLC with electrochemical detection, *LC-GC* **6**, 53–55 (1988).

D. B. Coultas, C. A. Howard, G. T. Peake, B. J. Skipper, and J. M. Samet, Salivary cotinine levels and involuntary tobacco smoke exposure in children and adults in New Mexico, *Am. Rev. Resp. Dis.* **136**, 305–309 (1987).

R. M. Dajani, J. W. Gorrod, and A. H. Beckett, Hepatic and extrahepatic reduction of nicotine-1'-N'-oxide in rats, *Biochem. J.* **130**, 88 (1972).

L. Davis, Tobacco use and associated health risks. In: *Tobacco Smoking and Nicotine, a Neurobiological Approach*, eds. W. Martin, G. Van Loon, E. Iwomoto, and L. Davis, Plenum Press, New York, pp. 15–24 (1988).

M. V. Djordjevic, K. D. Brunnemann, and D. Hoffmann, Identification and analysis of a nicotine-derived N-nitrosamino acid and other nitrosamine acids in tobacco, *42nd Tobacco Chemist Research Conference, Lexington, KY*, Abstr. 41 (1988).

E. F. Domino, Behavioral, electrophysiological, endocrine and skeletal muscle actions of nicotine and tobacco smoking. In: *Electrophysiological Effects of Nicotine*, eds. A. Remond and C. Izard, North Holland Biomedical Press, Elsevier, Amsterdam, pp. 133–146 (1979).

C. Feyerabend, *Environmental Carcinogens. Methods of Analysis and Exposure Measurement*, Vol. 9, *Passive Smoking*, IARC Sci. Publ. *81:* pp. 299–301.

P. G. Foiles, L. M. Miglietta, S. A. Akerkar, R. B. Everson, and S. S. Hecht, Detection of O^6-methyldeoxyguanosine in human placental DNA, *Cancer Res.* **48**, 4184–4188 (1988).

A. R. Folsom, T. Pechacek, R. DeGaudemaris, R. Leupker, D. Jacobs, and R. Gillum, Consumption of low yield cigarettes: Its frequency and relationship to serum thiocyanate, *Am. J. Public Health* **74**, 564–568 (1984).

K. Fuxe, K. Anderson, P. Eneroth, A. Härfstrand, A. M. Janson, G. von Euler, L. F. Agnati, B. Tinner, C. Köhler, and C. B. Hersh, Neuroendocrine and trophic actions of nicotine. In: *The Pharmacology of Nicotine*, eds. M. J. Rand and K. Thurau, IRL Press, Washington, D. C., pp. 293–320 (1988).

P. A. Gillies, B. Wilcox, C. Coates, F. Kristmundsdotir, and D. J. Reid, Use of objective measurement in the validation of self-reported smoking in children aged 10 and 11 years: Saliva thiocyanate, *J. Epidemiol. Community Health* **36**, 205–209 (1982).

K. H. Ginzel, Introduction to the effects of nicotine on the central nervous system, *Ann. New York Acad. Sci.* **142**, 101–120 (1967).

K. H. Ginzel, Muscle relaxation by drugs which stimulate sensory nerve endings. In: *The Effect of Nicotine Agents, Neuropharmacol.* **12**, 149–164 (1973).

K. H. Ginzel, The lungs as sites of origin of nicotine-induced skeletomotor relaxation and behavioral and electrocortical arousal in the cat. In: *The Pharmacology of Nicotine,* eds. M. J. Rand and K. Thurau, IRL Press, Washington, D. C., pp. 269–292 (1988).

J. W. Gorrod and P. Jenner, The metabolism of tobacco alkaloids. In: *Essays in Toxicology,* Vol. 6, ed. W. J. Hayes, Jr., Academic Press, New York, pp. 35–78 (1975).

R. A. Greenberg, N. J. Haley, R. Etzel, and F. Loda, Measuring the exposure of infants to tobacco smoke, *New Engl. J. Med.* **310**, 1075–1078 (1984).

N. J. Haley, C. M. Axelrad, and K. A. Tilton, Validation of self-reported smoking behavior. Biochemical analyses of cotinine and thiocyanate, *Am. J. Publ. Health* **73**, 1204–1207 (1983).

N. J. Haley, S. G. Colosimo, C. M. Axelrad, R. Harris, and D. W. Sepkovic, Biochemical validation of self-reported exposure to environmental tobacco smoke *Environ. Res.* **49**, 127–135 (1989a).

N. J. Haley, D. W. Sepkovic, and D. Hoffmann, Cigarette smoking as a risk for cardiovascular disease. VI. Compensation with nicotine availability as a single variable, *Clin. Pharmacol. Ther.* **38**, 164–170 (1985).

N. J. Haley, D. W. Sepkovic, and D. Hoffmann, Elimination of cotinine from body fluids: Nicotine disposition in smokers, and nonsmokers. *Am. J. Public Health* **79**, 1046–1048 (1989b).

N. J. Haley, D. Hoffman, and E. L. Wynder, Uptake of tobacco smoke components. In: Banbury Report 23: *Mechanisms in Tobacco Carcinogensis,* New York, pp. 3–19 (1987).

S. S. Hecht and D. Hoffmann, Tobacco-specific nitrosamines, an important group of carcinogens in tobacco and tobacco smoke, *Carcinog.* **9**, 875–884 (1988).

S. S. Hecht, A. Abbaspour, and D. Hoffmann, A study of tobacco carcinogenesis. XLII. Bioassay in A/J mice of some structural analogues of tobacco-specific nitrosamines, *Cancer Lett.* **42**, 141–145 (1988a).

S. S. Hecht, T. E. Spratt, and N. Trushin, Evidence for 4-(3-pyridyl)-4-oxobutylation of DNA in F344 rats treated with the tobacco specific nitrosamines 4-(methyl-nitrosamino)-1-(3-pyridyl)-1-butanone and N′-nitrosonornicotine, *Carcinog.* **9**, 161–165 (1988b).

S. S. Hecht, J. D. Adams, S. Numoto, and D. Hoffmann, Induction of respiratory tract tumors in Syrian golden hamsters by a single dose of 4-(methylnitrosamino)-1-(3-pyridyl)-1-butanone (NNK) and the effect of smoke inhalation, *Carcinog.* **4**, 1287–1290 (1983a).

S. S. Hecht, N. Trushin, A. Castonguay, and A. Rivenson, Comparative tumori-genicity and DNA methylation in F344 rats by 4-(methylnitrosamino)-1-(3-pyri-dyl)-1-butanone and N-nitrosodimethylamine, *Cancer Res.* **46**, 498–502 (1986a).

S. S. Hecht, A. Castonguay, A. Rivenson, B. Mu, and D. Hoffmann, Tobacco specific nitrosamines: Carcinogenicity, metabolism, and possible role in human cancer, *J. Environ. Health Sci. CI* **1**, 1–54 (1983b).

S. S. Hecht, A. Rivenson, J. Braley, J. DiBello, J. D. Adams, and D. Hoffmann, Induction of oral cavity tumors in F344 rats by tobacco-specific nitrosamines and snuff, *Cancer Res.* **46**, 4162–4166 (1986b).

S. S. Hecht, C. B. Chen, R. M. Ornaf, E. Jacobs, J. D. Adams, and D. Hoffmann, Chemical studies on tobacco smoke. LVII. Reaction of nicotine and sodium nitrite: Formation of nitrosamines and fragmentation of the pyrrolidine ring, *J. Org. Chem.* **42**, 72–76 (1978a).

S. S. Hecht, C. B. Chen, N. Hirota, R. M. Ornaf, T. C. Tso, and D. Hoffmann, A study of tobacco carcinogenesis. XVI. Tobacco specific nitrosamines: Formation from nicotine *in vitro* and during tobacco curing and carcinogenicity in strain A mice, *J. Natl. Cancer Inst.* **60**, 819–824 (1978b).

F. Henderson, R. Morris, H. Reed, P. C. Hu, J. Mumford, L. Forehand, R. Burton, J. Lewtas, S. K. Hammond, and N. J. Haley, Serum and urine cotinine as quantitative measures of passive smoke exposure in young children, *Proc. 4th Int. Conf. Indoor Air Qual. Climate* **2**, 18–21 (1987).

R. I. Herning, R. T. Jones, T. Bachman, and A. H. Mines, Puff volume increases when low-nicotine cigarettes are smoked, *Br. Med. J.* **282**, 187–189 (1981).

J. Hilfrich, S. S. Hecht, and D. Hoffmann, A study of tobacco carcinogenesis. XV. Effects of N'-nitrosonor-nicotine and N'-nitrosoanabasine in Syrian golden hamsters, *Cancer Lett.* **2**, 169–176 (1977).

P. Hill and E. L. Wynder, Smoking and cardiovascular disease. Effects of nicotine on the serum epinephrine and corticoids, *Am. Heart J.* **87**, 491–496 (1974).

P. Hill and E. L. Wynder, Nicotine and cotinine in human breast fluid, *Cancer Lett.* **6**, 251–254 (1979).

P. Hill, N. J. Haley, and E. L. Wynder, Cigarette smoking as a risk for cardiovascular disease. I. Biochemical analyses of carboxyhemoglobin, plasma nicotine, cotinine and thiocyanate versus self-reported smoking data, *J. Chron. Dis.* **36**, 439–449 (1983).

D. Hoffmann and J. D. Adams, Carcinogenic tobacco-specific N-nitrosamines in snuff and in the saliva of snuff-dippers, *Cancer Res.* **41**, 4305–4308 (1981).

D. Hoffmann and K. D. Brunnemann, Endogenous formation of N-nitrosoproline in cigarette smokers, *Cancer Res.* **43**, 5570–5574 (1983).

D. Hoffmann and S. S. Hecht, Nicotine-derived N-nitrosamines and tobacco-related cancer: current status and future directions, *Cancer Res.* **45**, 935–944 (1985).

D. Hoffmann, G. Rathkamp, and Y. Y. Liu, On the isolation and identification of volatile and non-volatile N-nitrosamines and hydrazines in cigarette smoke. In: *N-Nitroso Compounds in the Environment,* IARC Scientific Publication No. 9, eds. P. Bogovski and E. A. Walker, IARC, Lyon, France, pp. 159–165 (1974).

D. Hoffmann, K. D. Brunnemann, J. D. Adams, and S. S. Hecht, Formation and analysis of N-nitrosamines in tobacco products and their endogenous formation in tobacco consumers. In: *N-Nitroso Compounds: Occurrence, Biological Effects and Relevance to Human Cancer,* IARC Scientific Publication No. 57, eds. I. K. O'Neill, R. C. Von Borstel, C. T. Miller, J. E. Long, and H. Bartsch, IARC, Lyon, France, pp. 743–762 (1984).

D. Hoffmann, A. Castonguay, A. Rivenson, and S. S. Hecht, Comparative carcinogenicity and metabolism of 4-(methylnitrosamino)-1-(3-pyridyl)-1-butanone and N'-nitrosonor nicotine in Syrian golden hamsters, *Cancer Res.* **41**, 2386–2393 (1981).

D. Hoffmann, A. Rivenson, S. Amin, and S. S. Hecht, Dose-response study of the carcinogenicity of tobacco-specific nitrosamines in F344 rats, *J. Cancer Res. Clin. Oncol.* **108**, 81–86 (1984b).

B. Holmstedt, Toxicity of nicotine and related compounds. In: *The Pharmacology of Nicotine*, eds. M. J. Rand and K. Thurau, IRL Press, Washington, D.C., pp. 61–90 (1988).

IARC, *IARC Monographs on the Evaluation of the Carcinogenic Risk of Chemicals to Humans*, Vol. 38, *Tobacco Smoking*, IARC, Lyon, France, pp. 163–194 (1986).

P. Jenner, J. W. Gorrod, and A. H. Beckett, Factors affecting the *in vitro* metabolism of R(+) and S(−)nicotine in guinea-pig liver preparations, *Xenobiotica* **3**, 563–572 (1973a).

P. Jenner, J. W. Gorrod, and A. H. Beckett, The absorption of nicotine-1′-N′-oxide and its reduction in the gastrointestinal tract in man, *Xenobiotica* **3**, 344–349 (1973b).

P. Jacob, III, M. Wilson, and N. L. Benowitz, Improved gas chromatographic method for the determination of nicotine and cotinine in biologic fluids, *J. Chromatogr.* **222**, 61–70 (1981).

M. J. Jarvis, H. Tunstall-Pedoe, C. Feyerabend, C. Vesey, and Y. Saloojee, Biochemical markers of smoke absorption and self-reported exposure to passive smoking, *J. Epidemiol. Commun. Health* **38**, 335–339 (1984).

G. C. Kabat and E. L. Wynder, Lung cancer in nonsmokers, *Cancer* **53**, 1214–1221 (1984).

H. J. Klimisch and L. Stadler, Untersuchungen zur Bildung von N′-nitrosonornikotin aus Nicotin-N′-oxyd, *Talanta* **23**, 614–616 (1979).

L. C. Koo, J. H.-C. Ho, and R. Rylander Life-history correlates of environmental tobacco smoke: a study on nonsmoking Hong Kong Chinese wives with smoking vs nonsmoking husbands. *Social Sci Med* **26**, 751–760 (1988)

G. A. Kyerematen, M. D. Damiano, B. H. Dvorchek, and E. S. Vesell, Smoking induced changes in nicotine disposition. Application of a new HPLC method for nicotine and its metabolites, *Clin. Pharmacol. Ther.* **32**, 769–780 (1982).

G. A. Kyerematen, L. H. Taylor, J. D. deBethizy, and E. S. Vesell, Pharmacokinetics of nicotine and 12 metabolites in the rat, *Drug Metab. Disp.* **16**, 125–129 (1988).

K. F. Ladd, H. L. Newmark, and M. C. Archer, N-Nitrosation of proline in smokers and nonsmokers, *J. Natl. Cancer Inst.* **73**, 83–87 (1984).

J. J. Langone, H. B. Gjika, and H. Van Vunakis, Nicotine and its metabolites. Radioimmunoassays for nicotine and cotinine, *Biochem.* **12**, 5025–5030 (1973).

E. J. LaVoie, G. Prokopczyk, J. Rigotty, A. Czech, A. Rivenson, and J. D. Adams, Tumorigenic activity of the tobacco-specific nitrosamines 4-(methylnitrosamino)-1-(3-pyridyl)-1-butanone (NNK), 4-(methylnitrosamino)-4-(3-pyridyl)-1-butanol (iso-NNAL) and N′nitrosonornicotine (NNN) on topical application to Sencar mice, *Cancer Lett.* **37**, 277–283 (1987).

B. L. Lee, N. L. Benowitz, and P. Jacob III, Influence of tobacco abstinence on the disposition, kinetics and effects of nicotine, *Clin. Pharmacol. Ther.* **41**, 474–479 (1987).

D. A. Machacek and N-S. Jaing, Quantitation of cotinine in plasma and saliva by liquid chromatography, *Clin. Chem.* **32**, 979–982 (1986).

B. MacMahon, D. Trichopoulos, P. Cole, and J. Brown, Cigarette smoking and urinary estrogens, *New Engl. J. Med.* **307**, 1062–1065 (1982).

M. B. Mattammal, V. M. Lakshmi, J. V. Zenser, and B. B. Davis, Lung prostaglandin H. synthase and mixed-function oxidase metabolism of nicotine, *J. Pharm. Exp. Ther.* **242**, 827–832 (1987).

J. J. Michnovicz, R. J. Hershcopf, H. Nagonuma, H. L. Bradlow, and J. Fishman, Increased 2-hydroxylation of estradiol as a possible mechanism for the anti-estrogenic effect of cigarette smoking, *New Engl. J. Med.* **315**, 1305–1309 (1986).

H. Muronaka, E. Higashi, S. Itoni, and Y. Shemizu, Evaluation of nicotine, cotinine, thiocyanate, carboxyhemoglobin and expired carbon monoxide as biochemical tobacco smoke uptake parameters, *Int. Arch. Occup. Environ. Health* **60**, 37–41 (1988).

National Research Council, *Environmental Tobacco Smoke. Measuring Exposures and Assessing Health Risks,* National Academy Press, Washington, D.C. (1986).

G. B. Neurath, M. Duenger, D. Orth, and F. G. Peen, Trans-3'-hydroxycotinine as a main metabolite of nicotine in urine of smokers, *Int'l Arch. Occupat. Environ. Health* **59**, 199–201 (1987).

H. Ohshima, I. K. O'Neill, M. Friesen, J.-C. Béréziat, and H. Bartsch, Occurrence in human urine of new sulphur-containing N-nitrosamino acids N-nitrosothiazolidine 4-carboxylic acid and its 2-methyl derivative and their formation, *J. Cancer Res. Clin. Oncol.* **108**, 121–128 (1984).

G. Palladino, J. D. Adams, K. D. Brunnemann, N. J. Haley, and D. Hoffmann, Snuff-dipping in college students: a clinical profile, *Military Med.* **151**, 342–346 (1986).

S. F. Piazza, N. J. Haley, D. A. Clark, S. J. Caravella, C. M. Axelrad, and H. S. Dweck, Human milk contamination with nicotine and cotinine, *Pediatr. Res.* **21**, 401A (1987).

W. F. Pool, A. A. Houdi, L. A. Damani, W. J. Layton, and P. A. Crooks, Isolation and characterization of N-methyl-N'-oxonicotinium ion, a new urinary metabolite of R(+)-nicotine in the guinea pig, *Drug Metab. Disp.* **14**, 574–579 (1986).

P. Puustinen, H. Olkkonen, S. Kalonen, and J. Tuomisto, Microcomputer-aided measurement of puff parameters during smoking of low- and medium-tar cigarettes, *Scand. J. Clin. Lab. Invest.* **47**, 655–660 (1987).

A. Rivenson, D. Hoffmann, B. Prokopczyk, S. Amin, and S. S. Hecht, Induction of lung and exocrine pancreas tumors in F344 rtats by tobacco-specific and areca-derived N-nitrosamines, *Cancer Res.* **48**, 6912–6917 (1988).

J. B. Robertson and J. T. Bray, Development of a validation test for self-reported abstinence from smokeless tobacco products: preliminary results, *Prev. Med.* **17**, 496–502 (1988).

J. Rosenberg, N. L. Benowitz, P. Jacob, and K. M. Wilson, Disposition kinetics and effects of intravenous nicotine, *Clin. Pharmacol. Ther.* **28**, 517–522 (1980).

M. A. H. Russell, Nicotine replacement: the role of blood nicotine levels, their rate of change and nicotine tolerance. In: *Nicotine Replacement in the Treatment of Smoking: A Critical Evaluation,* eds. O. F. Pomerleau and C. S. Pomerleau, Alan Liss, New York, pp. 63–99 (1988).

M. A. H. Russell, C. Wilson, U. A. Patel, C. Feyerabend, and C. V. Cole, Plasma nicotine levels after smoking cigarettes with high, medium and low nicotine yields, *Br. Med. J.* **2**, 414–416 (1975).

M. Schiffman, N. J. Haley, J. S. Felton, A. W. Andrews, A. Kaslow, W. D. Lancaster, R. J. Kurman, L. A. Brinton, L. B. Lannom, and D. Hoffmann, Biochemical epidemiology of cervical neoplasia: Measuring cigarette smoke constituents in the cervix, *Cancer Res.* **47**, 3886–3888 (1987).

I. Schmeltz and D. Hoffmann, Nitrogen-containing compounds in tobacco and tobacco smoke, *Chem. Rev.* **77**, 295–311 (1977).

D. Schwartz-Bickenbach, B. Schulter-Hobein, S. Abt, C. Plum, and H. Nau, Smoking and passive smoking during pregnancy and early infancy: Effects on birth weight, lactation period and cotinine concentrations in mother's milk and infant's urine, *Toxicol. Lett.* **35**, 73–81 (1987).

D. W. Sepkovic and N. J. Haley, Biomedical applications of cotinine concentrations in biological fluids, *Am. J. Public Health* **75**, 663–636 (1985).

D. W. Sepkovic and N. J. Haley, Metabolism of nicotine in smokers and nonsmokers. In: *Tobacco Smoking and Nicotine*, eds. W. R. Martin, G. R. Van Loon, E. T. Iwamoto, and L. Davis, Plenum, New York, pp. 375–388 (1987).

D. W. Sepkovic, N. J. Haley, and D. Hoffmann, Elimination of cotinine in the plasma and urine of smokers and passively exposed smokers, *J. Am. Med. Assoc.* **256**, 863 (1986b).

D. W. Sepkovic, N. J. Haley, and E. L. Wynder, Thyroid activity in cigarette smokers, *Arch. Inter. Med.* **144**, 501–503 (1984c).

D. W. Sepkovic, N. J. Haley, C. M. Axelrad, and E. J. LaVoie, Thyroid hormone concentrations in rats after chronic nicotine metabolite administration, *Proc. Soc. Exp. Biol. Med.* **177**, 412–416 (1984b).

D. W. Sepkovic, N. J. Haley, C. M. Axelrad, A. Shigematsu, and E. J. LaVoie, Short-term studies on the *in vivo* metabolism of N-oxides of nicotine in rats, *J. Toxicol. Environ. Health* **18**, 205–214 (1986a).

D. W. Sepkovic, N. J. Haley, C. M. Axelrad, and E. L. Wynder, Cigarette smoking as a risk for cardiovascular disease. III. Nicotine regulation and compensation with increasing nicotine yield cigarettes, *Addict. Behav.* **8**, 56–66 (1983).

D. W. Sepkovic, M. Marshall, W. Rogers, P. Cronin, S. G. Colosimo, and N. J. Haley, Thyroid function and cigarette smoking in baboons, *Proc. Soc. Exp. Biol. Med.* **187**, 223–228 (1988).

D. W. Sepkovic, K. Parker, C. M. Axelrad, N. J. Haley, and E. L. Wynder, Cigarette smoking as a risk for cardiovascular disease. V. Biochemical parameters with increased and decreased nicotine content cigarettes, *Addict. Behav.* **9**, 255–263 (1984a).

M. Shaarawy and K. Z. Mahmoud, Endocrine profile and semen characteristics in male smokers, *Fert. Steril.* **38**, 255–257 (1982).

M. K. Shigenaga, A. J. Trevor, and N. Castagnoli, Jr., Metabolism-dependent covalent binding of (S)-[5-³H]nicotine to liver and lung microsomal macromolecules, *Drug Metab. Disp.* **16**, 397–402 (1988).

D. Umbenhauer, C. P. Wild, R. Montesano, R. Saffhill, J. M. Boyle, N. Huh, U. Kirstein, J. Thomale, M. F. Rajewsky, and S. H. Lu, O⁶-Methyldeoxyguanosine in oesophageal DNA among individuals at high risk of oesophageal cancer, *Int. J. Cancer* **36**, 661–665 (1985).

U. S. Surgeon General, *The Health Consequences of Smoking. Nicotine Addition*, U.S. Dept. of Health and Human Services, Public Health Service, National Institute of Health Publication No. 88-8406 (1988).

H. Van Vunakis, H. B. Gjika, and J. J. Langone, Radioimmunoassay for nicotine and cotinine. In: *Environmental Carcinogens: Methods of Analysis and Exposure Measurement*, Vol. 9, *Passive Smoking*, IARC Sci Publ. *81:* eds I. K. O'Neill, K. D. Brunnemann, B. Dodet, and D. Hoffmann, IARC, Lyon, France, pp. 317–330 (1987).

L. Wagenknecht, G. Cutter, C. Smoak, G. Friedman, and N. J. Haley, Black/white differences in cotinine levels among smokers in the United States. In: *The Pharmacology of Nicotine*, Vol. 9, eds. M. J. Rand and K. Thurau, IRL Press, Gold Coast, Queensland, Australia, pp. 30–31 (1988).

M. A. Wall, J. Johnson, P. Jacob, and N. L. Benowitz, Cotinine in the serum, saliva and urine of nonsmokers, passive smokers and active smokers, *Am. J. Publ. Health* **78**, 699–701 (1988).

P. Watson, Australian Aboriginal Exploitation of the Nicotine-Containing Plant, *Duboisia Hopwoodii*. In: *The Pharmacology of Nicotine*, Vol. 9, eds. M. J. Rand and T. Thurau, IRL Press, Gold Coast, Queensland, Australia, pp. 412–413 (1988).

E. L. Wynder and S. D. Stellman, The impact of long-term filter cigarette usage on lung cancer and larynx cancer risk. A case control study, *J. Natl. Cancer Inst.* **62**, 471–477 (1979).

Exposure Control Versus Risk Assessment: Lessons From the Study of Genotoxic N-Substituted Arenes

H.-G. Neumann
Institute of Pharmacology and Toxicology
University of Würzburg
Würzburg, Germany

1. INTRODUCTION

Arylamines were among the first chemicals shown to be carcinogenic in man. Since the time when the manufacturing of dyes was first associated with the occurrence of bladder tumors (Rehn, 1895), many more important industrial chemicals of this class were found to be carcinogenic in experimental animals or humans. As a consequence, health hazards from these compounds had to be considered over a long time, and basic research pioneered major developments in investigating the reaction mechanisms of chemical carcinogens. It is therefore not surprising that arylamines were among the first chemicals for which biological monitoring methods were developed. Arylamines or their metabolites were measured in urine in order to demonstrate the actual uptake, and the levels of methemoglobin in blood were determined in many cases of acute poisoning in order to select the appropriate treatment. These methods, however, are not satisfactory in controlling chronic, low-level exposures. One of the big disadvantages is that the results depend critically on the time at which the sample was collected. A major step forward was the proposal to analyze blood samples for specific hemoglobin-bound metabolites, as will be discussed in this chapter. Basic features of this method have been studied *in vitro* and in experimental animals. They will be discribed briefly as a basis for further discussion.

This method, which is based on protein binding and thus has widespread application, has also been applied to monitor humans and proved to be suitable for use in exposure control. Particularly two questions seem to inhibit the acceptance of this method: (1) How indicative is protein binding compared to DNA binding for determination of genotoxic properties of chemicals? (2) How can macromolecular binding be used to assess the tumor risk associated with the underlaying exposure? The broad experience with arylamines both in industrial hygiene and in basic cancer research makes them suitable examples to address these general questions in the field of biological monitoring.

2. THE REACTION OF *N*-SUBSTITUTED ARYL COMPOUNDS WITH BLOOD PROTEINS

2.1. The Generation of Nitroso-Arenes in the Metabolism of *N*-substituted Aryl Compounds

It is now generally accepted that *N*-substituted aryl compounds must be metabolically activated in order to exert their acute toxic and their genotoxic effects. There are numerous metabolic pathways that lead to reactive inter-mediates that bind covalently to macromolecules. The reactions of electro-philes with DNA have been studied in great detail, and many of the nucleic acid based adducts have been identified (Miller and Miller, 1969; Kriek *et al.*, 1984, Beland and Kadlubar, 1985). Although the reaction with pro-teins was detected first, far less is known about the adducts formed with amino acids. In addition to the early work on the reaction of hydroxamic acid esters with methionine (Poirier *et al.*, 1967), the reaction of nitroso-arenes with the SH group of cysteine drew some attention more recently (Dölle *et al.*, 1980). A sulfinic acid amide is formed (see below) that is stable in proteins *in vivo* but can readily be hydrolyzed under acidic or alkaline conditions to yield an arylamine. This can be extracted from a biological matrix and quantitatively determined (Fig. 1).

Thus nitroso-arenes became the key metabolites for biological monitor-ing of a great variety of chemicals collectively known as *N*-substituted aryl compounds (Neumann, 1984a). They are generated not only by *N*-oxidation of arylamines but also by the reduction of nitro-aromatic compounds (Fig. 2). These pathways extend the application to many environmental chemicals. In addition, aryl-azo compounds can be reduced by intestinal bacteria or reduc-tases in liver and other tissues to release two arylamine components that

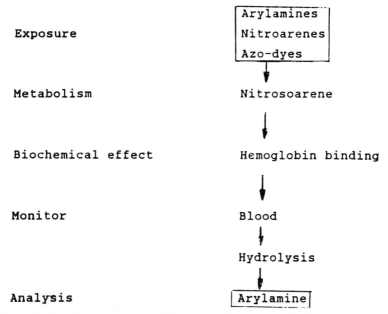

Figure 1. Control of exposure to *N*-substituted arenes by measuring the biochemical end point.

subsequently can be metabolically oxidized (Radomski and Mellinger, 1962). This allows one to monitor the bioavailability of carcinogenic compounds of a great number of azo dyes.

2.2. The Biological Role of Nitroso-Arene Metabolites

Arylhydroxylamines play a central role in the metabolism of *N*-substituted aryl compounds. They are considered penultimate metabolites of most biological effects. Either the *N*-hydroxy-amines as such or their *O*-esters give

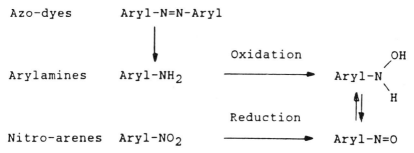

Figure 2. Nitroso-arenes are key metabolites for biomonitoring of different N-substituted aryl compounds.

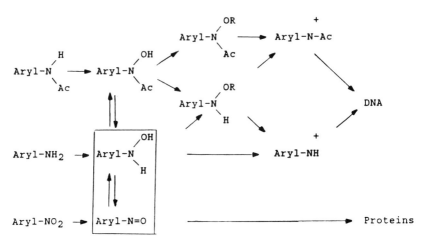

Figure 3. Pathways of metabolic activation of arylamines and nitroarenes. Ac = acetate, R = Ac or SO_3^-.

rise to nitrenium ions that react with DNA and produce promutagenic lesions. The esters are generated either by conjugation of *N*-hydroxy-amines with activated sulfate assisted by sulfotransferases or by conjugation with activated acetate. *N*-acetoxy-arylamines may be formed either directly by *N*-acetyltransferase or from hydroxamic acids by a *N,O*-transacetylase activity of *N*-acetyltransferase (Fig. 3).

N-hydroxy-amines and nitroso-arenes cannot be distinguished in many test systems because the former are readily oxidized by oxygen and the latter are readily reduced enzymatically. In bacterial test systems for mutagenicity both types of *N*-oxidation products are equally effective, and they exhibit the strongest direct mutagenicity among all metabolites of *N*-substituted aryl compounds (Glatt *et al.*, 1980). In experimental animals they constitute the core of *N*-oxidized metabolites, together with hydroxamic acids. These three metabolites are interconvertible *in vivo,* and the prevalence of one of them in different tissues depends on the acetylation status, i.e., the balance between acetylated and nonacetylated amines, and the redox state, i.e., the balance between oxidation and reduction (Neumann, 1986). In liver, for instance, nitroso-arenes are generated, but they are also reduced by aldehyde reductase such that only a small fraction is available for macromolecular binding. In erythrocytes, on the other hand, nitroso-arenes are generated by cooxidation of *N*-hydroxy-amines and hemoglobin in a cyclic process that provides a rather large fraction available for protein binding. This is a unique situation in which the availability of *N*-oxidized metabolites in circulation is amplified.

In summary, nitroso-arenes are not considered to be the ultimate genotoxic metabolites, i.e., those that react with DNA and generate promutagenic adducts, but they are closely linked to the activating pathway.

2.3. The Reaction of Nitroso-Arenes with Glutathione

The reaction of nitroso-arenes with mercaptanes was predominantly studied *in vitro* with glutathione as the nucleophile. A tentative scheme of the complex reaction was first proposed by Dölle *et al.* (1980). The essential features have been substantiated by further work in several laboratories (Eyer, 1979, 1985; Mulder *et al.*, 1982, 1984). The nucleophilic sulfur of the cystein moiety reacts with the electrophilic nitrogen of the nitroso group to give a hemimercaptal for which there are three options to stabilization (Fig. 4). The rates of the different reactions, and as a consequence the product pattern, depend on reaction conditions (pH, buffer concentration) and other substitutents at the aromatic ring (Eyer, 1985). First, the semimercaptal may rearrange to form a sulfinic acid amide. Electron-donating substituents (small Hammett constant) promote this reaction. Second, the hemimercaptal may be reduced to *N*-hydroxy-amine by glutathione. With a given glutathione concentration, reduction is favored by electron-withdrawing substituents (large Hammett constant). 4-Nitroso-acetophenone, for example, is almost completely reduced to the *N*-hydroxy-amine. Third, complete reduction to arylamine may occur, provided sufficient glutathione is available. This pathway is favored by electron-donating substituents. In the original study with nitrosobenzene, aniline was not detected, whereas *p*-toluidine always was a reaction product with nitrosotoluene.

The yield of a stable sulfinic acid amide under given reaction conditions, therefore, may vary considerably depending on the chemical structure of the aryl moiety.

2.4. The Reactions of Nitroso-Arenes *In vivo*

As soon as radioisotopes were introduced to study xenobiotics *in vivo,* erythrocytes were detected as targets for persistent binding. Phenylhydroxylamine labeled with iodine (Jackson, 1953) and 2-aminonaphthaline (Sommerville *et al.,* 1956; Goldblatt *et al.,* 1960) were found to be elimi-

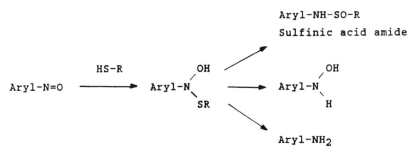

Figure 4. The reaction of nitroso-arenes with mercaptanes. The product pattern depends on substitution of the aryl moiety and reaction conditions.

nated from erythrocytes only in the course of their degradation. The key detection at that time was the finding that aniline is *N*-oxidized by liver microsomes and phenylhydroxylamine is further oxidized to nitrosobenzene in erythrocytes (Kiese and Uehleke, 1961). Side reactions were recognized, and a role for SH groups was assumed as early as 1962 (Boyland *et al.*). 2-Acetylaminofluorene became the key model compound to study metabolic activation and macromolecular binding of carcinogens (Miller and Miller, 1969), and it was soon recognized that metabolites circulating in the blood of rats were covalently bound to plasma proteins and erythrocytes (Weisburger *et al.*, 1966, 1969). *Trans*-4-dimethylaminostilbene, another arylamine model compound, was also found to accumulate in erythrocytes of rats (Rjosk and Neumann, 1971), and hemoglobin was identified as the major target molecule that covalently reacted with aminostilbene metabolites (Groth and Neumann, 1971).

Trans-4-nitrosostilbene was proposed to be the predominant reactant when it was found that 80% of the hemoglobin adduct could be hydrolyzed under mild conditions to give *trans*-4-aminostilbene (Töpner, 1975). This was supported by the observation that when *trans*-4-nitrosostilbene was intravenously injected, 20% of the dose rapidly bound to hemoglobin and *trans*-4-aminostilbene was obtained under the same conditions as the predominant cleavage product. The *in vitro* studies described above also support this proposal. Other minor covalent reaction products have not yet been identified. The strongest evidence for a reaction of nitroso-arenes with cysteine in hemoglobin to form a sulfinamide was provided by Ringe *et al.* (1988), who analyzed the structure of human hemoglobin modified by *N*-hydroxy-4-aminobiphenyl by x-ray crystallography and found the only binding position to be at cysteine 93β.

3. ANIMAL EXPERIMENTS

The idea to use hemoglobin-bound metabolites of arylamines as a dosimeter for reactive metabolites *in vivo* was pursued simultaneously with the pioneering work of Ehrenberg and his collaborators, who studied the covalent binding of alkylating agents to hemoglobin and the use of hemoglobin binding as a dose monitor (Ehrenberg *et al.*, 1974). A number of basic features were studied in animal experiments to lay the groundwork for the application of this technique to humans.

3.1. Binding of Aromatic Amines to Hemoglobin

Binding to hemoglobin in rats has been studied with a number of labeled aromatic amines, and in analogy to the DNA-binding index, (Lutz, 1979) a

TABLE 19–1. Hemoglobin-Binding Index of Aromatic Amines in Rats 24 h After a Single Dose of Labeled Compound (Total Binding, Neumann, 1988)

Compound	Binding Index*
4-Aminobiphenyl	230
Trans-4-dimethylaminostilbene	164
Nitrobenzene	74
Benzidine	60
2-Acetylaminofluoren	18
Acetanilide	12

*Binding index = [binding (mmol/mol Hb)]/[dose (mmol/kg)].

hemoglobin-binding index was calculated (Table 1). In all cases studied most of the bound material (75–90%) could be hydrolyzed under moderate conditions, and the parent amine was identified as the preponderant cleavage product (Fig. 5). A notable exception is benzidine, a bifunctional arylamine, which yielded more than one cleavage product (see below). The parent amine benzidine and mono-acetylbenzidine are obtained in a ratio 1:10 (Albrecht and Neumann, 1985; Birner and Neumann, 1988; see also Neumann, 1988). Nitrobenzene was the first example of a nitroarene. Comparison with acetanilide shows that more of the reduction product of nitrobenzene becomes available than the *N*-oxidation product of acetanilid.

A series of labeled carcinogens tested for hemoglobin binding in rats by Pereira and Chang (1981) also contained some arylamines. The binding values were lower than those in Table 1, probably because hemoglobin was processed under conditions that partly hydrolyzed the sulfinamides.

3.2. Methods Suitable for Analysis of Arylamines

So far, the results indicate that arylamines in general generate hydrolyzable hemoglobin adducts *in vivo*. As the next step, analytical procedures

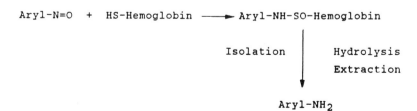

Figure 5. Hemoglobin binding of nitroso arenes.

TABLE 19–2. Hemoglobin-Binding Index of Monocyclic Aromatic Amines 24 h After Single Oral Administration of the Unlabeled Compounds to Rats. Hemoglobin was Hydrolyzed and the Extracted Parent Amine Analyzed by Capillary Gas Chromatography (Birner and Neumann, 1988)

Compound	Binding Index*
Aniline	22
N-Methylaniline	16
N,N-Dimethylaniline	11
o-Toluidine	4.0
m-Toluidine	4.9
p-Toluidine	4.3
2,4-Dimethylaniline	2.3
2,4,5-Trimethylaniline	0.7
p-Chloroaniline	569
3,4-Dichloroaniline	9
4-Chloro-o-toluidine	28
5-Chloro-o-toluidine	28
6-Chloro-o-toluidine	0.8

*Binding index = [binding (mmol/mol Hb)]/[dose (mmol/ kg)].

independent of a radioactive label had to be developed. With a series of monocyclic arylamines it was demonstrated that capillary gas chromatography (GC) with flame ionization detector (FID) (Albrecht and Neumann, 1985), later used with a nitrogen-sensitive detector (Birner and Neumann, 1988), was sufficient to analyze the underivatized amines extracted from processed blood samples of rats orally dosed with 0.6 mmol/kg of the arylamine. The detection limit was on the order of 5 pg/μl injected, or 10 μg amine bound to hemoglobin in 1 liter of blood. A wide range of hemoglobin-binding indices was observed (Table 2). In all cases, the parent amine was the only extractable product. Background levels were not detected with this method.

Following the identification of aminobiphenyl as a cleavage product from adducted hemoglobin by GC and gas chromatography–mass spectrometry (GC-MS) (Green *et al.*, 1984), a method developed in which the extracted amine was derivatized with pentafluorobenzoylchloride, precleaned by silica gel chromatography, and quantitated by GC using electron-capture detection. Better sensitivity and specificity, however, were obtained by analyzing a pentafluoropropionic acid derivative by negative-ion chemical ionization mass spectrometry, with a detection limit below 10 pg/10 ml blood (Bryant *et al.*, 1987).

3.3. The Dose Dependence of Hemoglobin Binding

Fundamental to the application of this technique is a linear dose–binding relationship *in vivo*. This was first demonstrated for the total binding of labeled *trans*-4-dimethylaminostilbene metabolites in rats after oral doses from 5×10^{-10} to 3.5×10^{-5} mol/kg (Neumann *et al.*, 1978; Neumann, 1980). In addition, hemoglobin samples of this experiment were hydrolyzed and *trans*-4-aminostilbene was quantitated in the extract. A perfectly linear dose dependence, in this case for a specific amino acid adduct, was obtained (Albrecht and Neumann, 1984; see Neumann, 1984b). The same linear relationship was obtained with labeled 4-aminobiphenyl in rats for doses ranging from 0.5 to 5000 μg/kg (Green *et al.*, 1984) and for total hemoglobin binding of labeled 2-acetylaminofluorene metabolites in mice and rats (Pereira and Chang, 1981).

These data are particularly important because within this wide range of doses studied, and down to extremely small doses, the same fraction of each dose is bound to hemoglobin, or, in other words, pharmacokinetics follow first-order behavior. This means that a linear pharmacokinetically determined relationship exists between hemoglobin binding and the reaction of metabolites with macromolecules, including DNA, in other tissues. The parallel dose–binding relationship in different compartments of rats, i.e., for DNA binding in liver, hemoglobin binding in blood, and DNA binding in kidney, provides experimental evidence for this conclusion (Neumann, 1980, 1984b). This does not mean that a constant ratio between hemoglobin and DNA binding exists at different time points, because the elimination rate differs for these two types of macromolecular adducts. On the other hand, for the initial extent of macromolecular modification a substance- and tissue-specific constant ratio between hemoglobin and DNA binding can be expected.

Deviations from linearity were seen only with very high dose levels approaching acute toxic doses (LD_{50}). This indicates that whenever processes become saturated, a linear dose dependence can no longer be expected. Delayed absorption, decrease of liver first-pass effects, and saturation of metabolic pathways may then be involved.

3.4. Accumulation of Adducts

Hemoglobin adducts accumulate during repeated administration to experimental animals. When 12 doses of *trans*-4-acetylaminostilbene (5×10^{-6} mol/kg) were orally administered to rats over a period of 6 weeks, steady-state levels were reached after 4 weeks (Neumann, 1983). With 4-aminobiphenyl (11.4 μg every other day) hemoglobin adducts accumulated to a level 30-fold higher than after a single dose within 8 to 9 weeks (Green *et al.*, 1984).

3.5. Elimination of Adducts

If the hemoglobin adducts are stable *in vivo,* their elimination should be linked to the degradation of erythrocytes. The average life span of an erythrocyte is 40 days in mice, 60–65 days in rats, and 120 days in humans. The adducts should therefore decrease linearly and should be eliminated after a single dose by the lifetime of the erythrocytes. This has been demonstrated for total binding of benzo(a)pyrene metabolites (Forsberg-Karlsson *et al.,* 1983) and for benzene metabolites (Segerbäck, 1985) to hemoglobin in mice.

With chronic administration of 4-aminobiphenyl to rats hemoglobin levels accumulated and after cessation of dosing declined at a rate of 2.5% per day. After 60–65 days the labeled adducts were cleared from the blood (Green *et al.,* 1984). After a single dose of *trans*-4-dimethylaminostilbene the hemoglobin adducts seemed to decrease exponentially over a period of 4 weeks with a half-life of 14 days (Neumann, 1979). An exponential decrease was also observed in a more extended study with benzidine, and measurable adduct levels were still found after 70 days in rats. The half-life was calculated in this case to be 11.5 days (Neumann, 1984b).

Thus conflicting data about the elimination kinetics of hemoglobin adducts exist, and more work is necessary to determine the extent to which the stability of adducts *in vivo* and the elimination kinetics of erythrocytes contribute. In addition, it has been proposed that the life span of erythrocytes may be altered by adduct formation. In all cases studied to date, the adducts were eliminated with a rate close to that of normal erythrocytes. This rate is slow enough to warrant high steady-state levels after repeated uptake. In other words, the hemoglobin adducts are sufficiently stable to be suitable as an integral dose monitor.

4. APPLICATIONS TO HUMANS

4.1. Occupational Exposures

The experience with this method in humans is still limited. Lewalter and Korallus (1985) were the first who analyzed hemoglobin adducts of aromatic amines after acute occupational exposures in accidents with aniline and *p*-chloroaniline. (These compounds are not classified as human or animal carcinogens.) They compared the levels of methemoglobin in blood, the excretion of the parent amines in urine, and the levels of hemoglobin adducts in erythrocytes. Methemoglobin levels and urinary excretion decreased much more rapidly than the hemoglobin adducts, which demonstrates a major advantage of determining the adducts: the data depend much less on the time of sample collection after exposure.

The authors also measured these parameters in workers under normal industrial exposure to aniline, i.e., air concentrations of aniline were below the MAK (maximum concentration at the workplace) value of 8 mg/m³. Two groups of subjects could be clearly distinguished as a result of the urinary excretion of acetanilide and were assigned as slow and rapid acetylators. Most interestingly, only the slow acetylators had measurable levels of hemoglobin–aniline adducts. It may therefore be concluded that under comparable exposure conditions all the subjects absorbed the compound leading to internal exposures, but only part of the individuals metabolized enough aniline to phenylhydroxylamine so that measurable levels of hemoglobin adducts could be formed. This clearly demonstrates the difference between internal exposure to the parent compound aniline, which was comparable in all individuals, and the availability of potentially hazardous metabolites, which was different in this case due to the genetically determined polymorphism of metabolic acetylation. This has important implications for risk assessment (see below).

Based on extensive studies in workers exposed to aniline a biological tolerance value (BAT value) of 100 μg liberated from hemoglobin adducts per liter of blood has been established in the Federal Republic of Germany (Bolt *et al.*, 1985). The original BAT (biological tolerance values for working materials) value of 1 mg aniline in urine is still listed and is largely based on the correlation with methemoglobin formation, i.e., methemoglobin levels of 5% were considered acceptable, and higher values were only observed in subjects excreting aniline concentrations in urine exceeding 1 mg/liter. In the above-mentioned study, the aniline concentrations in urine, however, were comparable in slow and rapid acetylators. Using the hemoglobin-adduct–derived BAT value means that different exposures to aniline are considered acceptable depending on the metabolic capacity of the individual.

Since phenylhydroxylamine and nitrosobenzene are also the metabolites viewed responsible for the toxic effects of nitrobenzene, the same BAT value of 100 μg aniline liberated from hemoglobin adducts in 100 ml blood was established for nitrobenzene. In both cases, aniline and nitrobenzene, observation of the tolerance value protects one from acute toxic effects but not from genotoxic effects and their possible contribution to the albeit weak carcinogenic potential of these chemicals.

Hydrolyzable hemoglobin adducts were also detected in workers handling the herbicides amitrole (aminotriazole) and diuron (Lewalter and Korallus, 1986). Whereas the internal exposure, measured as concentration of amitrole in urine after the working shift, closely followed the external exposure and thus varied considerably with time, the weekly determinations of hemoglobin adducts were less variable and considered sufficient to control alterations in thyroid function, an important biological end point for amitrole toxicity.

Diuron is a herbicide containing a 3,4-dichloroaniline moiety that can be metabolically released and yield the protein-binding *N*-oxidation product. Again, hemoglobin adducts were found with this compound particularly in weak acetylators. This example demonstrates in humans that the bioavailability of potentially hazardous arylamine components hidden in more complex structures can be measured.

4.2. Lifestyle Exposures

Cigarette smoke contains arylamines, and the level of 4-aminobiphenyl–hemoglobin adducts in the blood of smokers and nonsmokers has been compared. The mean value of 154 pg 4-aminobiphenyl/g hemoglobin was significantly higher in smokers (20–50 cigarettes/day) than in nonsmokers (28 pg/g hemoglobin; Bryant *et al.*, 1987). In this example, comparison on a group basis helped to detect environmental sources of carcinogens. Extending this principle, additional arylamines were considered and a trend to higher adduct levels in smokers was also observed for *o*- and *p*-toluidine, 2-naphthylamine, and 3-aminobiphenyl. *m*-Toluidine adducts were also present but clearly not associated with smoking. Differences in mean adduct levels seem to exist in populations from different geographical areas (Skipper *et al.*, 1988). The source of these amines in nonsmokers remains obscure. Passive smoking is one source; environmental exposure to the respective nitro-arenes is another possibility.

These data demonstrate the potential of measuring hemoglobin adducts for the detection of sources for nonoccupational exposure to carcinogens. Such background or lifestyle exposures must be accounted for when acceptable levels of occupational exposures are discussed. On the other hand, the variance of data must be investigated. Individual variation, analytical problems, and interlaboratory variation may contribute (see Farmer *et al.*, 1987).

5. EXPOSURE TO GENOTOXIC CHEMICALS AND THE ASSESSMENT OF CANCER RISK

The application of biological monitoring methods to genotoxic compounds is hampered by the difficulty in translating the obtainable data into cancer risk. The question of what the client should be told if macromolecular adducts are detected in his or her blood sample is difficult to answer. Therefore the prerequisites for employing these methods must be carefully defined (Ashby, 1988).

5.1. Protein Binding versus DNA Binding

With regard to the analysis of hemoglobin adducts, one of the issues to be discussed is the following: are DNA adducts better substitutes for the tissue dose of genotoxic chemicals than protein adducts? If blood samples are the source for analysis of adducts, neither lymphocytes (for DNA) nor erythrocytes (for hemoglobin) represent the target for tumor-initiating events. In both cases, the adducts serve as an indirect measure for the tissue dose of reactive metabolites. This tissue dose may or may not be representative for the dose in the target tissue of tumor formation. With ethylene oxide, a rather long-lived electrophile that is distributed rather equally in different compartments, it was possible to predict the level of DNA-adduct formation in other tissues rather precisely from the erythrocyte dose (Ehrenberg *et al.*, 1974). With other chemicals that generate more reactive and short-lived intermediates, this is less likely to be the case. The type of electrophile formed determines the reaction rate with the different soft and hard cellular nucleophiles. This not only influences the pattern of DNA adducts generated but also the relationship between DNA and protein binding, and must be considered for each individual chemical and its possibly different ultimate metabolites. Moreover, some of the ultimate metabolites may be formed within the target cell or within the target tissues so that distribution of proximate and ultimate reactants into the circulation and into different compartments becomes a rather complex problem. However, from the limited information available, it may be concluded that the initial DNA-adduct levels in different tissues vary less than one would expect. With benzo(a)pyrene (Stowers and Anderson, 1984) and *trans*-4-acetylaminostilbene (Hilpert and Neumann, 1985) DNA-adduct levels varied within 1 order of magnitude between the most- and the least-exposed tissue measured.

The advantage of measuring a specific DNA adduct in lymphocytes is that it may be identical with the promutagenic lesion generated in the target tissue and therefore most representative for the genotoxic effect. As outlined before, hemoglobin binding nitrosoarenes are not the ultimate metabolites reacting with DNA, but as long as first-order kinetics prevail, their concentration is directly linked to that of DNA-binding hydroxylamines or their conjugates. If these are genotoxic, the hemoglobin adduct in erythrocytes is as indicative for the presence of genotoxic electrophiles in target tissues as the DNA adduct in lymphocytes. Moreover, the persistence of hemoglobin adducts may be more predictable than that of DNA adducts. Repair does not interfere, and the fate of erythrocytes in the circulation is less variable. With some established carcinogens DNA adducts could not even be detected in lymphocytes; for example, 2-acetylaminofluorene (Willems *et al.*, 1987) and aflatoxin B_1.

It should be emphasized that proving the genotoxic potential of a chemical is not the purpose of hemoglobin-adduct measurements. This is rather a means of determining the tissue dose of a critical metabolite of more or less genotoxic N-substituted arenes *in vivo*. It will therefore soon become instead questions of technical feasibility, required sensitivity, and cost that may contribute to the decision about which analytical tool to use. The conclusions about the exposure to the chemical in question are ultimately the same with both types of adducts.

5.2. The Role of Macromolecular Damage in the Multistage Concept of Carcinogenesis

With ethylene oxide a linear relationship has been established between dose and (1) mutation rate, (2) tumor incidence in salivary glands and brains of female rats, (3) DNA binding in rat liver, a nontarget tissue, and (4) initiation of rat liver enzyme–altered foci in an initiation promotion experiment (Ehrenberg *et al.*, 1983, Denk *et al.*, 1988). These data strongly support the concept that the tissue dose of a genotoxic chemical correlates well with DNA damage and the incidence of tumor formation. They also provide a solid basis for the rad-equivalence concept and the prospective risk model proposed by Ehrenberg *et al.* (1983). And yet, it seems fair to say that risk estimates given on the basis of known amounts of ultimate carcinogens in the target tissue are still not possible (see Zielhuis and Henderson, 1986). This is not only due to the lack of knowledge about the specific type and the ultimate extent of genetically relevant lesions produced by a known tissue dose, but also about the multistage nature of tumor formation (Fig. 6).

The tumor incidence obtained with chemicals of low genotoxic potentials may depend strongly on secondary effects affecting, for instance, the promotion phase. Tissue-irritating and acute toxic effects of a chemical may govern the slope of the dose–response relationship (formaldehyde; Lutz, 1987). Complex correlations between DNA adducts and tumor incidence were reported for the generation of O^4-EtdT (ethyl-deoxythymidine) from diethylnitrosamine and rat liver tumors (Swenberg *et al.*, 1987) and DNA adducts generated by aminofluorene and mouse liver tumors (Beland *et al.*, 1988).

Moreover, there are chemicals that produce strong genotoxic effects related to measurable DNA adducts and yet have no (or very little) potential for producing tumors in these tissues unless supplemented by promoting effects (see Neumann, 1986). Thus the extent of initial or persistent DNA damage produced by a certain tissue dose of reactive metabolites does not correlate well with tumor formation in many cases, and DNA binding cannot be expected to be a reliable predictor of risk.

This complex situation is underlined by the difficulty in predicting the carcinogenic potential of monocyclic aromatic amines and nitro-arenes from *in vitro* and *in vivo* short-term tests. Most of them are genotoxic. However,

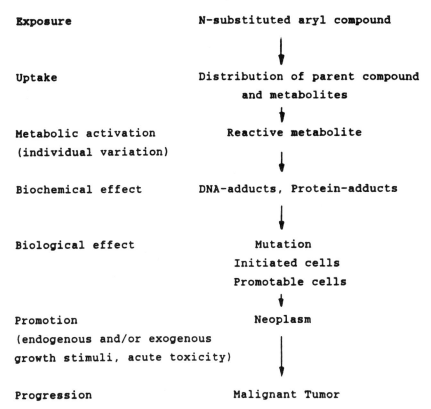

Figure 6. The position of biochemical effects in chemical carcinogenesis.

there seems to be no *in vitro* short-term test that predicts satisfactorily the genotoxic or carcinogenic properties. The tissue-, sex-, and species-specific effects are perplexing. Many of the inconsistencies may have pharmacokinetic reasons, i.e., they could possibly be explained if the target dose of the ultimate metabolites were known. However, in addition, acute toxic effects seem to play an important role in tumor development, and they may not all be related to genotoxic metabolites. The role of nongenotoxic effects with toxic or subtoxic doses on the biochemical level is only now being considered in a very preliminary stage.

The available information suggests that the tissue dose of genotoxic metabolites correlates well with the initiation of tumor cells but not necessarily with time to tumor and tumor incidence. In other words, DNA damage seems to be a necessary but not always sufficient prerequisite for tumor formation. This applies particularly to the human situation in which numerous unknown exogenous and endogenous factors may modify the development of tumors. There must be further developments in molecular epidemiology with respect to biological monitoring before the available data can be used sensibly for risk estimates.

5.3. The Role of Biochemical Effect Monitoring

In order to separate the end points under consideration, the term *biological monitoring* should be more clearly defined. Originally the term was introduced to indicate the difference between external exposure as measured by environmental monitoring and internal exposure as measured by analyzing the chemical in biological samples such as blood or urine. Later, it became necessary to distinguish between the actual uptake of the chemical (internal exposure) and the effects elicited by its metabolites, which are related to acute or chronic toxicity and which are subject to modification by numerous variables inherent in the individual. These can be biological effects, such as methemoglobinemia or point mutations in lymphocyte DNA or hemoglobin. If they are measured, the procedure should be called *biological* effect monitoring. The results can be interpreted directly in terms of health hazards. Protein or even DNA adducts represent primary biochemical effects and as such do not have disease character. If they are measured, the procedure should be called *biochemical* effect monitoring. This distinction should be very helpful until we know more about the relationship between macromolecular damage and tumor incidence.

5.4. The Potential of Improved Exposure Control by Biochemical Effect Monitoring

The limited experience with measuring arylamine–hemoglobin adducts clearly demonstrates progress in going from environmental monitoring to biological monitoring and then to biochemical effect monitoring in exposure control of compounds with established threshold limit values. In the occupational situation it means a change from control of the workplace to exposure control of the individual at the workplace. If the limits are exceeded, appropriate measures can be taken. For established carcinogens safe limits do not exist and therefore an all-or-none criterion for the test cannot be given if the nonthreshold hypothesis for genotoxic effects is accepted. This means that a yes or no answer cannot be given to the client. However, it seems possible to agree upon positive test criteria if exposure limits have been established, such as threshold limit values (TLV) (U.S.) or TRK (technical guiding concentrations) values (F.R.G.) from which exposure equivalence values can be delineated.

If such limits are exceeded on a group basis, external exposure must be reduced. Test results, however, should also be assessed on an individual basis with the aim of identifying increased susceptibility, for instance, due to genetic polymorphism, or to become aware of unknown exposures. The association of genetic polymorphism with increased susceptibility, however, must be treated with great care, and premature conclusions should be avoided. The increased risk of slow acetylators to develop bladder tumors from exposures to aromatic amines does not necessarily mean that rapid acetylators are less susceptible to tumor formation under these circumstances. There are reports indicating that rapid acetylators instead are at higher risk to develop colon tumors (Weber *et al.*, 1988).

Background and lifestyle exposures will frequently be observed. Since historical control data do not exist, control populations should be identified to account for these exposures. As such data become available, a more rational basis will develop for reducing environmental exposures and setting acceptable limits for occupational exposures. If applicable, the individual should be informed about an increased risk rather than a positive or negative test result, which requires certainly a good deal more of explanation and education.

It seems inevitable that this stage of improved exposure control, which should include well-planned prospective studies of molecular epidemiology, will proceed from limited statements about relative risk to satisfactory risk assessment.

REFERENCES

W. Albrecht and H.-G. Neumann, Covalent binding of benzidine to rat hemoglobin, *Naunyn-Schmiedeberg's Arch. Pharmacol. Suppl.* **325**, R21 (1984).

W. Albrecht and H.-G. Neumann, Biomonitoring of aniline and nitrobenzene, hemoglobin binding in rats and analysis of adducts, *Arch. Toxicol.*, **57**, 1–5 (1985).

J. Ashby, Comparison of techniques for monitoring human exposure to genotoxic chemicals, *Mutat. Res.* **204**, 543–551 (1988).

F. A. Beland and F. F. Kadlubar, Formation and persistence of arylamine DNA adducts *in vivo*, *Environ. Health Perspect.* **62**, 19–30 (1985).

F. A. Beland, N. F. Fullerton, T. Kinouchi, A. T. Lopez, W. A. Korfmacher, and M. C. Poirier, Macromolecular adduct formation in relation to tumorigenesis in mice continuously fed 2-acetylaminofluorene at multiple dose levels. *Proc. Ann. Meet. Am. Assoc. Cancer Res.* **29**, A377 (1988).

G. Birner and H.-G. Neumann, Biomonitoring of aromatic amines. II. Hemoglobin binding of some monocyclic aromatic amines, *Arch. Toxicol.*, **62**, 110–115 (1988).

H. M. Bolt, H.-G. Neumann, and J. Lewalter, Zur Problematik von BAT-Werten für aromatische Amine, *Arbeitsmed. Sozialmed. Präventivmed.* **20**, 197–201 (1985).

E. Boyland, D. Manson, and R. Nery, The reaction of phenylhydroxylamine and 2-naphthylhydroxylamine with thiols, *J. Chem. Soc.* 606–611 (1962).

M. S. Bryant, P. L. Skipper, S. R. Tannenbaum, and M. Maclure, Hemoglobin adducts of 4-aminobiphenyl in smokers and non-smokers, *Cancer Res.* **47**, 602–608 (1987).

B. Denk, J. G. Filser, D. Oesterle, E. Deml, and H. Greim, Inhaled ethylene oxide induces preneoplastic foci in rat liver, *J. Cancer Res. Clin. Oncol.* **114**, 35–38 (1988).

B. Dölle, W. Töpner, and H.-G. Neumann, Reaction of arylnitroso compounds with mercaptanes, *Xenobiotica* **10**, 527–536 (1980).

L. Ehrenberg, K. D. Hiesche, S. Osterman-Golkar, and I. Wennberg, Evaluation of genetic risk of alkylating agents: tissue dose in the mouse from air contaminated with ethylene oxide, *Mutat. Res.* **24**, 83–103 (1974).

L. Ehrenberg, E. Moustacchi, S. Osterman-Golkar, with an Appendix by G. Ekman, Dosimetry of genotoxic agents and dose-response relationships of their effects. *Mutat. Res.* **123**, 121–182 (1983).

P. Eyer, Reactions of nitrosobenzene with reduced glutathione, *Chem. Biol. Interact.* **24**, 227–239 (1979).

P. Eyer, Reactions of nitrosoarenes with sulphydryl groups: Reaction mechanism and biological significance. In: *Biological Oxidation of Nitrogen in Organic Molecules,* eds. J. W. Gorrod and L. A. Damani, Ellis Horwood, Chichester and VCH Verlagsgesellschaft, Weinheim, pp. 386–399 (1985).

P. B. Farmer, H.-G. Neumann, and D. Henschler, Estimation of exposure of man to substances reacting covalently with macromolecules, *Arch. Toxicol.* **60**, 251–260 (1987).

J. Forsberg-Karlsson, L. Ehrenberg, S. Osterman-Golkar, and S. Hussain, Estimation of risks of genotoxic effects of benzo(a)pyrene, *The Coal-Health-Environment Project Technical Report No. 80* (1983).

H. R. Glatt, F. Oesch, and H.-G. Neumann, Factors responsible for the metabolic formation and inactivation of bacterial mutagens from trans-4-acetylaminostilbene, *Mutat. Res.* **73**, 237–250 (1980).

M. W. Goldblatt, A. F. Henson, and A. R. Sommerville, Metabolism of bladder carcinogens. 3. The metabolic path of 2-(8-^{14}C)-naphthylamine in several animal species, *Biochem. J.* **77**, 511–516 (1960).

L. C. Green, P. L. Skipper, R. J. Turesky, M. S. Bryant, and S. R. Tannenbaum, *In vivo* dosimetry of 4-aminobiphenyl in rat via a cysteine adduct in hemoglobin, *Cancer Res.* **44**, 4254–4259 (1984).

U. Groth and H.-G. Neumann, The relevance of chemico-biological interactions for the toxic and carcinogenic effects of aromatic amines. V. The pharmacokinetics of related aromatic amines in blood, *Chem. Biol. Interact.* **4**, 409–419 (1971).

D. Hilpert and H.-G. Neumann, Accumulation and elimination of macromolecular lesions in susceptible and non-susceptible rat tissues after repeated administration of trans-4-acetylaminostilbene, *Chem. Biol. Interact.* **54**, 85–95 (1985).

H. Jackson, Studies with erythrocytes labelled with radioactive p-iodophenylhydroxylamine, *Biochem. J.* **57**, 619 (1953).

M. Kiese and H. Uehleke, Der Ort der N-Oxidation des Anilins im höheren Tier, Naunyn-Schmiedeberg's *Arch. Exp. Pathol. Pharmakol.* **242**, 117–129 (1961).

E. Kriek, L. DenEngelse, E. Scherer, and G. Westra, Formation of DNA modifications by chemical carcinogens, identification, localization and quantification, *Biochim. Biophys. Acta* **738**, 181–201 (1984).

J. Lewalter and U. Korallus, Blood protein conjugates and acetylation of aromatic amines. New findings on biological monitoring, *Int. Arch. Occup. Environ. Health* **56**, 179–196 (1985).

J. Lewalter and U. Korallus, Erythrocyte protein conjugates as a principle of biological monitoring for pesticides, *Toxicol. Lett.* **33**, 153–165 (1986).

W. K. Lutz, *In vivo* covalent binding of organic chemicals to DNA as quantitative indicator in the process of chemical carcinogenesis, *Mutat. Res.* **65**, 289–365 (1979).

W. K. Lutz, Quantitative evaluation of DNA-binding data *in vivo* for low-dose extrapolations, *Arch. Toxicol. Suppl.* **11**, 66–74 (1987).

E. C. Miller and J. A. Miller, The metabolic activation of carcinogenic aromatic amines and amides, *Prog. Exp. Tumor Res.* **11**, 273–301 (1969).

G. J. Mulder, F. F. Kadlubar, J. B. Mays, and J. A. Hinson, Reaction of mutagenic phenacetin metabolites with glutathione and DNA, possible implications for toxicity, *Molec. Pharmacol.* **26**, 342–347 (1984).

G. J. Mulder, L. E. Unruh, F. E. Evans, B. Ketterer, and F. F. Kadlubar, Formation and identification of glutathione conjugates from 2-nitrosofluorene and N-hydroxy-2-aminofluorene, *Chem. Biol. Interact.* **39**, 111–127 (1982).

H.-G. Neumann, Significance of metabolic activation and binding to nucleic acids of aminostilbene derivatives *in vivo, National Cancer Institute Monograph No. 58*, pp. 165–171 (1979).

H.-G. Neumann, Dose-response relationship in the primary lesion of strong electrophilic carcinogens, *Arch. Toxicol. Suppl.* **3**, 69–77 (1980).

H.-G. Neumann, The dose dependence of DNA interactions of aminostilbene derivatives and other chemical carcinogens. In: *Developments in the Science and Practice of Toxicology*, eds. A. W. Hayes, R. C. Schnell, and T. S. Miya, Elsevier Science Pubishers, Amsterdam, pp. 135–144 (1983).

H.-G. Neumann, Analysis of hemoglobin as a dose monitor for alkylating and arylating agents, *Arch. Toxicol.* **56**, 1–6 (1984a).

H.-G. Neumann, Dosimetry and dose-response relationships. In: *Monitoring Human Exposure to Carcinogenic and Mutagenic Agents*, eds. A. Berlin, M. Draper, K. Hemminki, and H. Vainio, IARC Scientific Publication No. 59, International Agency for Research on Cancer, Lyon, France, pp. 115–126 (1984b).

H.-G. Neumann, Toxication mechanisms in drug metabolism, *Advances in Drug Research*, ed. B. Testa, Vol. 15, Academic Press, London, pp. 1–28 (1986).

H.-G. Neumann, Biomonitoring of aromatic amines and alkylating agents by measuring hemoglobin adducts, *Int. Arch. Occup. Environ. Health* **60**, 151–155 (1988).

H.-G. Neumann, B. J. M. Gaugler, and W. Taupp, The metabolic activation of trans-4-dimethylaminostilbene after oral administration of doses ranging from 0.025 to 250 μmol/kg. In: *Proceedings of the First International Congress on Toxicology*, eds. G. L. Plaa and W. A. M. Duncan, Academic Press, New York, pp. 177–190 (1978).

M. A. Pereira and L. W. Chang, Binding of chemical carcinogens and mutagens to rat hemoglobin, *Chem. Biol. Interact.* **33**, 301–305 (1981).

L. A. Poirier, J. A. Miller, E. C. Miller, and K. Sato, N-Benzoyloxy-N-methyl-4-aminoazobenzene: its carcinogenic activity in the rat and its reactions with proteins and nucleic acids and their constituents *in vitro, Cancer Res.* **27**, 1600–1613 (1967).

J. L. Radomski and T. J. Mellinger, The absorption, fate and excretion in rats of the water-soluble azo-dyes FD&C red No.2, FD&C red No.4, and FD&C yellow No.6, *J. Pharmacol. Exp. Ther.* **136**, 259–266 (1962).

L. Rehn, Blasengeschwülste bei Fuchsin-Arbeitern, *Arch. Klin. Chirurgie* **50**, 588–600 (1895).

D. Ringe, R. J. Turesky, P. L. Skipper, and S. R. Tannenbaum, Structure of the single stable hemoglobin adduct formed by 4-aminobiphenyl *in vivo, Chem. Res. Toxicol.* **1**, 22–24 (1988).

H. K. Rjosk and H.-G. Neumann, Zur Bedeutung chemisch-biologischer. Wechsel-wirkungen für die toxische und krebserzeugende Wirkung aromatischer Amine. II. Verteilung der Radioaktivität nach Applikation des Tritium-markierten Carcinogens trans-4-Dimethylaminostilben und der beiden unwirksamen Vergleichssubstanzen cis-4-Dimethylaminostilbene und 4-Dimethylaminobibenzyl in der Ratte, *Z. Krebsforsch.* **75**, 209–220 (1971).

D. Segerbäck, *In vivo* dosimetry of some alkylating agents as a basis for risk assessment, Ph.D. Dissertation, University of Stockholm (1985).

P. L. Skipper, M. S. Bryant, and S. R. Tannenbaum, Determination of human exposure to carcinogenic aromatic amines from hemoglobin adducts in selected population groups. In: *Carcinogenic and Mutagenic Responses to Aromatic Amines and Nitroarenes,* eds. C. M. King, L. J. Romano, and D. Schuetzle, Elsevier, New York, pp. 65–71 (1988).

A. R. Sommerville, A. F. Henson, M. E. Cooke, M. E. Farquarson, and M. W. Goldblatt, Metabolism of bladder carcinogens, *Biochem. J.* **63**, 290–294 (1956).

S. J. Stowers and M. W. Anderson, Ubiquitous binding of benzo(a)pyrene metabolites to DNA and protein in tissues of the mouse and rabbit, *Chem. Biol. Interact.* **51**, 151–166 (1984).

J. A. Swenberg, F. C. Richardson, L. Tyeryar, F. Deal, and J. Boucheron, The molecular dosimetry of DNA adducts formed by continuous exposure of rats to alkylating hepatocarcinogens, *Prog. Exp. Tumor Res.* **31**, 42–51 (1987).

W. Töpner, Zur Aktivierung krebserzeugender aromatischer Amie im Stoffwechsel, Proteingebundene Metaboliten als indirektes Mass für die Konzentration reaktionsfähiger Stoffwechselprodukte in verschiedenen Geweben, Ph.D. Dissertation, University of Würzburg (1975).

W. W. Weber, S. S. Mattano, and G. N. Levy, Acetylation pharmacogenetics and aromatic amine-induced cancer. In: *Carcinogenic and Mutagenic Responses to Aromatic Amines and Nitroarenes,* eds. C. M. King, L. J. Romano, and D. Schuetzle, Elsevier, New York, pp. 115–123 (1988).

J. H. Weisburger, P. H. Grantham, and E. K. Weisburger, Transport of carcinogens: rat blood plasma and red cell binding of isotope after N-hydroxy-N-2-fluorenyl-lacetamide, *Life Sci.* **5**, 41–45 (1966).

J. H. Weisburger, P. H. Grantham, and E. K. Weisburger, The transport of chemical carcinogens by blood. Metabolites of N-2-fluorenylacetamide and N-hydroxy-N-2-fluorenylacetamide as a function of time. In: *The Jerusalem Symposia on Quantum Chemistry and Biochemistry,* eds. E. D. Bergman and B. Pullman, Vol. 1, *Physico-Chemical Mechanisms of Carcinogenesis,* Israel Acad. Sci. Humanities, Jerusalem, pp. 262–283 (1969).

M. I. Willems, W. K. deRaat, R. A. Baan, J. W. Wilmer, M. J. Lansbergen, and P. H. Lohman, Monitoring the exposure of rats to 2-acetylaminofluorene by the estimation of mutagenic activity in excreta, sister-chromatid exchanges in peripheral blood cells and DNA adducts in peripheral blood, liver and spleen, *Mutat. Res.* **176**, 211–223 (1987).

R. L. Zielhuis and P. T. Henderson, Definitions of monitoring activities and their relevance for the practice of occupational health, *Int. Arch. Occup. Environ. Health* **57**, 249–257 (1986).

Ethical, Economic and Legal Considerations

20

The Bioethics of Biomonitoring People for Exposure to Carcinogens

George J. Annas
Boston University Schools of Medicine and Public Health
Boston, Massachusetts

Traditionally, individuals seek medical care when they are sick or injured. Public health practitioners, on the other hand, attempt to prevent the injury or illness from occurring. One primary method is to actively seek out asymptomatic individuals, identify those at risk, and intervene to try to prevent harm. The most powerful tool in identifying at-risk individuals is screening. Even though they are usually performed with only beneficent intent and are widely accepted by the public, screening tests raise serious questions of autonomy, stigmatization, confidentiality, informed consent, and efficacy (Elias and Annas, 1987).

This book focuses on detecting and/or monitoring biological markers that can identify exposure to carcinogens at the molecular level prior to the manifestation of disease. Such measurements could be of the internal dose present in the cells, tissue, or bodily fluids, or a biologically effective dose reflecting interaction with cellular macromolecules or identifying an irreversible biologic effect that is known to be linked pathogenically to cancer (Perera, 1990). Although such testing does not involve locating potential markers of susceptibility, such as genetic traits affecting carcinogen metabolism or DNA repair, many of the ethical issues raised by genetic screening are similar to those raised by biological screening. This chapter explores the ethics of the major forms of genetic and biological screening tests with a view toward identifying some basic ethical guidelines that all human population screening programs should follow, and opening a discussion on the bioethics of biomonitoring people for exposure to carcinogens that should be amplified and agreed upon before large-scale population screening commences.

It should be noted at the outset that attention to the ethical dimensions of screening programs is generally relegated to a very low level of concern, and almost always becomes an issue only *after* a particular screening program has been developed and problems occur. For example, serious discussion of the ethics of genetic screening in this country did not really begin until after blacks objected to mandatory sickle-cell screening prior to admission to grade school (Elias and Annas, 1987). The black community saw clearly that since there was no treatment for sickle-cell disease, screening could only lead to discrimination against those with the disease.

Likewise, the acquired immunodeficiency syndrome (AIDS) epidemic has brought with it calls for mandatory screening of various populations (the military, hospital patients, surgeons, prisoners, etc.) on the grounds (among others) that now since there are at least some treatments available that may delay the onset of the disease for those with human immunodeficiency virus (HIV) infection, the asymptomatic but infected individuals should be identified so that they can take advantage of treatment. This argument has thus far failed, primarily because the gay rights community has persuasively argued that there is such a severe stigma, accompanied by discrimination in employment, housing, insurance, and other areas, that only voluntary, anonymous testing should be relied upon by governmental agencies. The public health community has so far concurred (Levine and Bayer, 1989). In addition, there seems to be general agreement that no widespread routine or mandatory testing of HIV is indicated unless and until: (1) there is an almost completely accurate screening test, with no false positives; (2) there is effective counseling before and after testing; (3) there is a meaningful intervention that can protect the individual; (4) the individual can give informed consent to testing; (5) the results can be kept strictly confidential (with a possible exception of the individual's spouse or regular sex partner); and (6) the screening program has community support (Gostin *et al.,* 1986).

On the other hand, it should be noted that most of these issues apply only to *individual* screening where the results can be linked to a specific individual. Anonymous screening of groups in a manner in which the results cannot be linked to the individual screened do not raise concerns about confidentiality and discrimination, and thus such screening (such as screening all newborns for HIV infection by sending anonymous samples of blood to a central lab) raises far fewer serious ethical issues. Screening workers for biological changes based on exposure to carcinogens in a way that made identification of individual workers impossible would not raise major ethical problems provided, of course, that the test is safe and noninvasive and the result were made known to the population screened.

1. DRUG SCREENING TESTS IN THE U.S. SUPREME COURT

HIV screening is the most publicized and debated screening test now in wide use, but it pales when compared to the more routine and pervasive biological screening of individuals to identify those who have been taking illicit drugs. This type of screening test is so pervasive that there are dozens of lawsuits now in the courts challenging its use, and two major cases on drug screening in the employment context have already been heard by the U.S. Supreme Court. These cases are worth reviewing not only because they show how little protection workers in the public sector can expect from the U.S. Constitution (workers in the private sector are not, of course, covered by the Constitution directly, since it only restricts government actions), but also because it dramatically illustrates how routine the justices of the U.S. Supreme Court see the screening of blood and urine collected from specific individuals to determine what activities they may have been engaged in. Of course, if collecting biological samples to find evidence to be used *against* individuals is acceptable, collecting such samples to benefit individuals will be even more acceptable.

The first case involved regulations promulgated by the Secretary of Transportation under the Federal Railroad Safety Act of 1970 [*Skinner* v. *Railway Labor Exec. Assoc.* 109 S.Ct. 1402 (1989)]. These regulations, announced in 1985, were based in part on a 1979 study that concluded that 23% of railroad operating personnel were "problem drinkers" and statistics that showed that from 1975 to 1983 there were 45 train accidents involving "errors of alcohol- and drug-impaired employees" resulting in 34 deaths and 66 injuries. The regulations required the collection of blood and urine samples for toxicological testing of railroad employees following any accident resulting in a fatality or in a release of hazardous materials or railroad property damage of $500,000, and any collision resulting in a reportable injury or damage to railroad property of $50,000. After such an event, all crew members are to be sent to an independent medical facility where both blood and urine samples will be obtained from them in an attempt to determine the cause of the incident. Employees who refuse testing may not perform their job for nine months. Under another provision, the railroad *may* perform breath and urine tests on individuals whom it has "reasonable suspicion" are under the influence of drugs or alcohol while on the job.

The second case involved the Customs Service. In 1986 the Commissioner of Customs announced the implementation of a drug-testing program for certain customs officials, finding that although "Customs is largely drug-free . . . unfortunately no segment of society is immune from the threat of illegal drug use" [*National Treasury Employee Union* v. *Von Rabb*, 109 S.Ct. 1384 (1989)]. Drug testing was made a condition of placement or employment for positions that met one of three criteria: positions that have direct involvement in drug interdiction or enforcement of drug laws, that require the carrying of firearms, or that require handling classified material. After an employee qualifies for a position covered by the rule, he or she is notified by letter that final selection is contingent upon successful completion of drug screening. An independent drug testing company contacts the employee and makes arrangements for the urine test. The employee is required to remove outer garments and personal belongings, but may produce the sample behind a partition or in the privacy of a bathroom stall. A monitor of the same sex, however, is required to remain close at hand "to listen for the normal sounds of urination," and dye is added to the toilet water to prevent adulteration of the sample. The sample is then tested for the presence of marijuana, cocaine, opiates, amphetamines, and phencyclidine. Confirmed positive results are transmitted to the medical review officer of the agency and can result in dismissal from the Customs Service. Test results may not, however, be turned over to any other agency, including criminal prosecutors, without the employees' written consent.

2. THE FOURTH AMENDMENT

The Fourth Amendment provides:

> The right of the people to be secure in the persons, houses, papers, and effects against unreasonable searches and seizures, shall issue not be violated, and no Warrants shall issue, but upon probable cause, supported by Oath or affirmation, and particularly describing the place to be searched, and the person or things to be seized.

Constitutional protections, of course, apply only to acts of the government. The Court had no trouble concluding, however, that the railroad was acting as an instrument of the government. Of course, tests carried out on Customs Service employees, done pursuant to government rule, are covered by the Constitution.

The Court, in opinions written by Justice Anthony Kennedy, easily found that blood and urine tests are "searches" under the Fourth Amendment. As to taking a blood sample, the court noted, "this physical intrusion, penetrating beneath the skin, infringes an expectation of privacy that society is prepared to recognize as reasonable." Moreover, the chemical analysis of the blood "is a further invasion of the tested employee's privacy interests." Breath tests, the court found, implicated "similar concerns about bodily integrity and . . . should also be deemed a search." And although urine tests do not entail a surgical intrusion, "analysis of urine, like that of blood, can reveal a host of private medical facts about an employee, including whether she is epileptic, pregnant, or diabetic." The collection of urine also intrudes upon the reasonable expectation of privacy, especially when accompanied by a visual or aural monitor, and the Court amazingly and perhaps prudishly seems to think that having one's urination monitored is a greater violation of privacy than having a needle penetrate one's skin to remove blood.

Is this the type of search that requires a warrant, and if not, is this type of search "reasonable"? The Court concluded that no warrant is required for two primary reasons: (1) there are virtually no facts for a neutral magistrate to evaluate (because both the circumstances justifying the search and the limits of the search are "defined narrowly and specified in the regulations that authorize them"), and (2) the delay needed to procure a warrant might result in the destruction of valuable evidence as the drugs and/or alcohol metabolize in the employee. Warrantless searches have traditionally required at least some showing of "individualized suspicion"; but the court made it clear that "a search may be reasonable despite the absence of such suspicion" at least if the interference is "minimal" and it takes place in the employment context. This is because "[o]rdinarily, an employee consents to significant restrictions on his freedom of movement where necessary for his employment, and few are free to come and go as they please during working hours."

Blood tests are not unreasonable because they are usually "taken by a physician . . . according to accepted medical practice" and are "safe, painless, and commonplace." Breath tests, although not done by medical personnel, involve no piercing of the skin and can be done "with a minimum of inconvenience or embarrassment." Urine testing was seen as the most difficult to justify, because it involves "an excretory function traditionally shielded by great privacy." However, the railroad regulations do not require direct observation, and the test is done in a medical environment "by personnel unrelated to the railroad employer, and is thus not unlike similar procedures encountered often in the context of a regular physical examination."

More important, however, seems to have been the Court's view of the "diminished" expectation of privacy employees have when they enter certain occupations. The railroad industry was described as "regulated pervasively to ensure safety . . . ." The court notes that an "idle locomotive" is harmless, but "it becomes lethal when operated negligently by persons who are under the influence of alcohol or drugs." Customs employees are involved in drug interdiction and so "reasonably should expect effective inquiry into their fitness and probity." This is especially true, the court opined, because "drug abuse is one of the most serious problems confronting society today [and] the almost unique mission of the Service gives the Government a compelling interest in ensuring that many of these covered employees do not use drugs even off-duty, for such use creates risks of bribery and blackmail against which the Government is entitled to guard."

Unlike the case with railroad employees, however, there was virtually no evidence that drug use is a problem in the Customs Service. Only 5 of 3,600 screened employees tested positive for drug use. The court, however, termed this finding a "mere circumstance," concluding that the Customs program was designed *to prevent* the promotion of drug users to sensitive positions, as well as to detect such employees. In the Court's words:

> The Government's compelling interests in preventing the promotion of drug users to positions where they might endanger the integrity of our Nation's borders or the life of the citizenry outweigh the privacy interests of those who seek promotion to these positions, who enjoy a *diminished expectation of privacy by virtue of the special, and obvious, physical and ethical demands of those positions* (emphasis added).

Nonetheless, the Court sent back to the lower court the section of the rules that dealt with individuals who had access to "classified" material for further proceedings because the list of positions covered (which ran from attorney to messenger) seemed overly broad to meet the goals of the rule.

The railroad decision was a 7-to-2 opinion, with Justice Thurgood Marshall dissenting. Justice Marshall said the opinion gutted the Fourth Amendment. He agreed that "declaring a war on illegal drugs is good public policy," but concluded that "the first, and worst, casualty of the war will be the precious liberties of our citizens." He was especially critical of Court's reading into the Fourth Amendment an exception to the probable cause requirement when "special needs" of non-law-enforcement agencies make either warrants or probable cause inconvenient requirements. The Court's reasoning, he noted, equated past cases that dealt with minimal searches of a person's possessions, with a search of the person's body itself, and thereby widened the "special needs" exception without requiring *any* evidence of wrongdoing on the part of the person.

The most powerful and surprising dissenting voice, however, was raised by Justice Antonin Scalia in the 5-to-4 Customs case. Like the railroad case, the Customs case was driven by the "war on drugs" and "the Government's compelling interests in safety and in the integrity of our borders.'' As Justice Scalia argued, however, whereas there was evidence of drug and alcohol abuse resulting in accidents and injury in the railroad case, in the Customs case "neither frequency of use nor connection to harm is demonstrated or even likely.'' He concluded therefore that the "Customs Service rules are a kind of immolation of privacy and human dignity in symbolic opposition to drug use.'' Justice Scalia noted, for example, that the record discloses not even one incident of the speculated harms actually occurring, nor even one incident "in which the cause of bribe-taking, or of poor aim, or of unsympathetic law enforcement . . . was drug use.'' Instead, the evidence is that the Customs Service is "largely drug free.'' The regulations thus expose "vast numbers of public employees" to the "needless indignity"of urine screening simply so the Customs Service can "set an important example" in the country's fight against drugs. Justice Scalia properly concluded that such a solely symbolic justification is "unacceptable" when the Fourth Amendment is violated:

> . . . liberties cannot be the means of making a point; . . . symbolism, even symbolism for so worthy a cause as the abolition of unlawful drugs, cannot validate an otherwise unreasonable search . . . Those who lose . . . are not just the Customs Service employees, whose dignity is thus offended, but all of us—who suffer a coarsening of our national manners that ultimately give the Fourth Amendment its content, and who become subject to the administration of federal officials whose respect for our privacy can hardly be greater than the small respect they have been taught to have for their own.

Justices Marshall and Scalia both seem correct. The Fourth Amendment now affords citizens far less protection against governmental searches, and this will inevitably result in a lessening of respect for the bodily integrity of all citizens.

As Justice Scalia quite correctly noted, the opinions can immediately be extended to all those employees who carry firearms, and from there, to all employees whose jobs, if performed under the influence of drugs, could harm the public, including: "automobile drivers, operators of potentially dangerous equipment, construction workers, school crossing guards.''

The cases raise a generic issue about screening. If the government can compel screening to determine if an individual is using a substance whose use is forbidden by the criminal law, is there anything for which the government cannot mandate screening to protect the public? For example, what would prevent the government from requiring biological screening for exposure to carcinogens or genetic screening for susceptibility to various diseases if and when such screening becomes available, assuming the screened-for disease rendered the individual ''less qualified'' for the job or potentially dangerous on the job? Indeed, the Court seems almost to eager to view any screening test done by a physician as routine medical practice as *de facto* reasonable, without any real analysis of how intrusive such testing can be (Annas, 1989a). And, of course, if the government can constitutionally mandate such testing, private employers can as well.

3. THE HUMAN GENOME PROJECT

Unlike biological screening, genetic screening has consistently raised publicly debated issues of ethics. In this regard, the proposed mapping and sequencing of the estimated 3 billion base pairs of the human genome (the 50,000 to 100,000 genes we are composed of) raises the ethical issues involved in screening on a scale never before contemplated. There is money to be made in the human genome project, and even ''ethicists'' are currently slated to have a share. How should the ethical issues raised by the human genome project be debated and resolved?

The *Wall Street Journal* summarized the case for the human genome project in early 1989 when it editorialized, ''The techniques of gene identification, separation, and splicing now allow us to discover the basic causes of ailments and, thus, to progress toward cures and even precursory treatments that might ward off the onset of illness ranging from cancer to heart disease and AIDS.'' All that is lacking ''is a blueprint—a map of the human genome.'' Noting that some members of the European Parliament had suggested that ethical questions regarding eugenics should be answered ''*before it proceeds*,'' the *Journal* opined that ''This, of course, is a formula for making no progress at all.'' The editorial concluded, ''The Human Genome Initiative . . . may well invite attack from those who are fearful of or hostile to the future. It should also attract the active support of those willing to defend the future.''

The National Institutes of Health (NIH) have created an Office of Human Genome Research headed by James Watson, and that office has issued a request for funding proposals to study ''the ethical, social, and legal issues that may arise from the application of knowledge gained as a result of the Human Genome Initiative.'' The initial announcement made it clear that such projects are to be about the ''immense potential benefit to mankind'' of the project and focus on ''the best way to ensure that the information is used in the most beneficial and responsible manner.''

Watson is perhaps the genome project's most prominent cheerleader, having said, among other things, that the project provides ''an extraordinary potential for human betterment . . . We can have at our disposal the ultimate tool for understanding ourselves at the molecular level . . . The time to act is now.'' And, ''How can we not do it? We used to think our fate is in our stars. Now we know, in large measures, our fate is in our genes'' (*Time,* 20 March 1989, p. 62).

Are there any difficult legal and ethical problems involved in mapping the human genome, or is everything as uncontroversial as the project's advocates paint it? NIH plans to devote only 1% to 3% of its genome budget to exploring social, legal, and ethical issues, and James Watson sees no dangers ahead. But Watson himself, reflecting on his own early career, wrote in 1967: ''Science seldom proceeds in the straightforward, logical manner . . . its steps are often very human events in which personalities and cultural traditions play major roles'' (Watson, 1968).

The human genome project has been frequently compared to both the Manhattan Project and the Apollo Project, and ''big biology'' is clearly happy to have its own megaproject of a size formerly restricted to physicists and engineers. But the sheer size of these two other projects obscures more important lessons. The Manhattan Project is familiar, but it still teaches us volumes about science and the unforeseen impact of technological change. In late 1945, Robert Oppenheimer testified before the U.S. Congress on the role of science in the development of the atomic bomb:

> When you come right down to it, the reason that we did this job is because it was an organic necessity. If you are a scientist, you cannot stop such a thing. If you are a scientist, you believe that it is good to find out how the world works; that it is good to find what the realities are; that it is good to turn over to mankind at large the greatest possible power to control the world . . . (Rhodes, 1986)

The striking thing in Oppenheimer's testimony is his emphasis on the notion that science is unstoppable with the simultaneous insistence that its goal is *control* over nature, irreconcilable concepts that seem to appear equally at the heart of the human genome project. Of course, with the atomic bomb, control quickly became illusory. The bomb, which carries with it the promise of the total annihilation of mankind, has made the nation state ultimately unstable and put it at the mercy of every other nation with the bomb. Necessity has forced all nuclear powers to move, however slowly, toward a transnational community.

The Apollo Project had its own problems. An engineering exercise, it was about neither the inevitability of scientific advancement nor the control of nature. Instead, it was about military advantage and commercialism, disguised as science and hyped as a peace mission. As Walter McDougall has persuasively documented, the plaque Astronaut Paul Armstrong left on the moon, which read, ''We came in peace for all mankind,'' was ironic:

The moon was not what space was all about. It was about science, sometimes spectacular science, but mostly about spy satellites, comsats, and other orbital systems for military and commercial advantage. "Space for peace" could no more be engineered than social harmony, and the UN Outer Space Treaty . . . drew many nations into the hunt for advantage, not integration, through spaceflight. (McDougall, 1985)

The *Wall Street Journal* seems more attuned to the commercial applications of gene mapping and sequencing than the NIH, although Congressional support of the project is based primarily on the hope that mapping the genome can help the U.S. maintain its lead in the biotechnology industry. Neither ethicists nor social planners played any real role in the Manhattan or Apollo projects. It appears they will at least play some minor role in the Genome Project. What should the role be, and how should it be structured?

4. A SUMMARY OF ETHICAL ISSUES IN GENETIC SCREENING

The basic legal and ethical issues implicit in the human genome project are, on the first level, the same issues involved in current genetic screening for various traits, such as carrier status for sickle-cell and Tay–Sachs disease (Macklin, 1986). Mapping and sequencing the human genome could, of course, lead to screening on an almost unimaginable scale—not only for certain diseases and traits, but also for tendencies toward certain diseases, such as cancer or manic depression. When all genetic traits can be deciphered in a genetic code (something that will require far more than a simple map of location), we will enter a new realm—taking not simply a quantitative step, but a qualitative one. Exactly what the consequences of such a step will be are not entirely foreseeable, just as it was unforeseeable that the most lasting impact of the Apollo Project would be the photographs of the planet Earth: a fragile blue jewel that helped energize environmental protection on a global scale.

There will be issues of information control and privacy. Employers, insurance companies, the military, and the government, among others, will want to have access to the information contained in our genome. Scientists may want such information restricted, but they will certainly have little influence over its use, as they had little influence over the use of the atomic bomb. Routine genetic screening will be easy to justify under current law (which already mandates newborn screening in most states, for example) in many settings. Although we are utterly unprepared to deal with issues of mandatory screening, confidentiality, privacy, and discrimination, we will likely tell ourselves that we have already dealt with them well, and so the genome project poses no threat here.

The second order of issues relates to what is generally termed "eugenics," the improvement of the species, either by weeding out genetic "undesirables" or by actually using genetic techniques (breeding or genetic engineering) to increase the number of desirable traits in offspring. Given the U.S.'s sad history of involuntary sterilization, we are unlikely to engage in a direct program of sterilization. Nonetheless, eugenics has its supporters, and the European Parliament is right to be worried about the dangers of repeating history. It was our own U.S. Supreme Court, after all, that wrote in 1927, approving involuntary sterilization of the mentally retarded:

> We have seen more than once that the public welfare may call upon the best citizens for their lives. It would be strange if it could not call upon those who already sap the strength of the State for these lesser sacrifices often not felt to be such by those concerned, in order to prevent our being swamped with incompetence. It is better for all the world, if instead of waiting to execute degenerate offspring for crime, or to let them starve for their imbecility, society can prevent those who are manifestly unfit from continuing their kind (*Buck* v. *Bell*, 274 U.S. 200, 207; 1972).

That may seem like ancient history, but in 1988 the U.S. Congress's Office of Technology Assessment (OTA), in discussing the "Social and Ethical Considerations" raised by the Human Genome Project, used strikingly similar language:

> Human mating that proceeds without the use of genetic data about the risks of transmitting diseases will produce greater mortality and medical costs than if carriers of potentially deleterious genes are alerted to their status and encouraged to mate with noncarriers or to use artificial insemination or other reproductive strategies (OTA, 1988).

The primary reproductive strategy, mentioned only in passing in the report, will be genetic screening of human embryos—already technically feasible, but not nearly to the extent possible once the genome is understood. Such screening need not be required, people can be made to *want* it, even to insist on it as their right. As OTA notes, "New technologies for identifying traits and altering genes make it possible for eugenic goals to be achieved through technological as opposed to social control" (OTA, 1988).

A third level of concern relates to the fact that powerful technologies do not just change what human beings can do, they change the very way we think—especially about ourselves. A map of the human genome could also lead to a more narrowly focused view of a "normal" gene complement, and how much deviation we permit before considering any individual genome "abnormal," deviant, or diseased. We haven't seriously begun to think about *how* to think about this issue, even though we know normalcy will be invented, not discovered.

It is easy to dismiss this concern, but as two leading social critics have rightly warned, it would be a serious mistake:

> If biological tests are used to conform people to rigid institutional norms, we risk increasing the number of people defined as unemployable, uneducable or uninsurable. We risk creating a biologic underclass (Nelkin and Tancredi, 1989).

At a Workshop on International Cooperation for the Human Genome Project held in Valencia in October, 1988, French researcher Jean Dausset suggested that the genome project posed great potential hazards that could open the door to Nazi-like atrocities. To attempt to avoid such results, he suggested that the conferees agree on a moratorium on genetic manipulation of germ line cells, and a ban on gene-transfer experiments in early embryos. Reportedly, the proposal won wide agreement among the participants, and was watered down to a resolution calling for "international cooperation" only after American participant Norton Zinder successfully argued that the group had no authority to enforce such a resolution.

Zinder was, of course, correct, and a moratorium and ban on research that no one wants to do at this point would have only symbolic value—and negative symbolic value at that. It would signal that the scientists could handle the ethical issues alone, and could monitor their own work. It would tend to quiet the discussion of both germ line research and gene transfers in early embryos—both subjects that deserve wide public debate. But Dausset also had a point. The Nazi atrocities grew out of the combination of a public health ethic that saw the abnormal as disposable, and a tyrannical dictatorship that was able to give the physicians and public health authorities unlimited authority to put their program into bestial practice (Proctor, 1988).

Ethics is generally taken seriously by physicians and scientists only when it either fosters their agenda or does not interfere with it. If it cautions a slower pace or a more deliberate consideration of science's darker side, it is dismissed as "fearful of the future," anti-intellectual, or simply uninformed. The genome project has been overhyped and oversold. It is the obligation of those who take legal and ethical issues seriously to ensure that the dangers, as well as the opportunities, are rigorously and publicly explored (Annas, 1989b).

5. LESSONS FROM OTHER SCREENING PROGRAMS

Screening programs are well-accepted public health interventions. They can also be used for personal reasons, such as prenatal screening for genetic abnormalities and for criminal law enforcement, such as drug screening. Massive potential screening programs, such as those implied in the human genome project, are proceeding rapidly before any serious work has been started on its ethical implications. These experiences suggest some lessons that are applicable to biological screening for carcinogen exposure where the results of the testing can be linked to the individual tested. The following guidelines are suggested to help protect the individual rights of those being screened.

1. Ethical issues should be fully explored *before* the screening program commences. In determining whether a screening program should be initiated on any population for which identifiable information will be collected, the following factors should be considered: (a) the frequency and severity of the condition, (b) the availability of an intervention with documented efficacy, (c) the extent to which detection by screening improves the outcome, (d) the validity and safety of the screening test, (e) the adequacy of the resources to ensure effective screening, prior informed consent, and follow-up treatment and counseling, (f) the cost-effectiveness of the screening, and (g) the acceptance of the screening by the group to be screened. For employees, screening should be restricted to job-related information. The optimal screening program will be initiated only after adequate education of those to be screened and the support of those to be screened (Holtzman, 1977; see also Ashford, 1986 and Rothstein, 1989).
2. To protect personal dignity and autonomy, screening should only be done on an individual with informed consent. Specifically, individuals should be informed of the nature of the screening test, what it is designed to detect, what the implications of detection or nondetection are, what can be done to help the individual if the test is positive, and what negative consequences, including loss of employment, insurance, etc., could follow a positive test result. This means that individuals should have a right to refuse to be screened (Elias and Annas, 1987). In exceptional cases where screening is mandatory, this should be spelled out in a clear statement prior to employment, and the screening procedures and tests should be agreed upon jointly by labor and management.
3. To protect individuals from discrimination, the results of the screening tests should be confidential, but should routinely be made available to the individual, and with the individual's consent, to the

individual's private physician. Information that can be linked to a specific individual should be made available to management only with the consent of the individual, unless the information indicates that the individual is a danger to other workers (Elias and Annas, 1987).

4. Counseling should be available both before and after screening, and the resources for any reasonable intervention that can benefit the screenee should be in place and available to the screenee before screening is offered. This means that sufficient qualified personnel and laboratory facilities for required testing and follow-up treatment must be available (Elias and Annas, 1987).

Taking ethics and human rights seriously demands that we address the major ethical and legal issues raised by population screening for exposure to carcinogens *before* screening programs are initiated. This is a novel idea, but we discount ethics and human rights at our peril.

REFERENCES

N. Ashford, Medical screening in the workplace: Legal and ethical considerations, *Sem. Occup. Med.* **1**, 67–79 (1986).

G. J. Annas, Crack, symbolism and the constitution, *Hastings Center Report* **19** (4), 35–37 (1989a).

G. J. Annas, Who's afraid of the human genome?, *Hastings Center Report* **19** (5), 19–21 (1989b).

S. Elias and G. J. Annas, *Reproductive Genetics and the Law,* Yearbook Medical Publication, Chicago (1987).

L. Gostin, W. Curran, and L. Clark, The case against compulsory casefinding in controlling AIDS—testing, screening and reporting, *Am. J. Law Med.* **7**, 22–45 (1986).

N. A. Holtzman, Newborn screening for genetic-metabolic diseases: progress, principles and recommendations, *U.S. Department of Health, Education and Welfare, Publication No. (HSTA) 78-5207* (1977).

C. Levine and R. Bayer, The ethics of screening for early interventions in HIV disease, *Am. J. Public Health* **79**, 1661–1667 (1989).

W. McDougall, *The Heavens and Earth,* Basis Books, New York, p. 413 (1985).

R. Macklin, Mapping the Human Genome: Problems of Privacy and Free Choice. In: *Genetics and the Law III,* eds. A. Milunsky and G. J. Annas, Plenum, New York, pp. 107–114 (1986).

D. Nelkin and L. Tancredi, *Dangerous Diagnostics,* Basic Books, New York (1989).

F. P. Perera, Validation of molecular epidemiologic methods. In: this volume, *Molecular Dosimetry and Human Cancer: Analytical, Epidemiological, and Social Considerations,* eds. J. Groopman and P. Skipper, Teldford Press, Caldwell, NJ (1990).

R. Proctor, *Racial Hygiene*, Harvard University Press, Cambridge, MA (1988).

R. Rhodes, *The Making of the Atomic Bomb*, Simon and Schuster, New York, p. 761 (1986).

M. A. Rothstein, *Medical Screening and the Employee Health Cost Crisis*, Bureau of National Affairs, Washington, D.C. (1989).

OTA, U.S. Congress, *Mapping Our Genes*, Government Printing Office, Washington, D.C. (1988).

J. Watson, *The Double Helix*, Macmillan, New York, p. 11 (1968).

Chapter

21

Genetic Testing in the Workplace: Costs and Benefits*

Leslie I. Boden
Environmental Health Section
Boston University School of Public Health
Boston, Massachusetts

1. WORKPLACE MEDICAL MONITORING AND SCREENING: AN OVERVIEW

Medical monitoring and screening are common components of corporate occupational health programs. Fitness for employment may be based on a medical examination that rules out existing illness or risk factors for future illness. Occupational Safety and Health Administration (OSHA) regulations require medical monitoring of workers exposed to cotton dust, lead, dibromochloropropane, acrylonitrile, asbestos, vinyl chloride, arsenic, coke-oven emissions, ethylene oxide, formaldehyde, and noise. Even where monitoring of current employees is not required by law, employers may decide that medical monitoring is justified by the potential costs of low back pain, silicosis, mercury poisoning, or other occupational diseases. Medical screening of prospective and current employees may uncover nonoccupational risk factors or illnesses as well, e.g., low back disease or cancer not related to workplace exposure. Employers may not be able to prevent such illnesses,

---

*A number of the ideas presented here were first developed in Boden, 1986.

but they can avoid paying their costs by refusing to hire prospective employees who are most at risk.

In 1972, 48% of all manufacturing workers were employed by companies requiring medical examinations before employees were hired or placed (Ratcliffe *et al.*, 1986). Ten years later, there was a higher level of preemployment screening, with manufacturing workplaces employing 59% of all manufacturing workers requiring preemployment or preplacement medical examinations (Ratcliffe *et al.*, 1986). In 1972, plants employing 21% of manufacturing workers provided periodic audiometric examinations (Ratcliffe *et al.*, 1986). This was prior to the full implementation of the Occupational Safety and Health Act of 1970, including the promulgation of OSHA regulations requiring medical screening. By the early 1980s, periodic audiometric examinations were provided in plants employing 40% of manufacturing workers. Similar growth is displayed for other periodic medical examinations. For example, plants employing 15% of manufacturing workers provided periodic blood tests in 1972. Ten years later, this proportion was 36% (Ratcliffe *et al.*, 1986).

2. WORKPLACE GENETIC TESTING

Genetic screening is testing for a preexisting genetic trait or for damaged genetic material. Genetic screening can be used for preemployment screening of prospective employees or preplacement screening of current employees. Genetic monitoring is performed periodically to determine whether occupational exposures have damaged genetic material. Both genetic screening and monitoring are used to determine the probability that an individual has or will develop a disease. For the purposes of this paper, genetic testing includes both screening and monitoring.

Currently, it appears that there is very little workplace genetic testing. In 1982, the National Opinion Research Center conducted a survey of genetic testing for the U.S. Office of Technology Assessment (OTA). The survey was sent to the 500 largest U.S. industrial companies, the 50 largest private utilities, and 11 major unions representing workers in these companies. The survey only found 6 companies that were performing genetic testing at that time (U.S. Office of Technology Assessment, 1983). Twelve companies had done genetic testing over the previous dozen years, but were no longer doing so. The most common testing programs were for sickle-cell trait, serum α_1-antitrypsin deficiency, and G6PD (glucose-6-phosphate dehydrogenase) deficiency. These genetic characteristics are, respectively, risk factors for anemia, obstructive airway disease, and hemolysis (a serious blood disease).

The OTA study is the most recent survey available. A more recent, if unrepresentative, source of information about workplace genetic testing is the research literature, where there are reports of genetic testing of workers exposed to benzene, B-naphthylamine, epichlorhydrin, ethylene oxide, vinyl chloride, epoxy resins, styrene, aromatic amines, propylene oxide, coke-oven emissions, and polycyclic aromatic hydrocarbons (Mason *et al.*, 1984; Uzych, 1986; de Jong *et al.*, 1988; Perera, 1990). Although this research literature has been growing, nothing suggests that workplace genetic testing as a decision tool has increased substantially since the OTA report was published.

Two OSHA regulations (coke-oven emissions and arsenic) require sputum cytology as an early indicator of lung cancer. However, OSHA has not generally required genetic monitoring as a part of its workplace standards. This is largely because currently available tests have not yet been demonstrated to be more effective predictors of cancer risk than more traditional screening methods.

There are three primary reasons for the limited use of genetic testing in industry. First, there is a dearth of studies confirming the relationship between genetic tests and increased risk of illness (Perera, 1990). Second, where genetic tests might be used to monitor exposure to occupational hazards, there is substantial variation in the relationship between genetic tests and exposures (Phillips *et al.*, 1988; Perera, 1988, 1990). This variation occurs both among individuals and over time for the same individuals. Third, currently available genetic tests are often expensive relative to the benefits that might accrue from using them.

By way of example, I have calculated the employer benefits per case of obstructive airway disease avoided using genetic screening for the *PiMZ* heterozygote of α_1-antitrypsin. The marker characteristics, relative risk, and disease frequency are taken from Newill *et al.* (1986). Hiring costs are assumed to be $1000 per additional worker screened and the screening test is assumed to cost $100. Workers who are screened out are assumed to have a 50% increased risk of obstructive lung disease. Calculations are based on a simple model of screening costs and benefits where the screening program does not lead to additional wage payments. In this case, costs to the employer of one case of obstructive lung disease must exceed $100,000 to make the test worthwhile. With hiring costs of only $200 per new hire, the cost per case must exceed $75,000. (See the Appendix for the model used.)

In summary, genetic testing is not currently in wide use in industry. However, continuing advances in our understanding of the relationship between genetic markers and the risk of illness may lead to more widespread use in the 1990s. This raises the issue of when such tests might be used and of the potential social benefits and costs of workplace genetic testing.

3. PRIVATE BENEFITS AND PUBLIC HEALTH IMPACTS

The net private benefits (benefits minus costs) of genetic testing to the employer determine the strength of the economic incentives to use it and are thus of considerable interest. If the costs of a test are greater than the perceived benefits to them, few employers would use it. However, the social benefits and costs of genetic testing may not be the same as the private benefits and costs to the employer. As a consequence, private incentives to screen can lead to employer actions that cannot be justified on social grounds. In some cases, testing will not reduce illness, but will diminish employment opportunities for workers with positive test results. In other cases, injuries and illnesses could actually increase as a consequence of employer screening policies. Where economic incentives encourage such outcomes, public regulation of testing should be considered.

4. BENEFITS TO EMPLOYERS OF GENETIC TESTING

Employers can benefit from genetic testing because it identifies workers who are at increased risk of disease. On the basis of that information, employers can decide whether to hire those at high risk of illness. If the work environment adds to that risk, employers can reduce exposure through environmental controls, with personal protective equipment, or by removing the worker from the risk. In addition, employers who control exposures stringently can use the results of both environmental monitoring and periodic genetic testing to support the contention that workers' disease was not caused by occupational exposures. This reduces the risk that the employer will be held liable to pay workers' compensation benefits. Genetic testing can help the employer avoid other disease costs, including:

- Time lost from work
- Decreased productivity
- Health care costs and medical insurance premiums paid by employers
- Worker turnover generated by ill health or death
- Tort suits for damage to third parties arising from worker illnesses

Constrained by law and collective bargaining, economic motivation will lead employers to choose the path with the greatest net benefits to them.

5. COSTS TO EMPLOYERS OF GENETIC TESTING

Balanced against the benefits of genetic testing are several types of costs, all of which will be weighed by the employer in deciding whether to initiate or continue a testing program. These include the direct costs of testing, higher wages needed to attract an adequate labor force from a smaller pool of acceptable workers, extra workers' compensation claims or tort suits incurred as a consequence of testing (Rothstein, 1984), and the costs of post-testing actions designed to reduce illness costs to the employer.

6. PUBLIC VERSUS PRIVATE COSTS AND BENEFITS

While employers will be motivated by the wish to maximize the net benefits *to them* of genetic screening, this goal will not always be consistent with maximizing the *social* benefits of screening. The social goals of workplace genetic screening include reducing the prevalence of disease, the severity of disease, the costs of treating disease, and the costs of testing. These goals are constrained by social policies, such as that of eliminating employment discrimination. In the cost–benefit framework, the four goals are integrated into the objective of minimizing the sum of the social costs of illness and screening. This can be contrasted with the goal of minimizing costs *to employers* of illness and screening. If private and social costs were identical, then employers, in pursuing their self-interest, would provide a socially efficient level of screening. However, private costs and benefits diverge from social costs and benefits for several reasons.

First, many employers do not pay the full costs of occupational disease and therefore do not have adequate incentives to provide genetic screening for these diseases. In principle, workers' compensation pays the wage loss, medical costs, and rehabilitation costs of occupational disease. However, many cases of occupational disease do not even enter the workers' compensation system (Boden, 1988). Workers who file claims for chronic occupational disease may find their claims barred by statutory barriers, such as restrictive definitions of occupational disease or statutes of limitations. When their claims are not barred by statute, it will be difficult for them to demonstrate that their diseases arose "out of and in the course of employment" (Boden, 1988). When they receive compensation, they recover, on average, less than one third of lost wages (Boden and Jones, 1987). Finally, only the largest firms are self-insured or retrospectively rated by their workers' compensation carriers, so that workers' compensation benefits paid to workers are fully reflected in the firms' expenses (Russell, 1974; Victor, 1982). Where payments to workers with occupational or nonoccupational diseases are made under company medical care and disability insurance, benefits vary widely and are often less adequate than under workers' compensation. As with workers' compensation, employers' costs for these programs often do not fully reflect payments to their employees because of the employers' insurance coverage.

Conversely, employers pay disability and medical costs for nonoccupational disease, providing incentives to perform excessive screening. Not hiring workers who are ill or at risk of chronic illness can reduce an employer's disability and medical insurance costs, but it will not reduce the social costs of disease (Rothstein, 1989). The burden will simply be shifted to the worker or to another employer. The social benefit of this kind of screening is zero, although the private benefit to the employer is positive. Its social costs (e.g., testing costs, unemployment, underutilization of worker skills) can be high.

The private benefits of preemployment genetic screening can be particularly high among firms that invest a great deal in worker training. These firms have a considerable investment in their workers, so that it is expensive for them to hire and train new workers when current employees become ill or injured.

If, through preemployment genetic testing of workers, the employer determines that some are at excess risk of developing an occupational disease, the employer may decide not to hire these job applicants. Similarly, if periodic genetic monitoring of workers exposed to a workplace hazard indicates excess risk on their part, the employer may decide to discharge them or otherwise remove them from injurious exposure. If firing workers, thus removing them from exposure, reduces the risk of disease, it may have some social benefit (in spite of the questionable ethics of this action). However, discharging workers and hiring others to take their place may only spread the risk and could conceivably increase the total disease burden (Bailar *et al.*, 1988). When the disease process is irreversible and there is a linear dose–response relationship, there will be no reduction in disease as a result of replacing workers without reducing exposure. However, the social costs of job loss may be significant.

Even where removal of workers with excess disease risk has no social benefit, it can reduce employers' costs. By reducing the excess risk of individual workers, removal makes it less likely that those with disease will be able to demonstrate that their illness arose ''out of and in the course of employment.'' They will thus be less likely to obtain workers' compensation benefits from the employer. While reducing the employer's costs, such actions place the burden of disease and job loss on the workers' shoulders. From a social perspective, incentives to monitor are excessive.

On the other hand, there may be inadequate incentives to monitor for early signs of increased occupational disease risk in workplaces with high employee turnover. Because a large proportion of the workforce is hired each year, preemployment screening costs will be large. Moreover, for work-related illnesses with delayed onset (like cancers with 5- to 30-year latent periods), the probability of workers becoming ill or injured before leaving the firm would be small for firms with short average job tenure. A long-latent disease may not manifest itself until after one or more intervening jobs. Because the worker is not likely to be aware of the cause of his or her disease, a workers' compensation claim will probably not be brought against the responsible employer. In this case, medical or disability benefits will be paid by the worker or the current employer. Among workplaces with high turnover, even if the social benefits of screening and reducing exposure are large, the private benefit will be small.

7. POTENTIAL MISALLOCATION OF GENETIC TESTING RESOURCES

It might seem that firms with the highest exposures to occupational health risks would be most likely to institute genetic screening programs. However, an analysis of employers' costs and benefits shows paradoxically that the firms with the highest risks may be less likely to find genetic screening beneficial than firms with lower risks. This may occur if the following conditions exist in the high-risk firms:

Wages are low. Low wages mean that workers' compensation and other disability payments will be smaller.

Turnover is high. High turnover increases the costs of screening by increasing the number of tests per year. It also reduces the benefits of screening for diseases with long latent periods by reducing the probability that workers will manifest disease during their tenure with the screening employer.

Private disability and medical insurance is minimal or nonexistent. This will reduce the costs of workers' illness to the employer.

Firms are small. For small firms, premiums for workers' compensation, health insurance, and disability insurance are not sensitive to changes in payments to workers (Russell, 1974; Victor, 1982). Thus the benefits of reducing illness rates would not accrue to small firms. Larger firms are also more likely to employ occupational health professionals, who may advocate genetic screening programs.

The characteristics likely to induce hazardous firms to screen less frequently than safer firms are often found together. Small nonunion firms offering jobs with low wages, poor working conditions, constant shifting from one task to another, and little if any access to training and increased

TABLE 21–1. Hiring Rate and Hourly Wages, by Major Industry Group, 1981 (Number of Industries)

Hires per 100 Employees	Average Hourly Wage Rate		
	Less than $7.00	$7.00–$8.49	More than $8.50
4.1–6.0	4	1	0
3.1–4.0	2	4	1
1.0–3.0	0	2	6

Source: U.S Department of Labor, Bureau of Labor Statistics (1983).

income are not uncommon (Piore and Doeringer, 1971; Graham *et al.*, 1983; Dickens and Lang, 1985). Similarly, it is routine for high-wage firms to also offer good working conditions, job security, substantial on-the-job training, and opportunity for advancement. Table 1 illustrates the relationship between hourly wages and turnover for manufacturing industries (U.S. Department of Labor Statistics, 1983). Because economic characteristics that increase the value of genetic screening are often found together, market forces may lead the firms with the best wages, working conditions, and employment security to adopt genetic screening programs. Workers with existing disease or at excess risk of work-related disease will be screened from employment in these good jobs, and directed toward employment in less attractive jobs, where job risks may well be higher.

The social benefits of genetic screening will tend to be greater at the most hazardous firms. However, if these firms do not capture the benefits of screening, they will not screen often enough. As a result, too many workers will develop occupational illnesses as a result of their exposures at these firms. In addition, workers with existing disease or an increased likelihood of developing disease will find themselves with limited job opportunities (Weinstock and Haft, 1974). The more widespread the hazard for which screening is done, the more severe will be the limitation on their economic opportunities.

8. ILLUSTRATIONS

Data on the frequency of genetic testing are sparse, possibly because very little genetic testing is currently done. However, data on other kinds of medical screening and monitoring illustrate the relationship between the frequency of testing and plant size, one of the factors discussed above.

1. As reported in the 1972 National Occupational Hazard Survey (NOHS) (Ratcliffe *et al.*, 1986), the proportion of plants requiring preemployment or preplacement physical examinations of new employees in-

creased with size in every manufacturing industry category (Table 2). For all manufacturing industries, 19% of small plants (employing fewer than 100 workers) required this form of medical screening. It was required in 49% of medium-sized plants (100 to 499 workers) and in 83% of large plants (greater than 500 workers). Data from the National Occupational Exposure Survey (NOES) conducted ten years later display the same relationship (Ratcliffe *et al.*, 1986). According to the NOES, medical screening of new or prospective employees was required in 27% of small manufacturing plants, in 56% of medium-sized plants, and in 88% of large plants.

2. Periodic medical monitoring also shows a strong correlation with plant size. The NOHS and NOES surveys provide information on periodic chest x-rays, eye exams, audiometric examinations, blood tests, urine tests, and pulmonary function tests. Table 3 demonstrates the strong relationship between plant size and the use of periodic monitoring as reported in the two surveys. Note, in addition, the fact that the proportion of plants doing medical monitoring dramatically increased substantially between 1972 and 1982 in almost every testing category.

9. THE WORKER'S PERSPECTIVE

Although both the employer and the worker want to decrease the incidence of worker illness, their goals are not identical. Employers want to reduce their costs associated with illness among workers in their own firms. Workers want to reduce the probability of their being ill *no matter whom they work for*. In addition, workers want jobs that offer high wages, security of tenure, and other positive attributes.

From the perspective of workers exposed to hazardous substances, genetic testing offers the hope of reducing the risk of illness and thus may be of substantial benefit. However, the workers at high risk of disease may not be given the opportunity to choose how to reduce that risk. The employer may decide to improve environmental controls. On the other hand, the workers may be removed from the risky jobs, reducing their wages or leaving them unemployed. For workers at risk of illnesses that are not occupational in origin, discharge provides no potential health benefits.

The risk of job loss is compounded by the fact that, given equal hazards, firms with better wages, working conditions, and job security are more likely to screen than those offering less attractive jobs. Excessive genetic screening by the firms with good jobs means that people with abnormal test results will not be able to work in the best jobs for which they are qualified. They may be relegated to work for firms with low-paying jobs and no opportunity

TABLE 21–2. Estimated Percentage of Workers With Screening Examinations by Size of Plant

Type of Examination	Estimated Percentage of Workers					
	Small Plants (<100 Workers)		Medium Plants (100–499 Workers)		Large Plants (>500 Workers)	
	1972–74	1981–83	1972–74	1981–83	1972–74	1981–83
Preplacement	19	27	49	56	83	88
Periodic	12	13	29	38	65	69

Source: National Institute for Occupational Safety and Health (NIOSH) National Occupational Hazard Survey 1972–1974; NIOSH National Occupational Exposure Survey 1981–1983; preliminary data (Ratcliffe *et al.*, 1986).

TABLE 21–3. Estimated Percentage of Workers With Specific Period Examinations by Size of Plant

Test	Small Plants (<100 Workers)		Medium Plants (100–499 Workers)		Large Plants (>500 Workers)	
	1972–74	1981–83	1972–74	1981–83	1972–74	1981–83
Chest Radiograph	9	8	20	21	50	65
Ophthalmologic	7	11	16	22	48	57
Audiometric	5	13	14	38	48	66
Blood	4	11	8	24	35	69
Urine	3	11	8	23	34	66
Pulmonary function	3	8	8	19	31	55

Estimated Percentage of Workers

Source: National Institute for Occupational Safety and Health (NIOSH) National Occupational Hazard Survey 1972–1974; NIOSH National Occupational Exposure Survey 1981–1983; preliminary data (Ratcliffe *et al.*, 1986).

for advancement. In some cases, they may actually face more hazardous working conditions in the firms that do not screen. Worse, they may be unemployed, facing not only reduced income but increased health risks (Cobb and Kasl, 1977). The impact of screening on job opportunities will be greatest for high-prevalence conditions like low-back pain, hearing impairment, and chronic obstructive lung disease.

10. POLICY IMPLICATIONS

Both the executive and legislative branches of the Federal government have recognized that workplace medical screening can have effects that are contrary to public policy. As a result, laws and regulations have been promulgated, limiting the application of employment screening (Rothstein, 1984; Ashford, 1990). OSHA provides employment protection to workers who are screened under the provisions of the lead standard. The Rehabilitation Act of 1973 and state handicap laws limit the legal use of medical screening to promote the public policy of hiring handicapped workers. Title VII of the Civil Rights Act of 1964 and the Age Discrimination in Employment Act prohibit the use of medical screening to support discriminatory employment practices.

This chapter raises a different public policy question. Incentives to screen may be too great for firms with good employment characteristics and too small for firms with poor employment characteristics. Equally as important, there may be excessive incentives to rely on voluntary and involuntary turnover to reduce the employer's costs of occupational disease. Turnover and employment screening take the place of environmental controls. As a result, the social costs of disease may remain high while the employer's costs are reduced. The burden of disease costs is shifted to former employees, other employers, and public expenditures. In addition, job opportunities are reduced for those with disease and for those at high risk of disease.

Where there are inadequate incentives to screen, OSHA might require employers to offer medical screening programs to workers, as is currently the case under several OSHA health standards. Where excessive screening is being done, the appropriate response of policy makers may be to impose more limits medical screening, or to restrict stringently the range of actions available to employers who screen, as OSHA's medical removal protection program currently does for lead-exposed workers. While screening and removal can provide important benefits to workers and employers, removal could be restricted to cases where the employer can demonstrate that disease, not just the employer's costs, will be reduced. Employers might also be required to demonstrate the health benefits of preemployment genetic screening prior to initiating such programs. The analysis in this chapter also provides support for OSHA's historic focus on environmental controls rather than medical screening, both from the standpoints of public health and of employment policy.

APPENDIX

This appendix develops a simple model of the costs and benefits of preemployment screening, by comparing annual labor costs with and without screening. In both cases, labor costs consist of the sum of wage costs, hiring costs, and disease costs attributable to the disease associated with the medical screening. Screening increases wage costs because some of the available pool of workers are eliminated from employment by the screening, thus leading to greater demand for the remaining workers. Assuming that the medical screening is done at the end of the hiring process, hiring costs are increased because more workers must apply to fill the available vacancies. On the other hand, disease costs among those hired should be reduced by screening out high-risk workers. Table 4 displays the parameters of screening test and the relevant disease. Annual labor costs without screening are the sum of hiring costs, wage costs, and disease costs. With testing, these costs are the sum of hiring costs, wage costs, disease costs, and testing costs.

TABLE 21–4. Parameters of Screening Tests*

Test Result	Will Develop Disease	Will Not Develop Disease	
Positive	bp	$(1-a)(1-p)$	m
Negative	$(1-b)p$	$a(1-p)$	$1-m$
Total	p	$1-p$	1

*Where m = marker frequency
p = disease frequency
b = sensitivity
a = specificity

Source: Adapted from Newill *et al.* (1986).

Annual Labor Costs without Screening

$$L(c_h h \ + \ w_0 \ + \ pc_d) \tag{1}$$

where c_h = cost per worker hired;
     h = turnover (new hire) rate per worker;
     L = number of employees;
     w_0 = wage rate;
     p = population disease frequency;
     c_d = employer's costs per case of disease among currently employed workers.

Annual Labor Costs with Screening

$$L \left[c_h h/(1-m) + (w_t + w_0) + pc_d(1-b)/(1-m) + c_t h/(1-m) \right] \qquad (2)$$

where m = marker frequency (proportion of population testing positive);
     c_t = cost per screening test;
     b = sensitivity;
     w_t = extra wages needed to attract additional hires.

Net Benefits of Screening

$$\text{NB} = L \left[-c_h hm/(1-m) - w_t + pc_d(b-m)/(1-m) - c_t h/(1-m) \right] \qquad (3)$$

where   NB = benefits − costs.

From Eq. (3), we can calculate the value of c_d that makes the employer's net benefit of screening positive.

REFERENCES

N. A. Ashford, In: *Molecular Dosimetry and Human Cancer: Analytical, Epidemiological, and Social Considerations,* eds. J. Groopman and P. Skipper, Telford Press, Caldwell, NJ, pp. 417–442 (1991).

J. C. Bailar, E. A. C. Crouch, R. Shaikh, and D. Spiegelman, One-hit models of carcinogenesis: Conservative or not?, *Risk Anal.* **8**, 485–497 (1988).

L. I. Boden, The impact of workplace characteristics on the costs and benefits of medical screening, *J. Occup. Med.* **28**, 751–756 (1986).

L. I. Boden, Workers' compensation. In: *Occupational Health: Recognizing and Preventing Work-Related Disease,* 2nd ed., eds. B. S. Levy and D. H. Wegman, Little Brown, Boston, pp. 231–246 (1988).

L. I. Boden and C. A. Jones, Occupational disease remedies: The asbestos experience. In: *Regulation Today: New Perspectives on Institutions and Policy,* ed. E. E. Baily, MIT Press, Cambridge, MA, pp. 231–246 (1987).

S. Cobb and S. V. Kasl, *Termination: The Consequences of Job Loss,* National Institute of Occupational Safety and Health, Cincinnati, OH (1977).

G. de Jong, N. J. van Sittert, and A. T. Natarajan, Cytogenetic monitoring of industrial populations potentially exposed, *Mutat. Res.* **204,** 451–464 (1988).

W. T. Dickens and K. Lang, A test of dual labor market theory, *Am. Econ. Rev.* **75,** 792–805 (1985).

J. Graham, D. M. Shakow, and C. Cyr, Risk compensation—In theory and practice, *Environment* **25,** 14 (1983).

T. J. Mason, P. C. Prorok, W. E. Neeld, and W. J. Vogler, Screening for bladder cancer at Dupont chambers workers, *J. Occup. Med.* **28,** 1011–1016 (1986).

C. A. Newill, M. J. Khoury, and G. A. Chase, Epidemiological approach to the evaluation of genetic screening in the workplace, *J. Occup. Med.* **28,** 1108–1111 (1986).

F. P. Perera, The significance of DNA and protein adducts in human biomonitoring studies, *Mutat. Res.* **205,** 255–269 (1988).

F. P. Perera, Validation of molecular epidemiologic methods. In: *Molecular Dosimetry and Human Cancer: Analytical, Epidemiological, and Social Considerations,* eds. J. Groopman and P. Skipper, Telford Press, Caldwell, NJ, pp. 53–76 (1991).

D. H. Phillips, K. Hemminki, A. Alhonen, A. Hewer, and P. L. Grover, Monitoring occupational exposure to carcinogens: Detection by ^{32}P-postlabeling of aromatic DNA adducts in white blood cells from iron foundry workers, *Mutat. Res.* **204,** 531–541 (1988).

M. Piore and P. Doeringer, *Internal Labor Markets and Manpower Analysis,* D. C. Heath, Lexington, MA (1971).

J. M. Ratcliffe, W. E. Halperin, T. M. Frazier, D. S. Sundin, B. S. W. Delaney, and R. W. Hornung, The prevalence of screening in industry, *J. Occup. Med.* **28,** 906–912 (1986).

M. A. Rothstein, *Medical Screening of Workers,* BNA Books, Washington, D.C. (1984).

M. A. Rothstein, *Medical Screening and the Employee Health Care Crisis,* BNA Books, Washington, D.C. (1989).

L. B. Russell, Safety incentives in workmen's compensation insurance, *J. Human Resources* **9,** 361–375 (1974).

U.S. Department of Labor, Bureau of Labor Statistics, *Handbook of Labor Statistics, 1983,* Bulletin 2175, U.S. Government Printing Office, Washington, D.C. (1983).

U.S. Office of Technology Assessment, *The Role of Genetic Testing in the Prevention of Occupational Disease,* U.S. Government Printing Office, Washington, D.C. (1983).

L. Uzych, Genetic Testing and Exclusionary Practices in the Workplace, *J Pub. Health Policy* **7,** 37–57 (1986).

R. B. Victor, *Workers' Compensation and Workplace Safety,* Rand Corporation, Santa Monica (1982).

M. Weinstock and J. I. Haft, The effect of illness on employment opportunities, *Arch. Environ. Health* **29,** 79–83 (1974).

Chapter 22

Medical Screening for Carcinogens: Legal and Ethical Considerations*

Nicholas A. Ashford
Massachusetts Institute of Technology
Cambridge, Massachusetts

The collection of articles in this book documents exciting new directions for molecular epidemiology. The identification of DNA adducts, protein–carcinogen adducts, and other biomarkers of exposure to and effects of carcinogens raises new and challenging questions for application of this emerging science in the workplace and community. While most attention for possible applications has been focused on the workplace, the eventual use of biomarkers in the context of contaminated communities or consumers exposed to carcinogen-containing products may also be anticipated. The new science may have relevance for carcinogen regulation, workers' compensation, and damage suits in the courts.

Human monitoring in the workplace, popularly referred to as medical screening, is a collection of practices that focuses on the workers as an indicator that (1) disease has resulted from exposure to toxic substances, radiation, or other traumas (medical surveillance), (2) toxic substances have been absorbed into the body (biologic monitoring), (3) a particular worker may be especially predisposed to disease (genetic screening or other sensitivity probes), and (4) preclinical disease states exist, indicating harmful

---

*Adapted from N. A. Ashford *et al.*, 1990.

exposure has occurred (genetic monitoring). These monitoring practices, especially when required or carried out by the government or the employer, raise serious and complex legal and ethical concerns (Ashford *et al.*, 1990, 1984).

However thorny the legal and ethical problems of disease detection and the communication of information are in the context of the patient and personal physician relationship, workplace medical screening complicates the inquiry even further. This chapter seeks to construct a philosophic framework for examining the adequacy of law as an embodiment of ethical values concerning medical screening and for identifying possible solutions to the attendant legal and moral dilemmas. The analysis necessarily focuses on three sets of activities involving distinct participants: workers, employers, corporations, in-house physicians or physicians under contract, and the government. The sets of activities deserving separate consideration are as follows: (1) requiring the worker to submit to monitoring tests, (2) disseminating the results of the tests, and (3) using the test results. Because the different kinds of monitoring address different stages of the process from exposure to disease and because what is monitored affects different groups of workers differently, specification of particular problems and a case-by-case analysis are essential, lest we face useless generalities at the end. It is hoped that the analysis presented will contribute to a more thoughtful process of inquiry rather than advocate particular solutions.

1. PURPOSES AND TYPES OF HUMAN MONITORING

The purposes of human monitoring are diverse and sometimes conflicting. One purpose is to reduce the incidence of occupational disease or injury in the working population as a whole by providing indicators of *average* harm or risk of harm from exposure to toxic substances. This type of monitoring need not involve all workers, but only a statistically informative sample, in order to ensure an adequate average level of toxic material control. Another purpose is to protect especially sensitive workers, workers exposed to toxic material by means other than inhalation, and workers for whom nonoccupational sources may add to occupational exposure. In these cases, an employer may institute human monitoring to remove workers from potentially harmful exposures and thus avoid liability for increased worker compensation premiums and tort or product liability suits. Employers may also use human monitoring results to avoid meeting increased worker demands for preventive technology or other health and safety measures.

Both workers and employers tend to be risk-averse, that is they err on the side of caution to avoid human and economic costs each might otherwise have to bear. The worker, however, would prefer to have the workplace as hazard-free as possible, whereas the employer would prefer to minimize legal and financial liability by removing problems (workers) before they arise. The resolution of conflicts between these goals fundamentally depends on one's sense of fairness about where the costs of preventing harm should lie.

The concept of human monitoring encompasses four practices: medical surveillance, genetic monitoring, sensitivity screening, and biologic monitoring. The scientific and legal literature commonly uses medical surveillance and biologic monitoring interchangeably, but the terms are not the same. This chapter employs the following definitions of these four concepts, consistent with regulations recently promulgated by the Occupational Safety and Health Administration (OSHA).

1.1. Medical Surveillance

Medical surveillance is designed in an occupational setting to detect adverse health effects (or health status) resulting from hazardous exposures in the workplace. Medical surveillance testing serves to obtain types of information, such as the identification of workers who are suffering from an occupational injury or illness, epidemiologic data on occupational disease, and general or specific data on categories or types of workers. Medical surveillance focuses on monitoring specific organ systems that may be affected by exposure to workplace hazards. Especially if a disease is reversible or arrestable, medical surveillance may be preventive insofar as it serves as a warning signal prompting timely action to avoid future exposures and continuing or progressive adverse health effects.

Medical surveillance is most useful in three situations: (1) if compliance with the permissible exposure limits established by OSHA will not adequately ensure worker health; (2) if air measurement cannot sufficiently monitor worker exposure (for example, if a significant route of entry is not inhalation); and (3) if high-risk groups are exposed. Medical removal may also be appropriate in these three situations.

1.2. Genetic Monitoring

One particular type of medical surveillance that has received much attention is genetic monitoring. This type of monitoring includes the periodic testing of employees working with or possibly exposed to certain substances (such as known or suspected carcinogens) thought to cause changes in chromosomes or in DNA. Blood or other body fluid samples are collected for this monitoring. Generally, such monitoring is conducted in an attempt to determine if environmental exposures of a specific population (for example, workers in the same job category) to particular substances cause changes in genetic material in statistically significant numbers above background levels.

1.3. Sensitivity and Genetic Screening

A third kind of human monitoring, sensitivity screening, is practiced on an employee only once, usually as part of a preemployment or preplacement examination. This screening determines whether an individual possesses certain inherited or acquired traits that may predispose him to an increased risk of disease if exposed to particular substances. If that trait is inherited, the term genetic screening applies. Laboratory tests on body fluids, commonly blood, usually identify these traits.

1.4. Biologic Monitoring

Biologic monitoring should be distinguished from medical surveillance. In contrast to the latter, which is used to determine effects of exposure, biologic monitoring focuses on the exposure itself. It is a collection of activities designed to determine whether body fluids or organs contain a particular substance or its metabolites. Note that while some authors would include genetic monitoring within biologic monitoring, here the latter term is reserved for detection of the chemical itself, e.g., lead in blood.

Thus different types of monitoring activities yield different kinds of data: a distribution of indicators of health or disease status, a distribution of uptake indicators in the working population, or a distribution of biomarkers of interaction with genetic material. Medical surveillance generally provides the first type of data, biologic monitoring the second, and genetic monitoring the third. Distributions of health status or toxin uptake have their origins in variations of current and past exposures, current health status, and biologic make-up.

Different workers exposed equally to the same substance may exhibit differences in the results of monitoring. A number of factors cause a wide distribution of testing results in a homogeneously exposed population, including those determined by nature, those somewhat in control of the employee, and those in the control of the employer. These include stochastic variability, genetic predisposition, age, sex, environmental factors (including behavioral and/or lifestyle components), and preexisting disease. These factors may be relevant in fashioning the strategies for protecting workers who exhibit monitoring results indicating they are at high risk. Employer actions to change the work environment, provide alternative employment, or suggest medical treatment ideally follow from the correct interpretation of monitoring results.

Conducting any or all of the four types of human monitoring tests is certain to be a complex activity. Even before one conducts the tests, decides how to interpret the results, and determines what possible action to take, one must assess the value of the tests. To determine whether a test can reliably detect a certain disease or abnormality, one must consider the test's sensitivity, specificity, and predictive value, the frequency (prevalence) of the condition within a population, and the reproducibility of the results. Unless one knows the frequency of the underlying condition in the population to be screened, one cannot estimate the test's predictive value. In fact, few frequencies are known for conditions related to work environments.

In sum, the decision to remove a worker from a job may be closely linked to the results of monitoring. Realizing that the decision to remove a worker depends strongly on these results, one must assess the appropriateness and validity of the test for the substance of concern. For example, in a situation in which testing may give false-positive results (low specificity), workers incorrectly diagnosed as sick may be inappropriately removed. Conversely, asymptomatic workers who have false-negative (low sensitivity) may remain on the job when, in fact, they should be removed. In addition, a false sense of security may develop from the unwarranted conclusions that because the risk to the removed individual is reduced, the risk to the work group as a whole is lessened. The rotation of workers can, in some instances, cause more total disease in the working population (Ashford *et al.*, 1990, 1984).

The technical considerations discussed in this section bear significantly on the legal and ethical conduct and use of medical screening.

2. TOWARD AN ETHICAL THEORY FOR HUMAN MONITORING

The moral and legal inquiry in the area of human monitoring addresses the behavior of particular actors engaged in the conduct of monitoring tests, the dissemination of the test results, and the use of the information in the context of the employment relationship. Ladd argues that it is important to distinguish ethics from law, custom, institutional practices, and positive morality (the body of accepted popular beliefs of a society about morality): Ethics is concerned with what ought to be (Ladd, 1987). Moral problems emanating from conflicts concerning human monitoring may be categorized as conflicts of interest; conflicts of moral demands, for example, conflicts of duties; and conflicts in perceptions of what is right or wrong, fair or unfair. Certain rights are possessed by individuals, and those rights impose (moral) obligations on others. Rights and obligations must be viewed together in the context of particular relationships (Ladd, 1978). John Ladd and others have argued that people have a *general duty* to support the fulfillment of the moral requirements of relationships, whether their own or those of others. Some of what is necessary to carry out this duty is embodied in rights and obligations. These rights and obligations may sometimes be given the force of law.

> . . . the concept of rights, as a cluster of claims on society and its institutions on the part of the individual, derives its principal moral warrant from the concept of moral integrity. This concept, unlike the concept of simple self-determination, focuses on the integrity of personal relationships, concerns, and responsibilities. Everyone in society has a duty, individually and collectively, to defend, support, and nourish these moral relationships both personally and in others. The concept of rights provides an effective social and conceptual instrument for carrying out this general duty (Ladd, 1978).

The delineation of rights and duties gives rise to certain expectations or hopes on the part of society concerning human behavior. In imposing rules or legal principles on individuals and institutions, the law often embodies societal attitudes, values, and expectations. This sometimes, but not always, occurs when a significant societal consensus has been reached on a particular moral question derived from human conflicts. The law establishes legal rights, whose violation may be illegal, and the law provides remedies to correct their violation. But the law also recognizes that conflicts of legitimate interests, conflicts of legal duties, and differences in perception of what is right or wrong, fair or unfair, require a balancing in the fashioning of remedies. Indeed, there are both legal remedies (usually of statutory origin) and equitable remedies that give great discretionary power to the courts or adjudicating institutions, such as the Equal Employment Opportunity Commission. Rules are embodied in legislation and regulations; legal principles guide, but do not unequivocally settle other conflicts. In examining the conflict questions, the law does indeed view behavior in the context of relationships. Justifiable expectations one party has of another are translated into the legal concept of reliance. Thus the law will sometimes find a physician–patient relationship between worker and company physician when none was intended by the physician, because it was reasonable that the worker expected certain behavior or transmission of information from the physician. Similarly, although the legal construct of the modern corporation bestows limited personal liability on corporate officers or employees (Ladd, 1984), the courts will "pierce the corporate veil" when corporate behavior violates the ethical norm. Discrimination law is replete with discretionary justice (Davis, 1971).

The law, of course, does not always serve the ethical interest of the society so nobly. Legislation and legal institutions can be compromised by powerful special interest. In addition, if there is a lack of societal consensus or interest about a moral issue, the law may either not address that issue or fail to give helpful guidance concerning the boundaries of fair or equitable behavior. Thus it is important to engage in both a legal and ethical inquiry concerning human and institutional behavior.

In the context of the transfer of medical information resulting from workplace monitoring and discrimination resulting from its use, the legal and ethical norms are in a great state of flux. Conflicts of interest and conflicts of duty (for example, for the company physician or government official) abound. Moreover, given the weak scientific validity of many of the screening tests and resulting data, questions of what actions to take or not to take reflect differences in perceptions of fairness and risk-averseness. The worker would rather be safe and keep his or her job; the employer wants to limit his or her legal and economic liability. In the face of great uncertainty, the participants prefer to take few chances, and very divergent solutions are pursued concerning both the transmission of uncertain information and its use.

On the other hand, some employers may be motivated to undertake medical screening solely out of a sincere concern for the health of their employees and may even feel a moral obligation to do so. In this case, there may be no conflict if care is taken concerning the dissemination and use of the monitoring results.

3. LEGAL AUTHORITY FOR HUMAN MONITORING

The Occupational Safety and Health Act (OSHAct) covers most private (i.e., nongovernmental) workplaces. The OSHAct grants both the Occupational Safety and Health Administration (OSHA) and the National Institute for Occupational Safety and Health (NIOSH) the authority to promulgate regulations that require employers to conduct human monitoring. The authority granted to NIOSH for monitoring is broader in scope than that vested in OSHA, but financial limitations on NIOSH in exercising that authority give OSHA the greater practical grant of authority.

While employers are mandated to make medical testing programs available (generally consisting of medical surveillance alone or coupled with biological monitoring in some circumstances), participation by employees in all OSHA health standards testing programs is not required as a matter of law. However, employers may be able to require participation as a condition of employment.

OSHA health standards fall into two groups: (1) some 450 permissible exposure limits (being currently updated to 626) adopted from a list of

TABLE 22-1. Chemical Substances for which OSHA Has Promulgated Health Standards; 1972-1988*

2-Acetylaminofluorine	4-Dimethylaminoazobenzene
Acrylonitrile	Ethylene oxide
4-Aminodiphenyl	Ethyleneimine
Arsenic (inorganic)	Formaldehyde
Asbestos	Lead (inorganic)[†]
Benzene	Methyl chloromethyl ether
Benzidine	α-Naphthylamine
bis-Chloromethyl ether	β-Naphthylamine
Coke-oven emissions	4-Nitrobiphenyl
Cotton dust[†]	N-Nitrosodimethylamine
1,2-Dibromo-3-chloropropane[†]	β-Propiolactone
3,3'-Dichlorobenzidine (and its salts)	Vinyl Chloride

*Adapted from 29 C.F.R. Sec. 1910, Subpart Z (1988). OSHA has initiated rulemaking for nine other substances, including 1,3-butadiene, cadmium dust and fume, 2-ethoxyethanol (Cellosolve), 2-Ethoxyethyl acetate, Methylene chloride and 4,4'-Methylenedianiline: 53 *Federal Register* **21**, 246 (June 7, 1988).
[†]Not regulated as a carcinogen.

recommended threshold limit values (TLVs) constructed by the American Conference of Governmental Industrial Hygienists and (2) some 24 standards adopted through a more formal, careful, and deliberate process. The former standards are not established on the basis of carcinogenicity and have no recommended human monitoring requirements, while the latter are mostly carcinogens and have requirements for medical screening, though they are vague for the most part. These latter standards are listed in Table 1.

OSHA is currently considering a generic exposure monitoring standard that may provide more definitive guidelines for the TLVs. Genetic screening has been specifically disavowed by OSHA, but genetic monitoring appears to be well within the kinds of investigations that could satisfy medical screening requirements.

4. LEGAL AND ETHICAL PROBLEMS ARISING FROM HUMAN MONITORING

4.1. Requiring Workers to Submit to Monitoring

4.1.1. Personal privacy

In the abstract sense, an employee may always refuse to be the subject of human monitoring. OSHA, NIOSH, and the employer have no authority to compel employees to cooperate. Refusal to participate, however, may mean loss of a job. Thus the relevant inquiry is the extent to which the employer may condition employment on such cooperation. For example, may an employer require a prospective employee to submit to genetic or biologic screening as a precondition to employment? May he or she require a current employee to submit to periodic biologic monitoring or medical surveillance? These questions raise important issues of confidentiality and discrimination. Apart from these issues, however, there remains a question about the employer's general authority to require human monitoring of employees.

Monitoring in response to agency directive. At the outset, a distinction must be made between the human monitoring that OSHA, NIOSH, or the Environmental Protection Agency (EPA) may require and monitoring that the employer implements on his own initiative. When a federal agency requires monitoring, the worker has a valid objection only if he asserts a statutory or constitutional violation. Congress was mindful of constitutional considerations in developing human monitoring programs. For example, it specifically acknowledged the need for a balancing of interests when an employee asserted a religious objection to a monitoring procedure. Human monitoring can also impinge on the worker's constitutional right to privacy. In the case of human monitoring, the privacy right may be articulated in two ways: the right to physical privacy and the right to withhold information likely to prove detrimental to one's self-interest.

If an employee does not wish to comply with a monitoring procedure required by agency regulation, imposing that procedure as a condition of employment may invade that employee's constitutional right to physical privacy. It may, depending on the nature of the procedure, infringe on the right to be free from unwelcome physical intrusions and on the right to make decisions regarding one's own body. Although these rights are obviously related, the former is grounded in the Fourth Amendment's proscription against unreasonable search and seizure, whereas the latter is closely associated with the rights of personal privacy commonly identified with the Ninth and Tenth Amendments. Although protected by the Constitution, these rights of privacy are not inviolate.

Courts have recognized a general need to balance the privacy interests of the individual with the public health interests of society. In certain situations, the former will be deemed to outweigh the latter, but in others, intrusion will be permitted in the name of public health. To date, no reported decision has mentioned an asserted constitutional right to refuse participation in human monitoring as a condition of employment. Nevertheless, one can identify the factors that would bear upon an evaluation of that right.

The public health significance of human monitoring, when properly used, is difficult to deny. Gathering information through human monitoring to develop standards for the protection of worker health, or for the enforcement or evaluation of existing standards, serves an important public health purpose. Furthermore, although the Constitution protects against government paternalism, the fact that this public health interest parallels the affected worker's own interest in a healthy workplace may make monitoring a less onerous invasion of privacy than it would be otherwise. [Bayer raises the question of whether workers may actually have a moral obligation to cooperate in such monitoring for the collective good. He argues that without safeguards, coercive monitoring is unfair (Bayer, 1986).] To the extent that monitoring serves a legitimate public health purpose, a limited intrusion of physical privacy appears constitutionally permissible. The less the accuracy, reliability, or predictability of a particular intrusion, however, the weaker the case for violating physical privacy.

At some point, the degree of risk or intrusiveness of monitoring may be sufficiently compelling to outweigh the public health interests. Some forms of human monitoring may simply be too risky or too intrusive to be constitutionally permissible. Furthermore, even if a monitoring procedure is not constitutionally impermissible per se, the worker may have a right to insist on an alternate, less intrusive procedure that adequately fulfills public health purposes. To survive constitutional challenge, a regulation requiring human monitoring should be reasonably related to a legitimate public health goal and should impose the least intrusive method necessary to achieve its goal.

An additional and critical question is whether the employee may refuse to participate in a program of agency-directed monitoring when he believes that his employer may use the resulting information as a basis for termination. For example, the worker who suffers chromosomal damage as a result of workplace exposure may fear that medical screening will reveal this condition to the employer and thus induce removal. Participation in a monitoring program can be tantamount to self-incrimination.

This form of self-incrimination conflicts with the right to personal privacy. If there is a constitutional right to preserve the confidentiality of information pertaining to one's health, there may also be a right to retain that information within one's body. Stated differently, there may be a limited constitutional right to refuse to comply with physical procedures that result in the initial disclosure of confidential information. Although this right is not absolute, damage to the employee can be substantial if health data are likely to affect employment status adversely. A worker's interest in preserving his employment status may rise to the level of property protected by the Fifth Amendment.

In developing monitoring requirements, an agency should seriously consider the constitutional dimensions of human monitoring. To avoid a challenge on a self-incrimination basis, OSHA and NIOSH might consider including mandatory Medical Removal Protection (MRP) programs as part of their human monitoring requirements. MRP provides earnings protection and employment security during medical removal. Properly used, an MRP program would thus help ensure employee cooperation with monitoring.

Monitoring in the absence of agency directive. Under common law, employers can require their employees to comply with reasonable programs of human monitoring. Congress did not intend the OSHAct to "preempt the field" by authorizing the implementation of human monitoring requirements. One of the OSHAct's express purposes is to "stimulate employers . . . to institute new and to perfect existing programs for providing safe and healthful working conditions." Congress intended that employers take the initiative on a number of fronts, including human monitoring, in developing health and safety programs. As long as it promotes "safe and healthful working conditions," employer-initiated human monitoring would appear to be welcome. Similarly, nothing in the Act precludes employers who are subject to OSHA monitoring requirements from implementing additional programs. Furthermore, it could be argued that employers have a moral obligation to initiate monitoring if they suspect their employees are at risk. To date, however, this moral obligation has not been translated into a legally enforceable duty to undertake medical screening.

If an employer institutes a human monitoring program in the absence of an agency directive, he or she is still subject to applicable restrictions under state common law, state statute, and federal labor law. Common law requires that human monitoring be implemented in a reasonable fashion. Determining reasonableness involves balancing the benefits gained by monitoring against the risk, discomfort, and intrusiveness of the monitoring procedure. The National Labor Relations Act may also require such balancing. In a given jurisdiction, the balance might be affected by a state statute defining a right of personal privacy.

4.1.2. Informed consent

Assuming that a human monitoring program is permissible, there are limitations on the manner in which an employer may implement the program. In general, one who undertakes the performance of monitoring procedures has a duty to perform those procedures properly and will face liability for damages caused by the negligent administration of a monitoring procedure.

A troublesome question arises, however, with regard to the applicability of the doctrine of informed consent. Strictly speaking, informed consent is a medicolegal concept, and stems from a belief that persons have a right to make decisions governing their bodies and health. Thus a medical professional is said to have a duty to inform the patient honestly and accurately of the potential risks and benefits of a proposed medical procedure so that the patient can make an informed choice whether to consent to that procedure.

All human monitoring procedures are medical or quasimedical in nature. Most commonly, they are performed by medical professionals: physicians, physician assistants, nurses, or nurse practitioners. Thus, the concept of informed consent appears at first glance to be applicable. The differences between human monitoring and medical treatment, however, are not insignificant, and they raise serious questions whether and to what extent the traditional doctrine of informed consent has meaning in the occupational setting.

Initially, one may inquire to what extent the relationship between the worker and the medical professional who administers the monitoring procedure can be characterized as a physician–patient relationship. Quite often, neither the employee nor the union selects the occupational physician. Rather, the employer selects and often directly employs the physician. Accordingly, some courts have held that the performance of a physical examination, which would clearly establish a physician–patient relationship in a purely medical context, does not create that relationship if it is a preemployment examination requested by the prospective employer. To the extent that the physician–patient relationship does not exist in the occupational setting, traditional notions of informed consent may not be applicable to human monitoring.

Similarly, the doctrine of informed consent is closely tied to the concept of medical treatment. It assumes that not only is the patient being requested to submit to a procedure designed for his own benefit, but also that the patient is in a position to make a voluntary choice to participate. Human monitoring calls both these assumptions into question. Monitoring may not be "treatment" in the conventional sense of the word. In many cases, monitoring benefits the employer more than the employee. Furthermore, monitoring is usually compulsory in that it is a condition of continued employment. It may be meaningless to speak of "informed consent" if the worker-patient is not free to reject the proffered procedure without jeopardizing his job. In this light, the applicability of informed consent appears particularly dubious in the case of agency-directed monitoring. Neither the employee nor the employer has the discretion to discontinue monitoring.

Regardless of the applicability of informed consent in the traditional sense, a complete and accurate disclosure of risks seems an advisable adjunct to a program of human monitoring. Whether or not a physician–patient relationship exists, imposing a medical procedure on a person not fully informed of the risks of that procedure may still be a battery and may give rise to liability in tort. In addition, prudent social policy requires full disclosure of risks. If the employer is required to disclose all risks inherent in a program of human monitoring, employee and union scrutiny will act as an incentive for the employer to develop programs that use the safest and least intrusive techniques possible. Indeed, unions may have a right to demand such information as a part of the collective bargaining process. Recognition of a duty to disclose material risks seems as appropriate in the area of human monitoring as it is in the area of medical treatment.

A final question concerns the scope of the required disclosure. The employer should, of course, disclose all material physical risks. The most significant risk of all, however, may be dismissal from employment. Should employers or occupational physicians be required to warn employees that one of the risks of submitting to a program of human monitoring may be a loss of job? The Code of Ethical Conduct adopted by the American Occupational Medical Association and the American Academy of Occupational Medicine states that physicians should:

> . . . treat as confidential whatever is learned about individuals served, releasing information only when required by law or by over-riding public health considerations, or to other physicians at the request of the individual according to traditional medical ethical practice; and should recognize that employers are entitled to counsel about the medical fitness of individuals in relation to work, but are not entitled to diagnoses or details of a specific nature (Code of Ethical Conduct for Physicians Providing Occupational Medical Services, 1976).

Under this formulation, although the physician may not disclose to the employer the specific results of human monitoring, the employee's job security may be endangered nonetheless. Employers are "entitled to counsel about the medical fitness of individuals in relation to work." A preferable alternative practice would involve the worker in such discussions between the physician and the employer (Whorton and Davis, 1978).

4.2. Dissemination of Monitoring Results

4.2.1. Employee's right of access

An employer may not limit or deny an employee access to his medical or exposure records. The OSHA regulation promulgated on May 23, 1980 grants employees a general right of access to medical and exposure records kept by their employer. Furthermore, it requires the employer to reserve and maintain these records for an extended period of time. [In the absence of OSHA regulation, employees would arguably still have a right of access under common law or state statute in many jurisdictions (Annas, 1976).] There appears to be some overlap in the definition of "medical" and "exposure" records, because both may include the results of biologic monitoring. The former, however, is generally defined as those records pertaining to "the health status of an employee," while the latter is defined as those pertaining to "employee exposure to toxic substances or harmful physical agents."

The employer's duty to make these records available is a broad one. The regulations provide that upon any employee request for access to a medical or exposure record, "the employer *shall* assure that access is provided in a reasonable time, place, and manner, but in no event later that fifteen (15) days after the request for access is made." In addition to the right of access, there are duties to inform workers of exposure to occupational hazards (Ashford and Caldart, 1985).

4.2.2. Employees' right to confidentiality

Of all of the issues raised by human monitoring, employee confidentiality may have received the most attention (Annas, 1990). An employee's right to maintain the confidentiality of information regarding his body and health places a significant limitation on the ways in which others can use that information. As programs of human monitoring are developed, mechanisms must be found that maximize both the employee's interest in privacy and society's interest in promoting general workplace health and safety. In the final analysis, this may be more a technologic challenge than a legal or ethical one.

In a broad sense, private citizens do have a right to protect the confidentiality of their personal health information. With regard to governmental invasions of privacy, this right is created by the Bill of Rights and is one component of the right of personal privacy discussed previously. With regard to private intrusions, the right is grounded in state law. In the medical setting, it grows out of the confidential nature of the physician–patient relationship, although rights of confidentiality exist outside this relationship as well. In essense, the recognition of a right of privacy reflects an ongoing societal belief in the need to protect the integrity of the individual.

This right to privacy, however, is not absolute and may be limited or waived. Courts nonetheless remain vigilant in their attempts to protect individual privacy. They generally look for a reasonable middle ground when faced with legitimate interests on both sides of the confidentiality question. They prefer an approach that permits both the use of health information for a socially useful purpose and the protection of the privacy of the individual. The key is the development of information-based technology that will make that approach more readily available.

4.2.3. Notification of workers at high risk

Caldart (1986) and Ashford and Caldart (1985) have addressed the worker's right to information and the employer's duty to provide information concerning occupational risks without request. Recently, there has been increasing attention to the government's responsibility to notify workers if they have been identified as being part of a high-risk group based on epidemiologic studies. (The reader is referred to the excellent works of Schulte and associates, 1984, 1985, 1986, which deal with the multitude of legal and ethical problems arising out of the right-to-know.)

4.3. Employer Use of Human Monitoring Results

Even if an employer obtains human monitoring data through a legitimate exercise of his right of access, the right to use such data is not absolute. Employers may not use health information to discriminate against employees on a basis deemed impermissible by federal or state law. Beyond discrimination, however, a more essential—and perhaps more difficult—question arises: To what extent may employers use health or exposure information to limit or terminate the employment status of individual employees or to deny employment to a prospective employee? Furthermore, to what extent and under what conditions does the employer have an obligation to remove the worker? If removing a worker and rotating another employee to take his place reduces each worker's individual risk but increases the total number of diseased workers, what is the employer to do?

4.3.1. Common law limitations

In early common law, an employer had the right to take an employee's health into account in determining whether to continue to employ that person. If the employment contract was "open," with no definite term, the employee could be discharged for any reason, including health status, at the will of the employer. If the contract of employment was for a definite term, the employee could be discharged for "just cause." Typically, significant illness or disability constituted "just cause." Although federal labor law, workers' compensation, and recent common law limitation on the doctrine of "employment at will" have profoundly affected the nature of employee–employer relations in this century, courts continue to recognize an employer's interest in discharging employees who cannot perform their work safely. Thus if the worker has no statutory or contractual protection, an employer may retain a general common law right to discharge the worker whose health status makes continued employment dangerous or whose health status prevents him from performing his job. Workers' compensation legislation, of course, facilitates the termination of permanently disabled workers.

Human monitoring, however, places the issue in a somewhat different light. Monitoring designed to reveal whether an employee has been, or in the future may be, harmed by the workplace raises the question of whether the employer may discharge an employee merely because the employee was, or may be, harmed by a situation created by the employer. The rights of the employer to discharge the employee might not be as broad then as in the general case.

Suppose an employer is complying with an existing OSHA standard for a particular toxic exposure and monitoring reveals that one of the firm's employees is likely to suffer serious and irreparable health damage unless he is removed from the workplace. In this situation, the employer is complying with public policy as enunciated by OSHA and, absent a mandatory MRP provision, is arguably free to discharge the employee. If an employer fails to comply with applicable OSHA standards, however, or if no standard exists, and the employer permits workplace exposure levels that violate state and federal requirements to maintain a safe place of employment, the employer is violating the public policy embodied in the OSHAct. Workers are only infrequently able to obtain compensation for disability due to occupational disease. In this case, to permit the employer to discharge the employee is to permit a further violation of public policy. An employer's use of human monitoring data for this purpose may well be impermissible as a matter of public policy, and employers may be obliged by common law to find safe assignments for the workers at comparable pay or bear the costs of their removal as part of doing business.

4.3.2. Limitations under the OSHAct general duty clause

The use of monitoring data to limit or deny employment opportunities raises other issues under the general duty clause of the OSHAct. When monitoring information reveals that an employee risks serious health damage from continued exposure to a workplace toxin, it may also indicate that the employer is in violation of the general duty clause. When a workplace exposure constitutes a recognized hazard likely to cause death or serious physical harm, an employer violates the general duty clause if he does not take appropriate steps to eliminate the hazard. In the case of toxic substances, this would appear to require reduction of the exposure, not mere removal of presumptively sensitive employees from the site of exposure.

The issue is amenable to regulatory solution. The implementation of mandatory MRP for toxic substances exposure in general, as OSHA has done with its lead standard, might be accomplished by a generic MRP standard. An employer's compliance with a mandatory MRP provision for a particular exposure would remove the threat of a general duty clause citation.

4.3.3. Limitations under the antidiscrimination laws

In addition to potential liability under the common law and the OSHAct general duty clause, an employer who uses monitoring information to limit employment opportunities may also face liability under antidiscrimination laws. Although not all workplace discrimination is prohibited, state and federal law forbid certain bases for discrimination. Many of these may apply to an employer's use of human monitoring information. A detailed discussion of the relevant discrimination laws is beyond the scope of this chapter, but an outline of their potential impact on human monitoring follows.

Section 11(c) of the OSHAct. Section 11(c)(1) of the OSHAct prohibits employers from discharging or otherwise discriminating against any employee ''because of the exercise by such employee on behalf of himself or others of any right afforded by this chapter.'' If an employee insists on retaining his job in the face of medical data indicating that continued exposure to a workplace toxin will likely pose a danger to health, the employee may well be asserting a right afforded by the OSHAct. The Act's general duty clause imposes on employers a duty to maintain a workplace that is free of ''recognized hazards'' likely to cause death or serious physical harm. Inferentially then, the Act vests employees with a concomitant right to insist that their workplace be free of such hazards. By insisting on retaining employment, the employee is asserting his right to a workplace that comports with the requirements of the general duty clause. Accordingly, an employer who discharges or otherwise discriminates against a worker because of perceived susceptibility to a toxic exposure arguably violates the Section 11(c) prohibition. When an employer asserts that an employee cannot work without injury to health, the employer admits that the workplace is unsafe. That admission triggers the remedial provisions of the OSHAct.

An OSHA regulation, issued under Section 11(c) and upheld in an unanimous Supreme Court decision (*Whirlpool Corp. v. Marshall,* 1980) gives individual workers a limited right to refuse hazardous work when there is a situation likely to cause ''serious injury or death'' (Ashford and Katz, 1977). The employer may not take discriminatory action against the employee by discharging the employee or by issuing a reprimand to be included in the employment file. According to the district court to which the issue was remanded for consideration, withholding the employee's pay during the period in which the employee exercises the right is also prohibited.

As a worker may be absent from a hazardous work assignment under certain conditions without loss of pay or job security, it seems anomalous to allow an employer to discharge or remove the employee without pay because of the same hazardous condition. This would make the employee's status depend on whether he asserted a right to refuse hazardous work before the employer took action to discharge him from employment.

Handicap discrimination. Employees may be able to assert further rights against discriminatory use of human monitoring data under laws protecting the handicapped. Congress and most states have passed laws barring discrimination against handicapped individuals in certain employment situations. The laws, which vary widely among the jurisdictions, all place potential limitations on the use of human monitoring data. Although the courts have adopted a case-by-case approach, the worker who is denied employment opportunities on the basis of monitoring results often falls within the literal terms of many handicap discrimination statutes. In general, two issues will be determinative: whether the workplace in question is covered by a state or federal handicap act, and if so, whether the worker in question is handicapped under the act.

At present, the general applicability of handicap discrimination statutes to the use of human monitoring information remains unclear. Examining the definitional criteria in the federal act, on which many of the state statutes are based, will illustrate the issues facing courts—and the potential range of logical interpretations. The Rehabilitation Act of 1973, as well as the 1990 Americans with Disabilities Act (ADA), defines a handicapped individual as "any person who (i) has a physical or mental impairment which substantially limits one or more of such person's major life activities, (ii) has a record of such an impairment, or (iii) is regarded as having such an impairment.

In the great majority of cases, the persons facing reduced employment opportunity as a result of human monitoring data do not presently have a substantially debilitating medical condition and thus do not satisfy either the first or second clauses of the federal definition. Rather, they are perceived as having an increased risk of developing such a condition in the future. The Acts seem designed primarily to protect the seriously handicapped, but their language is broad enough to cover discriminatory practices based on data obtained through human monitoring.

Even in cases in which handicap discrimination is established, an employer may escape liability if the discriminatory practice is reasonably necessary for efficient operation of the business. The Rehabilitation Act provides employers with no affirmative defense, but does require the handicapped individual to prove that he is "qualified" for the job. Thus, if a handicap prevents a worker from safely or effectively performing the job, an exclusionary practice may be permissible under the Act. The ADA and most state handicap statutes include some form of affirmative defense. Although these vary among jurisdictions, many appear analogous to the familiar defenses that have developed under Title VII of the Civil Rights Act.

Civil rights and age discrimination. Employers who exclude workers on the basis of monitoring information may also run afoul of the more general laws against discrimination. Title VII prohibits employment discrimination on the basis of race, color, religion, sex, or national origin. The scope of the Civil Rights Act is substantially broader than that of the federal Rehabilitation Act, and it affords protection for the great majority of American employees. In addition, many states extend similar protection to employees not covered by the federal act. The Age Discrimination in Employment Act and some state acts provide protection of comparable breadth against discrimination on the basis of age.

As with handicap discrimination, the applicability of these laws to the use of human monitoring information is not yet clear. The practical impact of an exclusionary practice, however, may fall disproportionately on a particular race, sex, ethnic or age group.

The Supreme Court has long held that a claim of disparate impact states a viable cause of action under the Civil Rights Act. A similar rationale has been applied in the area of age discrimination. In a 1975 decision, the Court held that job applicants denied employment on the basis of a preemployment screen establish a prima facie case of racial discrimination when they demonstrate that "the tests in question select applicants for hire or promotion in a racial pattern significantly different from that of the pool of applicants." Proof of disparate impact thus requires statistical analysis demonstrating a "significantly" disproportionate effect on a protected class. The cases provide no clear guidance, however, as to the level of disproportion that is required before an effect is deemed significant.

The potential for disparate impact inheres in many uses of human monitoring data. A genetic screen for sickle-cell anemia, for example, will disproportionally exclude blacks and certain ethnic groups because they have a much higher incidence of this trait than does the general population. Similarly, tests that consistently yield a higher percentage of positive results in one gender than the other may give rise to exclusionary practices that discriminate on the basis of sex. (The permissibility of fetus protection policies, which exclude fertile women from the workplace to avoid exposure to reproductive hazards, is beyond the scope of this article, as it does not involve discrimination on the basis of monitoring data. For a brief discussion of this issue, see Ashford and Caldart, 1983.)

Finally, a wide variety of exclusionary practices based on monitoring data may have a disparate impact on older workers. Older workers have been in the workforce longer and usually have been exposed to hazardous work environments much more often than their younger colleagues. Their prior exposure may have impaired their health or left them more vulnerable to current workplace hazards. They may, for example, have a preexisting illness as a result of previous workplace exposures. Their age alone may account for a certain degree of body deterioration.

When the plaintiff establishes a prima facie case of disparate impact, the employer will have an opportunity to justify the alleged exclusionary practice by showing that its use constitutes a "business necessity." If such a showing is made, the practice will withstand a charge of disparate impact discrimination. The Supreme Court has characterized the business necessity defense as requiring "a manifest relation to the employment in question." In the works of an often cited opinion, this means that the practice must be "necessary to the safe and efficient operation of the business." Furthermore, if the plaintiff can establish that another, less discriminatory practice will accomplish the same purpose, the business necessity defense will not stand.

The business necessity defense is available only in cases of disparate impact. If a practice is discriminatory on its face or involves disparate treatment, the employer may avoid liability only by demonstrating that the basis of the discrimination constitutes a bona fide occupational qualification (BFOQ). This defense is available under the Civil Rights Act for discrimination based on sex, national origin, or religion (but not for discrimination based on race or color) and under the Age Discrimination in Employment Act. The BFOQ defense requires the employer to establish that the discriminatory practice is "reasonably necessary to the normal operation of business." The Supreme Court has characterized the defense as an "extremely narrow" one.

There are two principal reasons why a business necessity may be difficult to establish for exclusionary practices based on human monitoring data. The first is that most of these practices are not designed to protect the health and safety of the public or of other workers. Instead, their business purpose is the protection of the excluded worker and, not incidentally, the protection of the employer from the anticipated costs associated with the potential illness of that worker (see Boden, 1991). That position may well encounter a chilly judicial reception. As noted in one recent analysis, "the courts are usually skeptical of an employer's argument that it refuses to hire qualified applicants for their own good, and they often require a higher level of justification in these cases than in cases in which public safety is at stake" (McGarity and Schroeder, 1981).

Another, potentially more serious obstacle to the successful assertion of a business necessity defense is the unreliability of the screening procedures themselves. If the exclusion of susceptible (high-risk) individuals truly is a business necessity, its rationale disappears if the test used as the basis for such exclusion cannot provide reasonable assurance that those excluded are actually susceptible (at high risk). Indeed, without such assurance, the test becomes little more than an instrument for arbitrariness and only adds to the discriminatory nature of the exclusionary practice. As many of the tests are currently far from reliable, the availability of the business necessity defense is questionable.

The foregoing discussion of discrimination has presupposed that the screened worker will be excluded from the workplace. Employers may, however, have another option. In many cases, they may be in a position to provide these workers with other jobs in workplaces that do not involve exposures to the substances from which they may suffer adverse health effects. If such alternative positions were supplied, at benefit levels comparable to those of the positions from which exclusion was sought, employers might avoid the proscriptions of the various discrimination laws. Providing an alternative position would certainly remove much of the incentive for filing a discrimination claim. Further, even if such a claim was filed, courts might find an adequate MRP program obviated the charge of discrimination. This could be one area in which good law and good social policy coincide.

4.4. The Use of Monitoring Data in Tort and Workers' Compensation Cases

From a legal perspective, the most important impact of human monitoring information may be its use as evidence in tort and workers' compensation cases. Although its potential in this area is still to be realized, the science of biological markers may eventually serve to answer the evidentiary question that has plagued most compensation cases involving human exposure to toxic substances: how do we kown whether a *particular* exposure caused a *particular person's* medical condition? At present, the problem remains a major one.

Obviously, to the extent that increased use of human monitoring adds to the existing database on the observed correlations between particular diseases and particular chemicals, it will provide increased evidence for use in compensation proceedings generally. More than this, though, human monitoring has the potential to bring about a change in the *nature* of the evidence used in these cases.

Typically, the evidence offered to prove causation in chemical exposure cases is premised on a *statistical* correlation between disease and exposure. Whether the underlying data are from epidemiologic studies, from toxicological experiments, or from the results of a complicated risk assessment model, they are usually *population*-based. This places the plaintiff at the mercy of the attributable risk (expressed as the percentage of cases of the disease attributable to the exposure) for the study population. Unless the attributable risk is greater than 50%—that is, unless the incidence rate among those exposed to the chemical is more than double the background rate— the plaintiff cannot prove *on the basis of the available statistical evidence,* that it is more likely than not that his or her particular case of the disease was caused by the chemical exposure.

The developing science of human monitoring may offer a way to distinguish individual claimants from the group. Conceivably, the data generated by various human monitoring procedures will:

- increase our knowledge of the ''subclinical'' effects of toxic substances, thus permitting us to track the effect of a chemical exposure over time, and also expanding the universe of ''medical conditions'' for which compensation may be provided;
- eventually enable us to establish that a particular person has been exposed to a particular chemical (or class of chemicals); and
- eventually enable us to establish that a particular person's medical condition (or subclinical effect) was caused by exposure to a particular chemical (or class of chemicals).

Already, human monitoring data are being used in some situations to show subclinical changes thought to be associated with particular chemical exposures. This can have many applications in toxic substance compensation cases. In the long term, evidence of subclinical changes occurring between the time of exposure and the time of disease may be a way of distinguishing those whose disease was caused by the exposure from those who contracted the disease because of other factors. Such evidence may also give rise to more immediate legal relief. There is a growing trend toward allowing those who can establish that they have been exposed to a toxic substance—and that they thus have been placed at increased risk of future harm—to recover the costs of medical surveillance from the responsible party. Proof of certain subclinical effects, such as DNA damage, would tend to support an allegation of increased risk, and would make the claim for medical surveillance all the more compelling. Furthermore, such evidence may support a separate claim for damages *for having been put at increased risk of future harm,* although it is not clear that such a claim would be viable in most states.

Finally, some evidence of "subclinical" effects may give rise to a right to recover compensation *for those effects themselves.* For example, human monitoring can detect certain changes in the immune system. There is a body of literature suggesting that chemical exposures can harm the immune system (Descotes, 1986), and evidence of immune system damage has been offered in recent cases involving toxic substance exposure (Rothman and Maskin, 1989). Thus far, allegations of immune system damage have met with mixed success in the courts, both because the relationship between chemical exposure and immune system damage is not yet clear, and because the evidence of immune system damage was not always considered persuasive. Although human monitoring may not be able to tie particular immune system deficiencies to particular exposures, it should be able to establish with greater certainty whether immune system damage has, in fact, occurred.

Looking farther to the future, it is quite possible that further developments in the science of biomarkers will permit the identification of "chemical footprints"—a distinctive change in the DNA that can be tied to a particular chemical or class of chemicals. At the very least, this should make it much easier to distinguish those who have been exposed to a particular chemical in the workplace from those who have not, and to identify which of the many potential defendants was responsible. More importantly, it should eventually permit the correlation of particular cases of diseases such as cancer with exposure to particular chemicals (or classes of chemicals). To the extent that this happens, it will narrow the scope of the evidence from the population to the individual, and will place the deliberations in these cases on firmer scientific footing.

5. POSITIVE USES OF HUMAN MONITORING

The strategies used for human monitoring must be fashioned on a toxin-specific basis because the state-of-the-art techniques differ from substance to substance. In general, medical surveillance and biologic or genetic monitoring for populations should be used only in combination with environmental monitoring. In the case in which a specific harmful substance cannot be identified, however, and the workplace is suspected to being unsafe, medical surveillance may indicate whether a problem exists. In the future, genetic monitoring may serve as an early indicator that exposure to certain chemicals has occurred in a worker population. However, the use of that screening for this purpose is in its infancy. Genetic screening focuses on removal of the worker before exposure and is preventive for that worker only.

Human monitoring should be used only if (1) given the specific workplace problem, monitoring serves as an appropriate preventive tool; (2) it is used in conjunction with environmental monitoring; (3) the tests are accurate and reliable and the predictive values are high; (4) it is not used to divert resources from reducing the presence of toxic substances in the workplace or from redesigning technology; and (5) medical removal protection for earnings and job security is provided. New solutions involving both technologic innovation and job redesign may obviate the necessity of human monitoring. Conflicts now arise only because, under existing technology, workers continue to be exposed to toxic substances.

If conditions (1)–(5) are met, the question remains whether the employer is obligated to remove the worker. It would seem so, since the employer owes that duty of care to his employees. But what if the tests are unreliable? The extent of the obligation would appear to be less, although the dissemination of the test results to the worker arguably would be required. Is the employer under a greater or lesser obligation to remove the worker if he provides no earnings and job security protection—greater because it costs him nothing to protect the worker, or lesser because the worker is economically disadvantaged?

Finally, if a particular workplace cannot be made safer and the removal (and rotation or replacement) of workers results in lower individual risk but greater total disease, is this morally defensible? (Exposure to radiation and some carcinogens can represent such a situation.) These are difficult questions.

6. SOLUTIONS TO THE LEGAL AND ETHICAL CONFLICTS

The extensive discussion in the previous section reveals the complexity and difficulty with which the law balances competing interests and equities.

In some cases, the law embodies the belief that *de facto* discrimination against certain protected classes of people (a minority race) is to be avoided affirmatively and cannot be justified by resulting health benefits. In other cases, such as those involving workers who are perceived as handicapped, the burdens to industry of providing protection may be relevant in deciding how much employment security to require. Discrimination law necessarily involves the exercise of discretion by courts and adjudicating institutions—with its attendant inconsistency and unpredictability.

Discriminatory practices and consequential tort suits, antidiscrimination suits, deterioration of labor–management relations, and agency sanctions may follow poorly conceived and poorly executed human monitoring programs. The weaker the scientific foundation for the tests, the less secure are the legal grounds and defenses available to the employer. In light of the sometimes preliminary, unreliable, and nonspecific nature of the techniques used in human monitoring, the practice is a problematic activity itself in most instances.

With legal and ethical norms in flux, it is important to examine the policy options for dealing with future and continuing ethical dilemmas. The possible strategies which deserve consideration include the following.

1. Encouragement of ethical enquiry in the conduct and use of medical screening, that is, educating workers, management, and health professionals to think more seriously about the problems. Perhaps ethicists should be consulted in designing the screening programs.
2. Use of legislative and regulatory means to clarify rights and duties, such as encouraging OSHA to promulgate a generic earnings and job security protection requirement for all medical removal or enacting legislation that requires workers to be notified of occupational risks and prohibits discrimination outright.
3. Encouragement of the use of self-help techniques by workers, such as the right to refuse hazardous work, union bargaining, and the filing of discrimination complaints.
4. Encouragement of better disposition of conflicts by improving procedural fairness in conflict resolution, such as full and complete disclosure of information to workers, the better maintenance of confidentiality of worker records, or the use of corporate ombudspersons.

These options differ in the extent to which rights and freedoms of some are diminished to protect those of others. Regulatory or legislative fiats define acceptable behavior and in the process decrease freedom of choice. On the other hand, freedom from harm and discrimination is preserved. Sharpening self-help mechanisms preserves choice, but fosters an adversarial solution. Education can persuade and enlighten; it can also sensitize the discriminated to assert their rights. Procedural fairness tends to right the imbalance of access to legal and political institutions. The choice of options at any one time reflects the seriousness with which society wishes to address the moral and legal dilemmas. Thinking about these problems is a first and necessary step (Atherley *et al.*, 1986; Derr, 1988; Lappe, 1986; Samuels, 1986).

REFERENCES

G. Annas, Legal aspects of medical confidentiality in the occupational setting, *J. Occup. Med.* **18**, 537 (1976).

G. Annas, The bioethics of biomonitoring people for exposure to carcinogens. In: this volume, *Molecular Dosimetry and Human Cancer: Analytical, Epidemiological, and Social Considerations,* Telford Press, Caldwell, NJ, pp. 385–399 (1991).

N. A. Ashford and J. I. Katz, Unsafe working conditions: Employee rights under the labor Management Relations Act and the Occupational Safety and Health Act, *Notre Dame Lawyer* **52**, 802 (1977).

N. A. Ashford and C. C. Caldart, The control of reproductive hazards in the workplace: A prescription for prevention, *Indus. Rel. Law J.* **5**, 523, 533–535, 541–547 (1983).

N. A. Ashford and C. C. Caldart, The "right to know," Toxics information transfer in the workplace, *Annu. Rev. Public Health* (6), 383–401 (1985).

N. A. Ashford, C. J. Spadafor, and C. C. Caldart, Human monitoring: Scientific, legal and ethical considerations, *Harvard Environ. Law Rev.* **8**, 263–363 (1984).

N. A. Ashford, C. J. Spadafor, D. B. Hattis, and C. C. Caldart, *Monitoring the Worker for Exposure and Disease: Scientific, Legal and Ethical Considerations in the Use of Biomarkers,* John Hopkins University Press, Baltimore, MD, (1990).

G. Atherley, N. Johnston, and M. Tennassee, Biological surveillance: Rights conflict with rights, *J. Occup. Med.* **28**, 958–965 (1986).

R. Bayer, Biological monitoring in the workplace: Ethical issues, *J. Occup. Med.* **28**, 935–939 (1986).

L. Boden, Genetic testing in the workplace: Costs and Benefits. In: this volume, *Molecular Dosimetry and Human Cancer: Analytical, Epidemiological, and Social Considerations,* Telford Press, Caldwell, NJ, pp. 401–416 (1991).

C. Caldart, Promises and pitfalls of workplace right to know, *Sem. Occup. Health,* **1**, 81 (1986).

Code of Ethical Conduct for Physicians Providing Occupational Medical Services, *J. Occup. Med.* **18**, 703 (1976).

K. C. Davis, *Discretionary Justice,* University of Illinois Press, Urbana, IL (1971).

P. Derr, Ethical considerations in fitness and risk evaluations. In: *Occupational Medicine: State of the Art Reviews,* eds. J. Himmelstein and G. Pransky, 3(2), 193–208 (1988).

J. Descotes, *Immunotoxicity of Drugs and Chemicals,* Elsevier, New York (1986).

J. Ladd, Legalism and medical ethics. In: *Contemporary Issues in Biomedical Ethics,* eds. J. Davis, B. Hoffmaster, and S. Shorten, Humana Press, Clifton, NJ, pp. 1–35 (1978).

J. Ladd, Corporate mythology and individual responsibility, *Int. J. Appl. Philos.* **2,** 1–17 (1984).

J. Ladd, The task of ethics. In: *Encyclopedia of Bioethics,* ed. W. T. Reich, The Free Press, New York, pp. 400–407 (1987).

M. Lappe, Ethical concerns in occupational screening programs, *J. Occup. Med.* **28,** 930–934 (1986).

T. McGarity and C. Schroeder, Risk-oriented employment screening, *Texas Law Rev.* **59,** 999 (1981).

R. Rothman and A. Maskin, Defending immunotoxicity cases, *Toxics Law Reporter (BNA)* **4,** 1219 (1989).

S. Samuels, Medical surveillance: Biological, social and ethical parameters, *J. Occup. Med.* **28,** 572–577 (1986).

P. A. Schulte, Problems in notification and screening of workers at high risk of disease, *J. Occup. Med.* **28,** 951–957 (1986).

P. A. Schulte and K. Ringen, Notification of workers at high risk: An emerging public health problem, *Am. J. Public Health* **74,** 485–490 (1984).

P. A. Schulte, K. Ringen, E. B. Altekruse, W. H. Gullen, K. Davidson, S. S. Anderson, and M. G. Patton, Notification of a cohort of workers at risk of bladder cancer, *J. Occup. Med.* **27,** 28 (1985).

Whirlpool Corp. v. *Marshall,* 455 U.S. 1,21 (1980).

D. Whorton and M. Davis, Ethical conduct and the occupational physician, *Bull. NY Acad. Med.* **54,** 733–740 (1978).

INDEX

A

Acetanilide, 369
Acetone, 31
N-Acetoxy-2-acetylaminofluorene, 20, 254
2-Acetylaminofluorene, 33, 368, 371
trans-4-Acetylaminostilbene, 375
Acetylation, 373, 378–379
N-Acetylation, 36, 55
N-Acetylation polymorphism, 18
Acetylbenzidine, 369
Acetylcholine, 329
Acetyl CoA-dependent O-deacetylation, 37
N,O-Acetyltransferase, 37
Acetyltransferases, 36–38, see also
    specific types
ACh, see Acetylcholine
Acid hydrolysis, 197–198, 200, 252, 351
Acquired immunodeficiency syndrome
    (AIDS), 386
Acrylamide, 192, 194, 201
Acrylonitrile, 401
Actinic keratosis, 22
Active oxygen, 20
Active smoking, 335–336
Adenocarcinomas, 18
AF, see Aflatoxin
Affinity chromatography, 271, 286, 318
Aflatoxin B_1, 29, 32, 36, 54, 123, 206,
    306, see also Aflatoxins
  activation of, 308
  detection of, 316
  dietary intake of, 312
  excretion of, 313
  formation of, 310
  metabolism of, 307
  reclassification of, 304
  in serum albumin, 318
Aflatoxin B_1-DNA adducts, 309–310, 315,
    321, see also Aflatoxin-DNA
    adducts
Aflatoxin B-lysine adducts, 319
Aflatoxin-DNA adducts, 223, 255, 271,
    see also Aflatoxin B_1-DNA adducts
Aflatoxin G_1, 29

Aflatoxin M_1, 307, 314
Aflatoxin P_1, 314
Aflatoxins, 29, 32, 303–322, see also
    specific types
  albumin-bound, 309
  dietary, 225, 312
  epidemiological studies of, 304–307
  excretion of, 303
  guanine and, 225
  intake of, 303
  metabolism of, 106, 303, 307–310
  metabolites of, 314
  in milk, 318
  in serum albumin, 318–321
  in tissue, 320–321
  total, 316
  in urine, 311–317
AGT, see O-Alkylguanine DNA
    alkyltransferase
AIDS, see Acquired immunodeficiency
    syndrome
Albumin, 19, 191, 269
  aflatoxins in, 309, 318–321
  plasma, 9
  serum, 318–321
Alcohol, 27, 31–32, 387, see also specific
    types
Aldehydes, 21, 31
Alkaline hydrolysis, 197, 207
Alkaline-phosphatase, 218, 223
Alkaloids, 36, see also specific types
Alkenylbenzenes, 132
Alkyl adducts, 293
N-Alkyl adducts, 285
3-Alkyladenines, 269, 282, 288
Alkylated DNA adducts, 263–276
  formation of, 264–265
  in human tissue, 273–274
  methods of detection of, 265–272
Alkylating agents, 21, 99, 263, 276, 281–
    294, see also specific types
  clinical studies and, 286–292
  epidemiological studies and, 286–292
  urine analysis in measurement of, 284–
    286